GREEN'S FUNCTIONS FOR SOLID STATE PHYSICISTS

GREEN'S FUNCTIONS FOR SOLID STATE PHYSICISTS

A REPRINT VOLUME WITH ADDITIONAL MATERIAL ON THE PHYSICS OF CORRELATED ELECTRON SYSTEMS

S. DONIACH
Stanford University, USA

E. H. SONDHEIMER
University of London, UK

ICP Imperial College Press

Published by

Imperial College Press
57 Shelton Street
Covent Garden
London WC2H 9HE

Distributed by

World Scientific Publishing Co. Pte. Ltd.
5 Toh Tuck Link, Singapore 596224
USA office: Suite 202, 1060 Main Street, River Edge, NJ 07661
UK office: 57 Shelton Street, Covent Garden, London WC2H 9HE

British Library Cataloguing-in-Publication Data
A catalogue record for this book is available from the British Library.

GREEN'S FUNCTIONS FOR SOLID STATE PHYSICISTS

First edition published in 1974 by W. A. Benjamin, Inc.

First published 1998
Reprinted 1999, 2004, 2008

ISBN-13 978-1-86094-078-1
ISBN-10 1-86094-078-1
ISBN-13 978-1-86094-080-4 (pbk)
ISBN-10 1-86094-080-3 (pbk)

Printed in Singapore by B & JO Enterprise

Contents

Preface to the Imperial College Press edition.

A lot has happened in the field of condensed matter physics since the original edition of "Green's functions for Solid State Physicists" was published in 1974. Nevertheless, the book has helped introduce several generations of condensed matter physics graduate students to the very powerful ideas of quantum many body theory and some of their applications, particularly those in the physics of itinerant magnetism and superconductivity that have nowadays come to be called "the correlated electron problem".

In preparing for the reprint edition, two new chapters have been added to the original text to provide an introduction to the recent developments in this branch of condensed matter physics. Chapter 11 focuses on the understanding of the Kondo problem which grew out of the exact solutions developed in the mid 1970's. The accompanying growth of experimental work culminating in the discovery of the heavy fermion superconductors gave substance to the idea that Coulomb repulsion between electrons in a narrow band metal can actually lead to attraction between the electrons and resulting instabilities at low temperatures to either a superconducting or an antiferromagnetic state.

Then in 1986, the discovery by Bednorz and Mueller of high T_c superconductivity in the cuprate compounds provided a bombshell in the field of correlated electron systems. For the first time it was possible to have materials in a superconducting state at temperatures well above that of liquid nitrogen. Nevertheless, in spite of more than 10 years of very intensive research by physicists in many countries, the mechanism of high T_c superconductivity remains a mystery at the fundamental level. Chapter 12 offers an introduction to some of the basic theoretical ideas of the physics of the cuprate compounds.

Although the theoretical concepts leading to the understanding of superconductivity, which resulted from the fundamental work of Bardeen, Cooper and Schrieffer in the 1950's, still provide some of the theoretical underpinnings for high T_c , there are still many aspects of the properties of these materials which do not fit in with the elementary quasiparticle ideas of Fermi liquid theory. Consequently it has become clear that new physical concepts need to be developed to explain these properties. A brief introduction to the physics of one-dimensional metals is included at the end of chapter 11 to serve as a basis for some of the new ideas in the physics of two-dimensional metals which may be applicable to high T_c. Their application in two dimensions is briefly introduced at the end of chapter 12.

The final chapter on understanding high T_c cannot be written at this time. Nevertheless it is our hope that this reprint edition, with the new material, will serve as an introduction and stimulus to the next generation of condensed matter physicists who seek to work on this challenging class of problems.

Sebastian Doniach
Stanford, California
Winter 1998

Preface

During the last 15 years the Green's function methods of quantum field theory have become generally recognized as a powerful mathematical tool for studying the complex interacting systems of solid state physics. In writing this elementary account we have tried to show the method in action, so that—without bothering about lengthy formal preliminaries—we use it from the start to discuss physical problems. The idea is to show in practice how the mathematics—in the form of the analytic properties of the Green's function in the complex energy plane—accounts for the physical effects (level shifts, damping, instabilities) characteristic of interacting systems. We concentrate on general physical principles and do not discuss experiments in detail, but we have included introductions to topics of current research interest such as the Mott metal-insulator transition and the singularities (x-ray, Kondo) associated with transient perturbations in an electron gas. We hope that the reader will feel compensated for any loss in generality of treatment by being kept in contact with real problems.

In the first three chapters the Green's function technique is illustrated on the exactly solvable example of the harmonic vibrating lattice. We go on to discuss scattering by random impurities in a gas of non-interacting fermions and show how to calculate the electrical conductivity of a metal. We then turn to the interacting fermi gas, devoting particular attention to magnetic instabilities. We finish with a short chapter on superconductivity. Two appendices deal briefly with second quantization and the fluctuation-dissipation theorem. We have also included a historical note on George Green. We have omitted several important subjects such as classical liquids, liquid helium, critical phenomena, and the details of the Landau theory of fermi liquids. Our choice of topics has been determined by our tastes and interests and should not be taken as canonical.

The book grew out of a course of intercollegiate graduate lectures given by S.D. in the University of London. We hope that it will appeal to beginning graduate students in theoretical solid state physics, as a first introduction to more comprehensive or more specialized texts, and also to experimentalists who would like a quick, if impressionistic, view of the subject. A basic knowledge of solid state physics and quantum mechanics, at undergraduate honors or first-year graduate level, is assumed.

We are indebted to a number of colleagues for their interest and helpful comments. We are particularly grateful to Dr. W. G. Chambers for showing us the treatment given in Appendix 1, and to the chairmen of our departments for facilitating consultation at close quarters.

S. DONIACH
E. H. SONDHEIMER

Introduction

The Theory of Condensed Matter

The science of condensed matter (thermodynamics, hydrodynamics, etc.) is in many ways much older than that of the atomic constituents. However, it is only in the last two or three decades that a systematic mathematical formulation of the many-body problem—with the 10^{23} or so degrees of freedom needed to describe a macroscopic sample—has become developed so that the properties of the simplest classes of condensed matter can be related back quantitatively to the properties of the constituent atoms.

There are two fundamental classes of properties possessed by condensed matter which belong essentially to its many-body character and do not occur for the individual constituent degrees of freedom. One is the existence of propagation—the notion of a sound wave, the transport of electronic charge in metals, the propagation of light in insulators; the other is the occurrence of phase transitions by which the matter changes its fundamental symmetry—for classical systems melting and freezing, for quantum systems phenomena such as magnetism, superconductivity and superfluidity. Both classes of effects involve phenomena of long range, spreading over distances much greater than the effective range of the basic atomic forces which mediate them.

What is the nature of the mathematical construction which links the atomic to the condensed description of matter? Its formulation takes on many guises, but the essential feature is that, even when the individual atomic interactions may be treated, in some sense, as "weak," the properties of the condensed system can only be treated correctly by taking them into account in infinite order. The most elementary example of such an infinite-order process leads to the occurrence of propagation through the "handing on" of excitation energy from one atom to the next. We shall see in Chaps. 1–3 that a natural formalism with which to set up the theory of the propagation phenomenon is the Green's function approach, which in a classical system reduces to the theory of correlation functions.

For a single degree of freedom the Green's function, or inverse differential operator, gives the amplitude of the degree of freedom at time t, given its amplitude at some previous time t'. This may refer either to a localized degree of freedom (e.g., a single atomic oscillator) or to a non-localized system (e.g., the amplitude of an electron wave function at position $\mathbf{x}$ at time t, given that at position $\mathbf{x}'$ at time t'). The many-body effects are then embodied in the repeated emission of Huyghens wavelets as the electron propagates through the medium, giving rise to an infinite series of multiply scattered waves which sums to provide the Green's function for an electron interacting with the medium (which may consist of other electrons). In this way one can obtain the response of complex interacting systems to simple forms of excitation without having to find the full eigenvalue spectrum, a task which is generally neither practicable nor of physical interest. The relation between the Green's function formalism and scattering theory will be studied in Chaps. 4 and 5.

A major simplification which occurs for homogeneous many-body systems is that the low-lying excited states with energies near the ground state can often be simply described in terms of the resulting propagating modes. Because of their mode-like nature (with rather well-defined excitation energy $\omega_{\mathbf{k}}$ as a function of the propagation wave-vector $\mathbf{k}$) the quanta of these elementary excitations are referred to as quasiparticles. The Green's function approach to the theory of elementary excitations is developed in Chaps. 1–6 through the study of a series of specific examples drawn from solid state physics. We shall see how the Green's function determines the excitation spectrum through its analytic properties in the complex energy plane. We also show how the excitations produced by applied external fields can be formulated in terms of Green's functions, leading to general expressions for measured quantities such as electrical conductivities and magnetic susceptibilities. An important property of Green's functions is that they are related, via the fluctuation-dissipation theorem, to time correlation functions which determine scattering cross-sections and which also give the averages needed to discuss properties of the ground state of the system (or, at finite temperature, the thermal equilibrium state). These relationships are illustrated a number of times in the text.

The phase transition phenomenon can also be reached via the elementary excitation concept. For some quantum systems this may be studied at zero temperature by seeing how an excited state, of lower symmetry than the ground state, becomes degenerate with the ground state as the interaction strength is increased. This "softening" of the excitation energy will show up as a singularity of the Green's function at the instability point of the system. In classical systems this singular behavior reduces to the Ornstein–Zernike theory of the two-particle correlation function. In Chaps. 7–10 we discuss the instability phenomenon in the context of the magnetic and superconducting instabilities of the interacting electron system. We also show how, once in the state of lower symmetry, the system acquires a new spectrum of elementary excitations which are no longer unstable.

Chapter 1

Lattice Dynamics in the Harmonic Approximation

We start our exploration of the Green's function approach by considering the problem of a lattice of interacting vibrating atoms. Within the harmonic approximation this represents perhaps the simplest type of many-body problem. By transforming to normal-mode coordinates the hamiltonian can be reduced to that of a set of independent oscillators and can thus be diagonalized exactly. But such exact closed solutions can only be obtained in quantum mechanics in exceptional cases, and it is therefore worth studying this problem by means of a more general formalism which can also be applied to more complicated cases, such as the important but much more difficult problem of the lattice dynamics of an *anharmonic* crystal. This is one reason for studying the harmonic problem by means of Green's functions. Another is that the Green's function approach provides a unified systematic method for calculating various quantities of physical interest. Thus we shall see that it gives the ground state energy of the system, or—more generally—the free energy at non-zero temperature, from which the thermodynamic properties such as the specific heat can be obtained.

In addition to these equilibrium properties, Green's functions also provide information on the excitation energies of a system. For example, scattering processes correspond to excitations in which one particle is added to the system, and we shall consider the scattering of thermal neutrons by the lattice vibrations as an example of this. There is also an important class of excitations in which the particle number is conserved; the theory of linear response to an externally applied field describes such excitations and, as we shall see in later chapters, provides expressions for dielectric response functions, electrical conductivities and magnetic susceptibilities in terms of appropriately defined Green's functions. Green's functions thus make it possible to evaluate measurable thermodynamic and transport properties by studying the response of a system to simple perturbations. This approach is particularly important for interacting many-particle systems, where the complete set of wave functions and energy levels is highly complex but is not in fact needed for studying properties related to experiment.

In the first three chapters we use the harmonic lattice as an exactly soluble example to study and compare the principal methods for calculating Green's functions. The physical phenomena produced by the interaction between atoms in this case are the propagating modes—excitation energy is handed on from one atom to the next so as to produce traveling sound waves (quantized as phonons). We shall find later that in more complicated cases also there exist excitations which take the form of sets of coupled oscillators—plasmon excitations in the case of an electron gas with Coulomb interactions (Chap. 6), and spin waves in the case of an insulating magnet (Chap. 8). Thus the phonon Green's function serves also as a prototype for studying a number of other interacting systems of interest in solid state physics.

1.1 THE GROUND STATE ENERGY

We consider the hamiltonian

$$H = \sum_i \frac{\mathbf{p}_i^2}{2M} + \frac{1}{2} \sum_{i \neq j} V(\mathbf{X}_i - \mathbf{X}_j), \tag{1.1.1}$$

which describes a simple lattice composed of N identical interacting atoms of mass M situated at the points $\mathbf{X}_i$ ($i = 1, 2, \ldots, N$). It is assumed that the potential energy V is a two-body potential which depends only upon the relative positions of pairs of atoms. We write $\mathbf{X}_i = \mathbf{R}_i + \mathbf{u}_i$, where $\mathbf{R}_i$ is an undisplaced lattice point and the lattice displacement $\mathbf{u}_i$ is assumed to be small. In the harmonic approximation V is expanded in powers of the $\mathbf{u}_i$ as far as second-order terms. Since the expansion is about the equilibrium configuration the coefficients of the linear terms are zero, and the expansion is

$$H = \sum_i \frac{\mathbf{p}_i^2}{2M} + \frac{1}{2} \sum_{i \neq j} \sum_{\alpha\beta} \frac{1}{2!} (u_i^\alpha - u_j^\alpha)(u_i^\beta - u_j^\beta)\, \nabla^\alpha \nabla^\beta V, \tag{1.1.2}$$

where u_i^α is a cartesian component of $\mathbf{u}_i$.

To separate out the interaction between different atoms we rewrite the potential energy term as

$$\frac{1}{4} \sum_{i \neq j} \sum_{\alpha\beta} \{(u_i^\alpha u_i^\beta + u_j^\alpha u_j^\beta) - (u_j^\alpha u_i^\beta + u_j^\beta u_i^\alpha)\} \nabla^\alpha \nabla^\beta V,$$

where we have separated the terms which involve only the displacements of a single atom i from the interaction terms which depend on the displacements of two different atoms i and j. If the interaction terms are neglected we have the *Einstein model* of lattice vibrations, in which each atom vibrates independently with constant frequency (Ω_0, say) in the potential well of its neighbors' force fields. Because of the interactions there is a tendency for the motion of adjacent atoms to be correlated, and this leads to a change in the Einstein frequencies of the system.

We treat the Einstein oscillators as the unperturbed system with hamiltonian H_0, and the interactions between atoms as producing a perturbation H_1. Thus we have $H = H_0 + H_1$, with

$$H_0 = \sum_i \left\{ \frac{\mathbf{p}_i^2}{2M} + \tfrac{1}{2} M \Omega_0^2 \mathbf{u}_i^2 \right\}, \tag{1.1.3}$$

$$H_1 = -\frac{1}{2} \sum_{i \neq j} \sum_{\alpha\beta} u_i^\alpha u_j^\beta \nabla^\alpha \nabla^\beta V(\mathbf{R}_i - \mathbf{R}_j). \tag{1.1.4}$$

If H is regarded as a classical hamiltonian, the basic theoretical problem is the calculation of the change in the vibration frequency Ω_0 brought about by the interaction H_1. In quantum-mechanical terms each Einstein oscillator has zero-point energy $\frac{1}{2}\hbar\Omega_0$, and the classical problem can be reformulated as the problem of calculating the change in the zero-point (or ground state) energy of the system. It is convenient for this calculation to consider the more general hamiltonian

$$H(\lambda) = H_0 + \lambda H_1, \tag{1.1.5}$$

where $0 \leqslant \lambda \leqslant 1$; this can be interpreted as a gradual "switching on" of the interaction H_1 with parameter λ. The hamiltonian $H(0)$ thus corresponds to the unperturbed problem and $H(1)$ is the full hamiltonian of the original problem. We shall in future drop explicit reference to the superscripts α, β and replace (1.1.4) by

$$H_1 = \tfrac{1}{2} M \sum_{i \neq j} D_{ij} u_i u_j, \qquad \text{with} \qquad D_{ij} = D_{ji}. \tag{1.1.6}$$

Thus we ignore the vector character of the displacements $\mathbf{u}_i$. This is an oversimplification of the original problem which keeps the notation as simple as possible but retains the main features of the formalism.

The exact solution of the classical problem is obtained by writing down the equations of motion of the coupled system and determining the normal modes of vibration. This establishes the propagating modes. From the hamiltonian equations of motion for the hamiltonian $H(\lambda)$ we find that the displacements u_i satisfy the N coupled differential equations

$$\ddot{u}_i + \Omega_0{}^2 u_i = -\lambda \sum_j D_{ij} u_j. \tag{1.1.7}$$

In fact, because of translational invariance—the fact that every lattice site is equivalent to every other site—these N equations reduce to a single equation. [With the full hamiltonian (1.1.4) we have instead $3N$ coupled equations which can be reduced to 3.]

The mathematical expression of translational invariance is Bloch's theorem, according to which two displacements u_i and u_j differ only by a phase factor:

$$u_j = e^{i\mathbf{k}\cdot(\mathbf{R}_j - \mathbf{R}_i)} u_i.$$

In other words, any solution u_i must be of the form

$$u_i(t) = e^{i\mathbf{k}\cdot\mathbf{R}_i} \xi_{\mathbf{k}}(t), \tag{1.1.8}$$

where $\mathbf{k}$ is some wave-vector in the reciprocal lattice. The equation of motion thus reduces to

$$\ddot{\xi}_{\mathbf{k}} + \Omega_0{}^2 \xi_{\mathbf{k}} = -\lambda \xi_{\mathbf{k}} \sum_j D_{ij} e^{i\mathbf{k}\cdot\mathbf{R}_j}.$$

The sum on the right-hand side must be independent of the origin $\mathbf{R}_i$, and we write

$$\sum_j D_{ij} e^{i\mathbf{k}\cdot\mathbf{R}_j} = \sum_j D_{ij} e^{-i\mathbf{k}\cdot(\mathbf{R}_i - \mathbf{R}_j)} = D_{\mathbf{k}} \text{ say}, \tag{1.1.9}$$

where $D_{\mathbf{k}}$ is a Fourier transform of the force function D_{ij}. The inverse formula is

$$D_{ij} = \frac{1}{N} \sum_{\mathbf{k}} D_{\mathbf{k}} e^{i\mathbf{k}\cdot(\mathbf{R}_i - \mathbf{R}_j)}, \tag{1.1.10}$$

where the sum over $\mathbf{k}$ is over all allowed values in the Brillouin zone (determined by periodic boundary conditions). [In the three-dimensional case (assuming one atom per unit cell) of the hamiltonian (1.1.4) $D_{\mathbf{k}}$ becomes a 3 x 3 matrix $D_{\mathbf{k}}^{\alpha\beta}$, called the *dynamical matrix*.]

Now, for the normal mode of frequency $\Omega_{\mathbf{k}}$, the normal coordinate $\xi_{\mathbf{k}}(t)$ contains a time factor $\exp(i\Omega_{\mathbf{k}} t)$. Putting this into the equation of motion, we immediately obtain our result

$$\Omega_{\mathbf{k}}^2 = \Omega_0^2 + \lambda D_{\mathbf{k}}. \tag{1.1.11}$$

We thus have one normal mode of frequency $\Omega_{\mathbf{k}}$ for each allowed value of $\mathbf{k}$ in the Brillouin zone and, with our choice of (collective) coordinates, the strongly interacting system is reduced to a set of independent simple harmonic oscillators. $\Omega_{\mathbf{k}}$ is the frequency of an *acoustic* mode of vibration. It is easy to prove the relation $\Omega_0^2 + \sum_j D_{ij} = 0$ (this is the condition that the force on any atom must vanish when all atoms are displaced equally); hence, when $\lambda = 1$, we have $\Omega_{\mathbf{k}} = 0$ at $\mathbf{k} = 0$. The coupling between oscillators represented by the hamiltonian H_1 is thus a *large* perturbation at $\mathbf{k} = 0$, reducing the unperturbed frequency Ω_0 to zero. For small values of $|\mathbf{k}|$ $\Omega_{\mathbf{k}}$ is linear in $|\mathbf{k}|$ and hence represents sound-wave propagation.

When the same problem is formulated in quantum-mechanical terms, the solution is equally simple. The displacements u_i and conjugate momenta p_i are now operators, and the classical hamiltonian can be re-interpreted as the hamiltonian of a set of independent harmonic quantum oscillators. Each simple harmonic oscillation of frequency $\Omega_{\mathbf{k}}$ corresponds to a quantum oscillator whose ground state energy is the zero-point energy $\frac{1}{2}\hbar\Omega_{\mathbf{k}}$. When we wish to emphasize the particle aspect of these quantized acoustic excitations, we speak of *phonons* with which can be associated crystal momentum $\hbar\mathbf{k}$ and energy $\hbar\Omega_{\mathbf{k}}$. The total ground state energy of the lattice is obtained by summing over all modes $\mathbf{k}$, and the *change* in the ground state energy E_G due to the interaction H_1 is therefore

$$\Delta E_G = \tfrac{1}{2}\hbar \sum_{\mathbf{k}} (\Omega_{\mathbf{k}} - \Omega_0), \tag{1.1.12}$$

where $\Omega_{\mathbf{k}}$ is given by (1.1.11) with $\lambda = 1$. (We choose units such that $\hbar = 1$ in future.)

1.2. THE GROUND STATE ENERGY AS AN INTEGRAL OVER COUPLING CONSTANT

An exact solution in closed form, such as can be obtained foı the harmonic oscillator by direct diagonalization of the hamiltonian, is only possible in exceptionally simple cases. We now want to rederive the ground state energy of the harmonic lattice by more general quantum-theoretical methods which, although more complicated, can also be applied to a wide range of problems for which the above type of closed solution cannot be obtained. The method consists of a calculation of ΔE_G using H_1 as a perturbation. Although the strength of the perturbation may be small, it connects together all the particles in the system. It is then not sufficient, as in elementary applications of perturbation theory to one- or two-particle systems, to calculate only the first few terms of the perturbation series and to assume that the higher terms are negligible. In fact the essential features of the perturbation (in the present case the existence of propagating modes) come from the high-order terms in the perturbation series, and we must therefore work to all orders in H_1. For this purpose we use a general formula for ΔE_G which is particularly suitable for perturbation theory. (The formula seems to be due to Pauli, and has been rederived many times.) When the hamiltonian is $H_0 + \lambda H_1$, the ground state energy E_G and the ground state eigenfunction $|\Psi_G\rangle$ are functions of λ, and $E_G(\lambda)$ is the expectation value

$$E_G(\lambda) = \langle \Psi_G(\lambda)|H(\lambda)|\Psi_G(\lambda)\rangle. \tag{1.2.1}$$

Hence

$$\frac{\partial E_G}{\partial\lambda} = \langle \Psi_G(\lambda)|H_1|\Psi_G(\lambda)\rangle + \left\langle \frac{\partial\Psi_G}{\partial\lambda}\,|H(\lambda)|\,\Psi_G\right\rangle + \left\langle \Psi_G|H(\lambda)|\,\frac{\partial\Psi_G}{\partial\lambda}\right\rangle. \tag{1.2.2}$$

The last two terms reduce to

$$E_G(\lambda)\left\{\left\langle\frac{\partial\Psi_G}{\partial\lambda}\middle|\Psi_G\right\rangle+\left\langle\Psi_G\middle|\frac{\partial\Psi_G}{\partial\lambda}\right\rangle\right\} = E_G(\lambda)\,\frac{\partial}{\partial\lambda}\,\langle\Psi_G|\Psi_G\rangle = 0,$$

assuming $|\Psi_G\rangle$ to be normalized. Therefore, integrating with respect to λ,

$$\Delta E_G = E_G(1) - E_G(0) = \int_0^1 d\lambda \langle \Psi_G(\lambda) | H_1 | \Psi_G(\lambda) \rangle. \tag{1.2.3}$$

The change in ground state energy is thus expressed entirely in terms of a matrix element of the perturbation H_1, but it is necessary to know this matrix element for all values of the coupling constant λ.

The formula (1.2.3) is general. For the lattice problem H_1 is given by Eq. (1.1.6), and thus

$$\Delta E_G = \tfrac{1}{2} M \sum_{i \neq j} D_{ij} \int_0^1 d\lambda \langle \Psi_G(\lambda) | u_i u_j | \Psi_G(\lambda) \rangle. \tag{1.2.4}$$

1.3. THE NEUTRON SCATTERING CROSS-SECTION

The matrix element in (1.2.4), which gives the change in the ground state energy, is a measure of the *correlation* brought about by the interaction between the displacements of two atoms, u_i and u_j. This is an example of a quantum-mechanical *correlation function*. We come across a more complicated, time-dependent, correlation function, which is more directly related to observable phenomena, when we consider the scattering cross-section for plane waves of the lattice of phonons. We shall discuss specifically the coherent inelastic scattering of slow neutrons, but a similar discussion can be given of the scattering of light (x-rays) by a crystal. Our discussion follows that given in Kittel (1963, Chap. 19), where further details can be found. The formulas for the scattering cross-sections are due to van Hove (1954).

We work in the first Born approximation, in which the incident and scattered particles are represented by plane waves $\exp(i\mathbf{K}\cdot\mathbf{x})$ and $\exp(i\mathbf{K}'\cdot\mathbf{x})$. The inelastic differential scattering cross-section per unit solid angle per unit energy range is

$$\frac{d^2\sigma}{d\Omega\, d\omega} = \sum_F \frac{K'}{K} \left(\frac{M}{2\pi}\right)^2 |\langle \mathbf{K}'\psi_F | H' | \mathbf{K}\psi_G \rangle|^2 \, \delta(\omega + E_G - E_F), \tag{1.3.1}$$

where H' is the interaction between particle and target, ω is the energy transfer to the target, and M is the reduced mass of the particle. It is assumed here that the target is initially in its ground state $|\psi_G\rangle$ with energy E_G; $|\psi_F\rangle$ is the final state with energy E_F. Thus we are assuming that the equilibrium state of the target corresponds to zero temperature; the extension to finite temperatures is developed in Chap. 2. If we assume further that H' is the sum of two-particle interactions between the incident particle and the target particles:

$$H' = \sum_j V(\mathbf{x} - \mathbf{X}_j), \qquad j = 1, 2, \ldots, N, \tag{1.3.2}$$

we have

$$\langle \mathbf{K}'\psi_F | H' | \mathbf{K}\psi_G \rangle = \langle \psi_F | \int d^3x\, e^{i\mathbf{q}\cdot\mathbf{x}} H' | \psi_G \rangle$$

$$= V_{\mathbf{q}} \sum_j \langle \psi_F | e^{i\mathbf{q}\cdot\mathbf{X}_j} | \psi_G \rangle, \tag{1.3.3}$$

where $\mathbf{q} = \mathbf{K} - \mathbf{K}'$ is the change in wave-vector of the incident particle and

$$V_{\mathbf{q}} = \int d^3x\, e^{i\mathbf{q}\cdot\mathbf{x}}\, V(\mathbf{x}) \tag{1.3.4}$$

is the Fourier transform of the scattering potential.

We now have

$$\frac{d^2\sigma}{d\Omega\, d\omega} = \sum_F \frac{K'}{K} \left(\frac{M}{2\pi}\right)^2 |V_{\mathbf{q}}|^2 \sum_{jl} \langle \psi_G | e^{-i\mathbf{q}\cdot\mathbf{X}_j} | \psi_F \rangle$$

$$\times \langle \psi_F | e^{i\mathbf{q}\cdot\mathbf{X}_l} | \psi_G \rangle \delta(\omega + E_G - E_F). \tag{1.3.5}$$

This can be simplified by introducing the Fourier representation of the delta-function:

$$\delta(\omega) = \frac{1}{2\pi} \int_{-\infty}^{\infty} e^{i\omega t}\, dt,$$

where t can be interpreted as a time variable. We can now write

$$e^{i(E_G - E_F)t} \langle \psi_G | e^{-i\mathbf{q}\cdot\mathbf{X}_j} | \psi_F \rangle = \langle \psi_G | e^{iHt}\, e^{-i\mathbf{q}\cdot\mathbf{X}_j}\, e^{-iHt} | \psi_F \rangle, \tag{1.3.6}$$

where $H = H_0 + \lambda H_1$ is the hamiltonian describing the target. The operator exponential is to be interpreted as an infinite series, and thus Eq. (1.3.6) may be rewritten with the time factors in the exponent:

$$\langle \psi_G | e^{-i\mathbf{q}\cdot \tilde{\mathbf{X}}_j(t)} | \psi_F \rangle, \tag{1.3.7}$$

where the time-dependent quantity

$$\tilde{\mathbf{X}}_j(t) = e^{iHt}\mathbf{X}_j\, e^{-iHt} \tag{1.3.8}$$

is a *Heisenberg operator* (see Sec. 1.4 below).

It is now possible to sum over the final states $|\psi_F\rangle$ in (1.3.5), using the completeness relation

$$\sum_F |\psi_F\rangle\langle\psi_F| = 1. \tag{1.3.9}$$

The differential cross-section is thus given by

$$\frac{d^2\sigma}{d\Omega\, d\omega} = \frac{K'}{2\pi K}\left(\frac{M}{2\pi}\right)^2 |V_\mathbf{q}|^2$$

$$\times \int_{-\infty}^{\infty} e^{i\omega t}\, dt \sum_{jl} \langle \psi_G | e^{-i\mathbf{q}\cdot\tilde{\mathbf{X}}_j(t)}\, e^{i\mathbf{q}\cdot\tilde{\mathbf{X}}_l(0)} | \psi_G\rangle. \tag{1.3.10}$$

Note that the operator exponentials do not commute at different times and cannot be combined into a single exponential. We thus see that the cross-section depends upon the two-body potential and upon the time Fourier transform of the time correlation function

$$\sum_{jl} \langle \psi_G | e^{-i\mathbf{q}\cdot\tilde{\mathbf{X}}_j(t)}\, e^{i\mathbf{q}\cdot\tilde{\mathbf{X}}_l(0)} | \psi_G\rangle.$$

Writing $\mathbf{X}_i = \mathbf{R}_i + \mathbf{u}_i$, where $\mathbf{R}_i$ is as before the undisplaced lattice position (not an operator), the correlation function is

$$F(\mathbf{q}, t) = \sum_{jl} e^{i\mathbf{q}\cdot(\mathbf{R}_l - \mathbf{R}_j)} \langle \psi_G | e^{-i\mathbf{q}\cdot\tilde{\mathbf{u}}_j(t)}\, e^{i\mathbf{q}\cdot\tilde{\mathbf{u}}_l(0)} | \psi_G\rangle. \tag{1.3.11}$$

An important application of this expression is to the discussion of recoilless (Mossbauer) emission of γ-rays from nuclei embedded in a crystal [see Kittel (1963), Chap. 20].

We now make an important approximation. We shall see that $F(\mathbf{q}, t)$ describes inelastic scattering in which phonons are emitted or absorbed by the scattered neutron. Usually the one-phonon processes, in which a single phonon is emitted or absorbed, dominate. We shall confine attention to these. This allows us to expand the exponentials in Eq. (1.3.11) and to keep only terms in the matrix element which involve products of two u operators. (The leading term, in which the exponentials are replaced by unity, represents elastic scattering and is of no interest here.) Furthermore, we are only interested in terms which involve u-operators at *different* times (the other terms are time-independent and lead to an overall factor—the "Debye–Waller factor"—multiplying the magnitude of the scattering cross-section. It is also easily seen that the expectation values of *single* u-operators are zero.) The higher-order terms in the expansion of (1.3.11), involving products of more than two u-operators, correspond to multi-phonon processes and will be neglected. Finally, we suppose for simplicity that the polarization of the phonon mode of interest is parallel to $\mathbf{q}$, so that the product $\mathbf{q} \cdot \mathbf{u}_j$ can be replaced by the scalar qu_j.

With all these approximations (and omitting the Debye–Waller factor), the correlation function (1.3.11) reduces to

$$F(\mathbf{q}, t) \simeq q^2 \sum_{jl} e^{i\mathbf{q}\cdot(\mathbf{R}_l - \mathbf{R}_j)} \langle \psi_G | \tilde{u}_j(t) \tilde{u}_l(0) | \psi_G \rangle, \tag{1.3.12}$$

where $\tilde{u}(t)$ is a Heisenberg operator as in (1.3.8).

1.4. THE GREEN'S FUNCTION AND ITS EQUATION OF MOTION

In Eqs. (1.2.4) and (1.3.12) we now have two examples, one time-independent and the other time-dependent, of correlation functions of interest in the lattice problem. These, and similar correlation functions, are conveniently calculated by means of Green's functions.

First we recall some general quantum-theoretical formulas [see Messiah (1961, Chap. VIII)]. In the Schrödinger picture or "representation" of quantum mechanics the vector $|\Psi_S(t)\rangle$ representing the state of a system at time t evolves in time in accordance with the Schrödinger equation

$$i \frac{\partial |\Psi_S\rangle}{\partial t} = H |\Psi_S\rangle. \tag{1.4.1}$$

This has the formal solution (assuming that H does not depend explicitly on the time)

$$|\Psi_S(t)\rangle = e^{-iH(t-t_0)}|\Psi_S(t_0)\rangle, \tag{1.4.2}$$

which describes the evolution of the system from time t_0 to time t. The physical observables are time-independent operators $\mathcal{O}_S$ in the vector space of the Schrödinger vectors $|\Psi_S\rangle$. The Heisenberg picture or "representation" is an entirely equivalent formulation, obtained from the Schrödinger picture by a unitary transformation to new state vectors $|\Psi_H\rangle$ and observables $\mathcal{O}_H$, defined by

$$|\Psi_H\rangle = e^{iH(t-t_0)}|\Psi_S(t)\rangle = |\Psi_S(t_0)\rangle, \tag{1.4.3}$$

$$\mathcal{O}_H(t) = e^{iH(t-t_0)}\,\mathcal{O}_S\,e^{-iH(t-t_0)}. \tag{1.4.4}$$

(It is immediately verified that this transformation leaves all matrix elements invariant.) Thus the state vectors are now stationary, and the time dependence has been transferred to the observables. Note that the hamiltonian H is unchanged by the transformation (1.4.4). Instead of the Schrödinger equation we now have the Heisenberg equation of motion which determines the time development of the Heisenberg operators. This is obtained by differentiating (1.4.4):

$$i\frac{d\mathcal{O}_H}{dt} = [\mathcal{O}_H, H], \tag{1.4.5}$$

where the square bracket denotes the commutator $\mathcal{O}_H H - H\mathcal{O}_H$. This is to be solved with the initial condition $\mathcal{O}_H(t_0) = \mathcal{O}_S$.

We now define the single-particle *Green's function* or *propagator* for the phonon problem as

$$G_{ij}(t-t') = -i\langle T[\tilde{u}_i(t)\tilde{u}_j(t')]\rangle_\lambda. \tag{1.4.6}$$

Here the Heisenberg picture is used for operators and state-vectors: thus $\tilde{u}_i(t)$ is a time-dependent operator defined as in Eqs. (1.3.8) or (1.4.4) (with $t_0 = 0$), H being the hamiltonian $H_0 + \lambda H_1$ of the lattice with coupling constant λ, and the expectation value $\langle\cdots\rangle_\lambda$ refers to the exact ground state of the system described by the time-independent Heisenberg vector $|\Psi_G(\lambda)\rangle$ which is assumed to be normalized. The symbol T is Dyson's time-ordering operator which rearranges the product of two time-dependent operators so that the operator which refers to the later time always stands on the left:

$$T[A(t)B(t')] = A(t)B(t') \qquad (t > t')$$
$$= B(t')A(t) \qquad (t < t'). \tag{1.4.7}$$

In the language of second quantization (see Appendix 1) the $\tilde{u}_i$ are field operators, and the Green's function is an expectation value of a product of such operators; as such it is simply (in general) a function of space and time coordinates. In fact, in the present problem the $\tilde{u}_i$ are localized operators containing no space dependence, and G_{ij} is thus a function only of time. Moreover, if H is independent of t, G is a function of the difference $t - t'$ of its arguments, as indicated in Eq. (1.4.6); this follows at once from the analysis of Appendix 2. The presence of the T symbol in the definition of G introduces the discontinuity at $t = t'$ which is characteristic of a Green's function. More generally, if the Green's function is formed from field operators which are expanded in terms of space-dependent basis functions such as plane waves [see Appendix 1, Eq. (A.1.13)], G will be a function of space coordinates $\mathbf{x}$, $\mathbf{x}'$ as well as of t and t'.

G_{ij} as defined in Eq. (1.4.6) is an example of a *time-ordered* or *causal* Green's function. It describes the propagation of disturbances in which a single particle is added to the many-particle equilibrium system at some time instant and removed again at a later time. In the present example G represents the propagation of phonon modes through the crystal lattice. Other Green's functions may de defined, and examples of these will be encountered later: in particular, the *retarded* single-particle Green's function is, for the present problem, given by

$$G_{ij}^R(t - t') = -i\theta(t - t')\langle[\tilde{u}_i(t), \tilde{u}_j(t')]\rangle_\lambda, \tag{1.4.8}$$

and the *advanced* Green's function is

$$G_{ij}^A(t - t') = i\theta(t' - t)\langle[\tilde{u}_i(t), \tilde{u}_j(t')]\rangle_\lambda, \tag{1.4.9}$$

where $\theta(t)$ is the *unit function* ($\theta(t) = 1$ for $t > 0$, $\theta(t) = 0$ for $t < 0$). Thus, for fixed t', the retarded function exists only for times t later than t', and the advanced function exists only for times t earlier than t'. We shall find when we consider the theory of linear response in later chapters that response functions such as electrical conductivity (Chap. 5), dielectric response (Chap. 6), and magnetic susceptibility (Chap. 7) are obtained in the form of retarded Green's functions. These are directly measurable quantities. Furthermore, retarded and advanced Green's functions have the advantage of possessing particularly simple analytical properties. On the other hand, we shall see in Chap. 3 that the time-ordered functions are particularly well adapted to evaluation by means of diagrammatic perturbation theory. In the present problem, where everything can be calculated exactly, we could use any of these Green's functions for our analysis, and we shall continue to work with the time-ordered form. There are general relations between the different Green's functions, some of which are derived in Appendix 2.

We now see that the expectation value required for the ground state energy (1.2.4) is the limit, as t tends to zero from positive values, of $\langle T[\tilde{u}_i(t)\tilde{u}_j(0)]\rangle_\lambda$, and thus we can express ΔE_G in terms of the Green's function as

$$\Delta E_G = \lim_{t\to 0+} \tfrac{1}{2}iM \sum_{i\neq j} D_{ij} \int_0^1 d\lambda G_{ij}(t). \tag{1.4.10}$$

Furthermore, we see that the time correlation function (1.3.12) required for the neutron scattering cross-section can, for $t > 0$, be expressed in terms of G as

$$F(\mathbf{q}, t) = iq^2 \sum_{jl} e^{i\mathbf{q}\cdot(\mathbf{R}_l - \mathbf{R}_j)} G_{jl}(t) \tag{1.4.11}$$

(the expression so obtained then holds for all t).

We are now ready to consider the calculation of G_{ij}. Several general methods are available. In this chapter we discuss one of these methods, where we proceed by setting up and solving a differential equation (the "equation of motion") for G. This equation is essentially an inhomogeneous form of the Schrödinger equation. The equation-of-motion approach emphasizes the similarity between G and the Green's functions familiar in classical vibration problems.

Since G depends on $t - t'$ there is no loss of generality in putting $t' = 0$. In differentiating, we must be careful to treat correctly the discontinuity at $t = 0$. We write

$$T[\tilde{u}_i(t)\tilde{u}_j(0)] = \theta(t)\tilde{u}_i(t)\tilde{u}_j(0) + \theta(-t)\tilde{u}_j(0)\tilde{u}_i(t),$$

thus splitting the time-ordered Green's function into the sum of retarded and advanced terms. The derivative $d\theta(t)/dt$ is the delta-function $\delta(t)$, and thus

$$i\frac{dG_{ij}(t)}{dt} = \delta(t)\langle \tilde{u}_i(t)\tilde{u}_j(0) - \tilde{u}_j(0)\tilde{u}_i(t)\rangle + \left\langle T\left[\frac{d\tilde{u}_i(t)}{dt}\,\tilde{u}_j(0)\right]\right\rangle.$$

The first term on the right vanishes, since, at equal times $t = 0$, $[u_i, u_j] = 0$. The time derivative in the second term can be obtained from the Heisenberg equation of motion

$$i\frac{d\tilde{u}_i}{dt} = [\tilde{u}_i, H] = e^{iHt}[u_i, H]\; e^{-iHt}.$$

Using $[p_i, u_j] = -i\delta_{ij}$, this is

$$e^{iHt}[u_i, \sum_j p_j^2/2M]\; e^{-iHt} = e^{iHt}(ip_i/M)\; e^{-iHt} = i\tilde{p}_i(t)/M,$$

and thus

$$i\frac{dG_{ij}(t)}{dt} = \frac{1}{M}\langle T[\tilde{p}_i(t)\tilde{u}_j(0)]\rangle; \tag{1.4.12}$$

we see that a new Green's function has appeared on the right. Differentiating again,

$$i\frac{d^2G_{ij}(t)}{dt^2} = \frac{\delta(t)}{M}\langle \tilde{p}_i(t)\tilde{u}_j(0) - \tilde{u}_j(0)\tilde{p}_i(t)\rangle$$

$$+\frac{1}{M}\left\langle T\left[\frac{d\tilde{p}_i(t)}{dt}\,\tilde{u}_j(0)\right]\right\rangle. \tag{1.4.13}$$

The first term on the right is

$$\frac{\delta(t)}{M}\langle [p_i, u_j]\rangle = -\frac{i}{M}\,\delta_{ij}\,\delta(t),$$

and to evaluate the second term we require the commutator

$$[p_i, H] = [p_i, \tfrac{1}{2}M\Omega_0^2u_i^2 + \tfrac{1}{2}\lambda M \sum_{jk} D_{jk}\,u_ju_k]$$

$$= -iM\Omega_0^2u_i - i\lambda M \sum_k D_{ik}\,u_k.$$

Since this involves only terms *linear* in the u_i, we see on substitution in Eq. (1.4.13) that the original Green's function now reappears on the right-hand side, and we obtain the closed set of coupled differential equations

$$\left(-\frac{d^2}{dt^2}-\Omega_0{}^2\right)G_{ij}(t)=\frac{1}{M}\,\delta_{ij}\,\delta(t)+\lambda\sum_k D_{ik}G_{kj}(t). \tag{1.4.14}$$

In more general cases, differentiation of a Green's function will not reproduce the original function on the right-hand side, and instead of a closed set of equations we obtain an infinite chain of equations involving Green's functions of successively higher orders. We then have to approximate at some stage to break off the chain, usually by some sort of "decoupling" procedure in which a high-order Green's function is expressed approximately as a product of Green's functions of lower order. The simple closed result obtained in the present problem depends on the special oscillator form of the hamiltonian.

1.5. THE ITERATION SOLUTION FOR G

If we know the solution of the equation of motion (1.4.14) for $\lambda = 0$ (corresponding to the unperturbed hamiltonian), we can obtain the general solution by iteration in powers of λ. For $\lambda = 0$, Eq. (1.4.14) reduces to a differential equation, with a delta-function term representing a unit applied impulse, of a form familiar in elementary discussions of Green's functions. The solution of this is clearly diagonal in the suffices i and j, of the form $G^0(t)\,\delta_{ij}$. The differential equation

$$\left(\frac{d^2}{dt^2}+\Omega_0{}^2\right)G^0(t)=-\frac{1}{M}\,\delta(t) \tag{1.5.1}$$

by itself does not, however, determine $G^0(t)$ completely. It tells us that, in each sub-interval $t < 0$ and $t > 0$, G^0 is some linear combination of the elementary solutions $\exp(\pm i\Omega_0 t)$ of the homogeneous equation. It also follows from Eq. (1.5.1) that G^0 is continuous at $t = 0$ and that its first derivative has a discontinuity there of magnitude $(-1/M)$. The time-ordered, retarded and advanced Green's functions all satisfy Eq. (1.5.1) and have these properties, and the distinction between them comes from initial (and final) conditions at $t = \pm\infty$.

In fact, to obtain $G^0(t)$ explicitly it is simplest not to use the differential equation, but to go back to the original definition of the Green's function. Thus, to obtain the time-ordered function, we have to evaluate

$$G^0(t) = -i\langle T[\tilde{u}_i(t)\tilde{u}_i(0)]\rangle_{\lambda=0}, \tag{1.5.2}$$

with

$$\tilde{u}_i(t) = e^{iH_0 t}u_i\, e^{-iH_0 t}, \qquad H_0 = \sum_i \left(\frac{p_i^2}{2M} + \tfrac{1}{2}M\Omega_0^2 u_i^2\right).$$

This is a standard harmonic oscillator problem, and we introduce *annihilation* and *creation operators* B_i, $B_i^\dagger$ in the usual way [Messiah (1961, Chap. XII)]. Put $\xi_i = u_i\sqrt{(M\Omega_0)}$, $p_{\xi_i} = p_i/\sqrt{(M\Omega_0)}$, so that

$$H_0 = \sum_i \tfrac{1}{2}\Omega_0(\xi_i^2 + p_{\xi_i}^2),$$

and write

$$B_i = \frac{1}{\sqrt{2}}(\xi_i + ip_{\xi_i}), \qquad B_i^\dagger = \frac{1}{\sqrt{2}}(\xi_i - ip_{\xi_i}). \tag{1.5.3}$$

These operators satisfy the boson commutation rules

$$[B_i, B_j^\dagger] = \delta_{ij}, \qquad [B_i, B_j] = [B_i^\dagger, B_j^\dagger] = 0, \tag{1.5.4}$$

and the hamiltonian becomes

$$H_0 = \sum_i \Omega_0(B_i^\dagger B_i + \tfrac{1}{2}). \tag{1.5.5}$$

The time dependence of the B_i, $B_i^\dagger$ is obtained from the Heisenberg equations of motion; thus

$$i\frac{dB_i}{dt} = [B_i, H_0] = \Omega_0 B_i, \quad \text{giving } B_i(t) = B_i\, e^{-i\Omega_0 t}, \tag{1.5.6}$$

and similarly

$$B_i^\dagger(t) = B_i^\dagger\, e^{i\Omega_0 t}, \tag{1.5.7}$$

where $B_i = B_i(0)$, $B_i^\dagger = B_i^\dagger(0)$. In terms of the B_i operators,

$$u_i = \frac{1}{\sqrt{(2M\Omega_0)}} (B_i + B_i^\dagger), \tag{1.5.8}$$

and thus

$$G^0(t) = -\frac{i}{2M\Omega_0} \langle 0 | T[(B_i(t) + B_i^\dagger(t))(B_i(0) + B_i^\dagger(0))] | 0 \rangle.$$

Here $|0\rangle$ is the normalized ground state of H_0, with the property $B_i|0\rangle = 0$ and $\langle 0|B_i^\dagger = 0$ for all i. Hence, for $t > 0$,

$$G^0(t) = -\frac{i}{2M\Omega_0} e^{-i\Omega_0 t} \langle 0 | B_i B_i^\dagger | 0 \rangle$$

$$= -\frac{i}{2M\Omega_0} e^{-i\Omega_0 t} \langle 0 | 1 + B_i^\dagger B_i | 0 \rangle = -\frac{i}{2M\Omega_0} e^{-i\Omega_0 t}. \tag{1.5.9}$$

Similarly, for $t < 0$,

$$G^0(t) = -\frac{i}{2M\Omega_0} e^{i\Omega_0 t}, \tag{1.5.10}$$

so that, for all t, we have the result

$$G^0(t) = -\frac{i}{2M\Omega_0} e^{-i\Omega_0 |t|}. \tag{1.5.11}$$

This undamped traveling wave, of frequency Ω_0, is just the Green's function representing the response of an undamped harmonic oscillator of natural frequency Ω_0 driven by a unit impulse at time $t = 0$.

For comparison we quote the expressions for the unperturbed retarded and advanced Green's functions, which may similarly be calculated from the definitions (1.4.8) and (1.4.9). We obtain the standing waves

$$G^{0,R}(t) = -\frac{\theta(t)}{M\Omega_0} \sin \Omega_0 t, \quad G^{0,A}(t) = \frac{\theta(-t)}{M\Omega_0} \sin \Omega_0 t. \tag{1.5.12}$$

Having obtained the unperturbed solution $G^0(t)\,\delta_{ij}$ of the equation of motion (1.4.14), we can now proceed to derive the complete solution. We could of course solve Eq. (1.4.14) directly in closed form, but as we are illustrating a general technique we prefer to use a more general approach in which the equation is solved by iteration, as a power series in the parameter λ.[1]

The Green's function $G_{ij}(t)$ is now to be obtained as the solution of the set of differential equations (1.4.14) which satisfies the boundary (or "initial") condition that $G_{ij}(t) = G^0(t)\,\delta_{ij}$ when $\lambda = 0$, where $G^0(t)$ is given by Eq. (1.5.11). This defines a boundary-value problem which can equally well be formulated in terms of an equivalent set of *integral equations.* These integral equations incorporate the boundary condition at $\lambda = 0$ and are thus particularly well adapted to solution by iteration in λ. To set up the integral equations we need only observe, as is clear from (1.5.1), that the function $G^0(t - t')$ is just the Green's function for the unperturbed operator $M(d^2/dt^2 + \Omega_0{}^2)$. The solution of the full equations (1.4.14) can thus be expressed as

$$G_{ij}(t) = G^0(t)\,\delta_{ij} + \lambda M \int_{\infty}^{\infty} dt'\, G^0(t - t') \sum_k D_{ik} G_{kj}(t'). \qquad (1.5.13)$$

It is easily checked directly that these integral equations are equivalent to the differential equations (1.4.14) together with the boundary condition at $\lambda = 0$.

The iterative solution of Eq. (1.5.13), obtained by successive approximation, is

$$G_{ij}(t) = G^0(t)\,\delta_{ij} + \lambda M \int_{-\infty}^{\infty} dt'\, G^0(t - t') D_{ij} G^0(t')$$

$$+ (\lambda M)^2 \sum_{i'} \int_{-\infty}^{\infty} \int_{-\infty}^{\infty} dt'\, dt''\, G^0(t - t') D_{ii'}\, G^0(t' - t'')$$

$$\times D_{i'j} G^0(t'') + \cdots \qquad (1.5.14)$$

[1] This is a more general method, in the sense that it will work in cases where no closed exact solution exists. On the other hand, the iteration series will converge only for sufficiently small values of λ, although the sum of the series defines a function which, by a process of analytic continuation, exists and is unique for general values of λ. In the present, exactly soluble, case the analytic properties of the solution are easily investigated. In more complicated problems, the justification of formal summations of perturbation series and their analytic continuation is not easily made rigorous, and the mathematical argument must usually be supported by physical reasoning.

The different terms in this series show how propagation develops through interaction between successively greater numbers of atoms, with excitation energy being handed on from site to site. To visualize this process, and to represent the terms in the series in a clear and simple way, we draw diagrams as shown in Fig. 1.1. These are Feynman diagrams

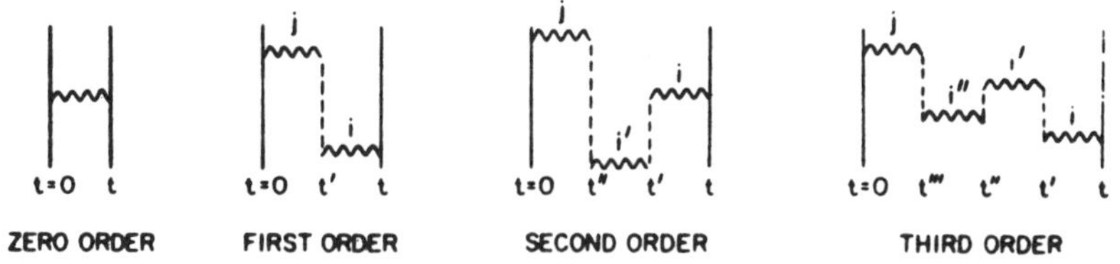

Fig. 1.1. Diagrammatic representation of the iteration solution for $G_{ij}(t)$.

of a specially simple kind, representing propagation (in time only) between times $t = 0$ and t (in general we have to deal with propagation in both time and space). We imagine (for $t > 0$) the time to be increasing from left to right. Each factor $G^0(t^\alpha - t^\beta)$ in any term of the perturbation series, which represents unperturbed propagation between successive scattering processes, is represented by a wavy line in a diagram connecting two times t^α and t^β and is labeled with the index of the atom carrying the excitation energy. At each time t^α at which a wavy line begins or ends, and excitation energy is handed on, there is an interaction factor D_{kl} represented by a dotted line in the diagram connecting two atoms k and l. The nth-order diagram contains n intermediate times between $t = 0$ and t, $(n + 1)$ wavy lines, n dotted lines and $(n - 1)$ intermediate atoms between i and j; a summation over all possible intermediate atoms and intermediate times is implied. Thus we can read off at once (reading from right to left) that the third-order diagram in Fig. 1.1 corresponds to the term

$$(\lambda M)^3 \sum_{i'i''} \int_{-\infty}^{\infty} \int_{-\infty}^{\infty} \int_{-\infty}^{\infty} dt'\, dt''\, dt'''\, G^0(t - t')D_{ii'}G^0(t' - t'')D_{i'i''}$$

$$\times\, G^0(t'' - t''')D_{i''j}G^0(t''' - 0)$$

in the iteration series.

1.6. SUMMATION OF THE ITERATION SERIES

Having seen how to obtain all the terms in the iteration series, we now show that in the present case the series can be summed exactly in closed form. It is simplest to return to the integral equation (1.5.13). The equation is exactly soluble because of the *convolution* form of the last term: if we introduce the time Fourier transforms of the functions involved, defined by the reciprocal formulas

$$G(t) = \frac{1}{2\pi} \int_{-\infty}^{\infty} G(\omega)\, e^{-i\omega t}\, d\omega, \qquad G(\omega) = \int_{-\infty}^{\infty} G(t)\, e^{i\omega t}\, dt, \tag{1.6.1}$$

and take the Fourier transform of the integral equation, the last term splits into a product of factors. The transformed equation is

$$G_{ij}(\omega) = G^0(\omega)\, \delta_{ij} + \lambda M G^0(\omega) \sum_k D_{ik} G_{kj}(\omega). \tag{1.6.2}$$

This can again be solved by iteration, the terms in the iteration series being the Fourier transforms of the terms in Eq. (1.5.14). Thus (1.5.14) transforms to

$$\begin{aligned} G_{ij}(\omega) = {} & G^0(\omega)\, \delta_{ij} + \lambda M G^0(\omega) D_{ij} G^0(\omega) \\ & + (\lambda M)^2 \sum_{i'} G^0(\omega) D_{ii'} G^0(\omega) D_{i'j} G^0(\omega) + \cdots. \end{aligned} \tag{1.6.3}$$

Because of the translational invariance of the lattice the terms in this series can be simplified by making a further transformation to **k** space. We introduce a Fourier expansion of $G_{ij}(\omega)$ in the lattice space, analogous to the Fourier expansion, Eqs. (1.1.9) and (1.1.10), of D_{ij},

$$G_{ij}(\omega) = \frac{1}{N} \sum_{\mathbf{k}} G_{\mathbf{k}}(\omega)\, e^{i\mathbf{k}\cdot(\mathbf{R}_i - \mathbf{R}_j)},$$

$$G_{\mathbf{k}}(\omega) = \sum_i G_{ij}(\omega)\, e^{-i\mathbf{k}\cdot(\mathbf{R}_i - \mathbf{R}_j)}. \tag{1.6.4}$$

This uncouples the summations in Eq. (1.6.3), and the transformed series is

$$G_{\mathbf{k}}(\omega) \doteq G^0(\omega) + \lambda M G^0(\omega) D_{\mathbf{k}} G^0(\omega) + (\lambda M)^2 G^0(\omega) D_{\mathbf{k}} G^0(\omega) D_{\mathbf{k}} G^0(\omega) + \cdots . \tag{1.6.5}$$

This is now a simple geometric series, with sum to infinity

$$G_{\mathbf{k}}(\omega) = \frac{G^0(\omega)}{1 - \lambda M G^0(\omega) D_{\mathbf{k}}} . \qquad (1.6.6)^2$$

In the three-dimensional case, when $D_{\mathbf{k}}$ is the 3 x 3 matrix $D_{\mathbf{k}}^{\alpha\beta}$, Eq. (1.6.6) involves a matrix inversion which can be performed by diagonalizing $D_{\mathbf{k}}^{\alpha\beta}$ with normalized eigenvectors $\epsilon^\alpha(\mu, \mathbf{k})$ (μ a polarization index)

$$\sum_\beta D_{\mathbf{k}}^{\alpha\beta} \epsilon^\beta(\mu, \mathbf{k}) = D_{\mathbf{k}}{}^\mu \epsilon^\alpha(\mu, \mathbf{k}). \tag{1.6.7}$$

Thus the three-dimensional analog of Eq. (1.6.6) is

$$G_{\mathbf{k}}^{\alpha\beta}(\omega) = \sum_{\mu=1}^{3} G_{\mathbf{k}}{}^\mu(\omega)\, \epsilon^\alpha(\mu, \mathbf{k})\, \epsilon^\beta(\mu, \mathbf{k}),$$

where

$$G_{\mathbf{k}}{}^\mu(\omega) = G^0(\omega)/(1 - \lambda M G^0(\omega) D_{\mathbf{k}}{}^\mu). \tag{1.6.8}$$

We continue to discuss the simpler form (1.6.6).

We now have to evaluate the Fourier transform $G^0(\omega)$. This requires some care if we are to obtain well-defined results. The function $G^0(t)$, Eq. (1.5.11), is a wave of constant amplitude, and the Fourier transform of this does not exist for real ω. We therefore work with the *complex* Fourier transform [Titchmarsh (1937)], and regard ω as a complex variable with a non-zero (but infinitesimal) imaginary part. The imaginary part must be chosen negative for $t < 0$ and positive for $t > 0$, and thus we write

$$G^0(\omega) = -\frac{i}{2M\Omega_0} \int_{-\infty}^{0} e^{i(\omega + \Omega_0 - i\eta)t}\, dt - \frac{i}{2M\Omega_0} \int_0^{\infty} e^{i(\omega - \Omega_0 + i\eta)t}\, dt,$$

[2] We can easily check in this case, by direct solution of Eq. (1.6.2) in closed form, that this result is in fact valid for all λ.

where η is a positive infinitesimal. These integrals now converge, and we obtain

$$G^0(\omega) = -\frac{1}{2M\Omega_0}\frac{1}{\omega+\Omega_0-i\eta}+\frac{1}{2M\Omega_0}\frac{1}{\omega-\Omega_0+i\eta}$$
$$=\frac{1}{M(\omega^2-\Omega_0^2+i\eta)} \qquad (\eta > 0). \tag{1.6.9}$$

We note that the singularities of $G^0(\omega)$ in the complex ω plane are two poles at the unperturbed frequencies $\omega = \pm(\Omega_0 - i\eta)$, one lying just above and the other just below the real axis, as indicated in Fig. 1.2.

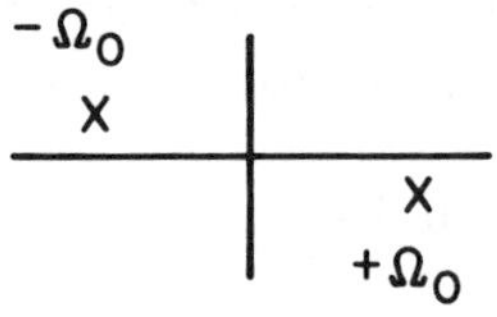

Fig. 1.2. Poles of the Fourier transform of the time-ordered Green's function in the ω plane.

To regain the t-dependent Green's function from Eq. (1.6.9), the Fourier integral is evaluated by contour integration. For $t > 0$, the contour consists of the real axis and a large semicircle in the lower half-plane. By Jordan's lemma, the integral over the semicircle tends to zero as the radius tends to infinity. The contour is described in the negative sense, and by the residue theorem the required integral is thus equal to $-2\pi i$ times the residue of the integrand at the pole $\omega = \Omega_0 - i\eta$ which lies inside the contour. This leads at once to the expression (1.5.9). For $t < 0$ the contour must be closed in the upper half-plane; we then pick up the pole at $\omega = -\Omega_0 + i\eta$ and regain Eq. (1.5.10).

With the value (1.6.9) for $G^0(\omega)$ we now at once have the exact result

$$G_{\mathbf{k}}(\omega) = \frac{1}{(G^0)^{-1}-\lambda MD_{\mathbf{k}}} = \frac{1}{M(\omega^2-\Omega_0^2-\lambda D_{\mathbf{k}}+i\eta)}$$
$$=\frac{1}{M(\omega^2-\Omega_{\mathbf{k}}^2+i\eta)}, \tag{1.6.10}$$

so that to obtain $G_{\mathbf{k}}$ from G^0 we simply have to replace Ω_0 by $\Omega_{\mathbf{k}}$, where $\Omega_{\mathbf{k}}^2 = \Omega_0^2 + \lambda D_{\mathbf{k}}$. This agrees with the result (1.1.11) given by the classical normal mode analysis.

A number of comments are worth making on the calculation which led to Eq. (1.6.10) and on the result:

(i) We see that the effect of the interaction on $G(\omega)$ is to shift the position of the poles from the unperturbed frequencies $\pm\Omega_0$ to the frequencies $\pm\Omega_{\mathbf{k}}$ of the phonon modes. This is a special case of a general property of the function $G_{\mathbf{k}}(\omega)$, valid also for fermion systems: $G_{\mathbf{k}}(\omega)$ is a meromorphic function with simple poles at the excitation frequencies (or energies) of the interacting system corresponding to wave-vector **k**. In the case of the phonon Green's function these excitation energies are of magnitude $\hbar\Omega_{\mathbf{k}}$ because the matrix elements connecting excited states with the ground state vanish unless the number of phonons changes by one.

(ii) The t-dependent Green's function $G_{\mathbf{k}}(t)$, obtained by contour integration from Eq. (1.6.10), is clearly an undamped wave

$$G_{\mathbf{k}}(t) = -\frac{i}{2M\Omega_{\mathbf{k}}}\, e^{-i\Omega_{\mathbf{k}}|t|} \qquad (1.6.11)$$

at the new frequency $\Omega_{\mathbf{k}}$. We see from the series (1.6.5) that, if it is broken off after any *finite* number of terms, $G_{\mathbf{k}}(\omega)$ will be a *polynomial* in $G^0(\omega)$ instead of a rational function, and its poles are then still at $\pm\Omega_0$. Thus, in any finite order (however large) of perturbation theory, $G_{\mathbf{k}}(t)$ still has the unperturbed frequency Ω_0. This demonstrates that, in order to obtain the propagating lattice modes and the phonon frequencies $\Omega_{\mathbf{k}}$, the perturbation theory must be taken to infinite order.

(iii) In the present problem the poles of $G_{\mathbf{k}}(\omega)$ lie just off the real axis in the ω plane, and $G_{\mathbf{k}}(t)$ represents undamped propagation. In more general cases (for example the problem of anharmonic phonons) a new phenomenon appears: the poles of $G_{\mathbf{k}}(\omega)$ may be situated at a *finite* distance from the real axis. Suppose there is a pole at $\Omega_{\mathbf{k}} - i\Gamma_{\mathbf{k}}$ in the lower half-plane ($\Gamma_{\mathbf{k}} > 0$): inverting the Fourier transform it is then seen that for $t > 0$ $G_{\mathbf{k}}(t)$ acquires a real exponential factor $\exp(-\Gamma_{\mathbf{k}}t)$. [If there are several poles in the lower half-plane, the dominant term in $G_{\mathbf{k}}(t)$ for large t arises from the pole which lies nearest to the real axis.] Thus $G_{\mathbf{k}}(t)$ has acquired a damping factor, and the excitation of frequency $\Omega_{\mathbf{k}}$ now has a finite lifetime $\Gamma_{\mathbf{k}}^{-1}$. In this way the Green's function can describe damping phenomena and finite lifetime effects. We shall encounter such phenomena in later chapters in connection with the electrical conductivity (Chap. 5) and the magnetic susceptibility (Chap. 7). The way in which the theory can lead to a modified analytic form of $G_{\mathbf{k}}(\omega)$ possessing finite damping is discussed further in Chap 5.

(iv) The poles of $G_{\mathbf{k}}(\omega)$, like those of $G^0(\omega)$, lie one just above and one just below the real axis. This analytic behavior corresponds to our definition of G as a causal (time-ordered) Green's function. We could equally well have worked with the retarded or advanced Green's functions, Eqs. (1.4.8) and (1.4.9), and we would then have obtained instead of Eq. (1.6.10) the expressions

$$G_{\mathbf{k}}{}^R(\omega) = \frac{1}{M(\omega^2 - \Omega_{\mathbf{k}}{}^2 + i\eta\omega)}, \qquad G_{\mathbf{k}}{}^A(\omega) = \frac{1}{M(\omega^2 - \Omega_{\mathbf{k}}{}^2 - i\eta\omega)}, \tag{1.6.12}$$

where η is again a positive infinitesimal. (We note that, for real ω, G^A is the complex conjugate of G^R.) These functions also have poles at the frequencies $\pm\Omega_{\mathbf{k}}$, but these poles now lie, in the case of $G^R(\omega)$, both just *below* the real axis, and, in the case of $G^A(\omega)$, both just *above* the real axis. Thus $G^R(\omega)$ is regular (free from singularities) in the whole upper half of the ω plane, and $G^A(\omega)$ is regular in the whole lower half of the ω plane. This is a general property of retarded and advanced Green's functions. It has the advantage in general that, if the function $G^R(\omega)$ is known in the lower half-plane, it can be obtained for all complex ω by the process of analytic continuation; similarly for $G^A(\omega)$.

(v) In the present problem the k-dependent Green's functions $G_{\mathbf{k}}(\omega)$, $G_{\mathbf{k}}(t)$ are given by simple expressions, of the same form as the unperturbed Green's functions but containing the phonon frequencies $\Omega_{\mathbf{k}}$. We do not calculate the lattice Green's function

$$G_{ij}(t) = \frac{1}{N} \sum_{\mathbf{k}} G_{\mathbf{k}}(t)\, e^{i\mathbf{k}\cdot(\mathbf{R}_i - \mathbf{R}_j)} \tag{1.6.13}$$

explicitly; evaluation of the sum over **k** would require knowledge of the phonon dispersion law giving the frequencies $\Omega_{\mathbf{k}}$ as functions of **k**. In fact explicit knowledge of the form of $G_{ij}(t)$ is not needed for the discussion of the physical quantities which will be considered in the next section.

1.7. CALCULATION OF THE GROUND STATE ENERGY AND THE NEUTRON CROSS-SECTION IN TERMS OF THE PHONON GREEN'S FUNCTION

We can now use the exact Green's function (1.6.10) to calculate the correlation functions (1.4.10) and (1.4.11) which determine the change in ground state energy and the neutron scattering. Using the **k**-space transformation (1.6.4) for $G_{ij}(t)$ we have

$$\Delta E_G = \lim_{t\to 0+} \tfrac{1}{2} iM \sum_{i\neq j} D_{ij} \int_0^1 d\lambda \frac{1}{N} \sum_{\mathbf{k}} G_{\mathbf{k}}(t)\, e^{i\mathbf{k}\cdot(\mathbf{R}_i - \mathbf{R}_j)},$$

and

$$\frac{1}{N} \sum_{i\neq j} D_{ij}\, e^{i\mathbf{k}\cdot(\mathbf{R}_i-\mathbf{R}_j)} = \frac{1}{N} \sum_{j} D_{\mathbf{k}} = D_{\mathbf{k}};$$

hence

$$\Delta E_G = \lim_{t\to 0+} \tfrac{1}{2} iM \int_0^1 d\lambda \sum_{\mathbf{k}} D_{\mathbf{k}} G_{\mathbf{k}}(t). \qquad (1.7.1)$$

But, from Eq. (1.6.11),

$$\lim_{t\to 0+} G_{\mathbf{k}}(t) = -\frac{i}{2M\Omega_{\mathbf{k}}};$$

hence

$$\Delta E_G = \tfrac{1}{4} \int_0^1 d\lambda \sum_{\mathbf{k}} D_{\mathbf{k}}/\Omega_{\mathbf{k}}$$

$$= \tfrac{1}{4} \sum_{\mathbf{k}} D_{\mathbf{k}} \int_0^1 \frac{d\lambda}{\sqrt{(\Omega_0^2 + \lambda D_{\mathbf{k}})}}$$

$$= \tfrac{1}{2} \sum_{\mathbf{k}} \{\sqrt{(\Omega_0^2 + D_{\mathbf{k}})} - \Omega_0\}, \qquad (1.7.2)$$

in precise agreement with the result (1.1.12) obtained from the normal mode analysis.

The time correlation function (1.4.11), whose Fourier transform gives the differential neutron scattering cross-section according to Eq. (1.3.10), is [using Eqs. (1.6.13) and (1.6.11) for $t > 0$]

$$F(\mathbf{q}, t) = \frac{iq^2}{N} \sum_{jl} e^{i\mathbf{q}\cdot(\mathbf{R}_l - \mathbf{R}_j)} \sum_{\mathbf{k}} G_{\mathbf{k}}(t)\, e^{i\mathbf{k}\cdot(\mathbf{R}_j - \mathbf{R}_l)}$$

$$= \frac{q^2}{2NM} \sum_{jl} \sum_{\mathbf{k}} \frac{1}{\Omega_{\mathbf{k}}}\, e^{-i\Omega_{\mathbf{k}} t}\, e^{i(\mathbf{q}-\mathbf{k})\cdot(\mathbf{R}_l - \mathbf{R}_j)} \,. \tag{1.7.3}$$

This expression describes an inelastic scattering process in which one phonon of frequency $\Omega_{\mathbf{k}}$ and wave vector $\mathbf{k}$ is emitted by the incident neutron. The time Fourier transform of Eq. (1.7.3) contains a delta function $\delta(\omega - \Omega_{\mathbf{k}})$, showing that the neutron loses energy $\hbar\Omega_{\mathbf{k}}$, and the target gains energy $\hbar\Omega_{\mathbf{k}}$, in the process. Also the lattice sum in Eq. (1.7.3) is zero unless $\mathbf{q} - \mathbf{k} = \mathbf{G}$, where $\mathbf{G}$ is a reciprocal lattice vector; this relation determines the change in the neutron's momentum and can be interpreted as representing conservation of crystal momentum in the scattering process.

Since the target is at zero temperature, there are no phonons present in equilibrium, and therefore Eq. (1.7.3) contains no terms which correspond to phonon *absorption* by the incident neutron. Such processes appear at non-zero temperature, as will be seen in Chap. 2.

The scattering of slow neutrons from crystals permits a direct experimental determination of the phonon dispersion law. The peaks in the differential cross-section as a function of energy transfer correspond to the frequencies of the phonon modes and, by measuring the cross-section for different values of the momentum transfer $\mathbf{q}$, one can determine the relation between $\Omega_{\mathbf{k}}$ and $\mathbf{k}$ for the phonons. The lattice vibration spectra of many substances have been determined in this way [see Brockhouse (1964)].

For the present problem, in which only harmonic terms were included in the expansion of the lattice potential energy, the iteration series could be summed in closed form because the repeated time integrals in Eq. (1.5.14) are convolutions, corresponding to simple chain diagrams of the form shown in Fig. 1.3. In practice the anharmonic terms, which arise from the higher terms in the expansion of V in powers of the lattice

Fig. 1.3. Chain diagram for the harmonic lattice.

displacements, are not negligible (there are indeed features such as the thermal expansion which do not exist at all within the harmonic approximation), and it is therefore an important task of solid-state theory to describe the lattice dynamics of an anharmonic crystal. When anharmonicity is included the perturbation hamiltonian contains products of 3 or more u operators. We then have to consider more complicated diagrams such as Fig. 1.4 which cannot be summed

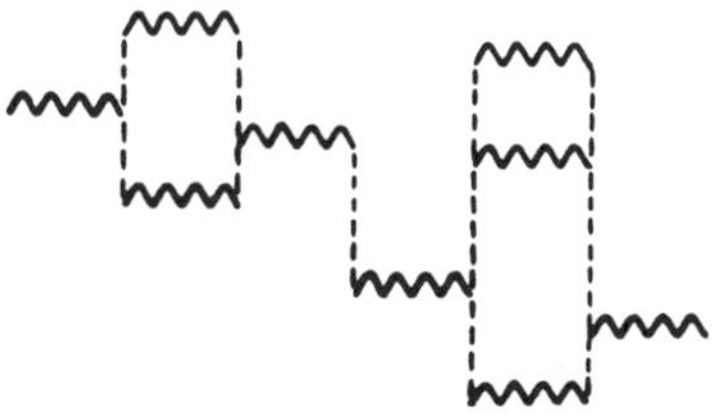

Fig. 1.4. A diagram for the anharmonic lattice.

exactly. It is, however, possible to perform *partial* summations by picking out subsets of diagrams similar in form to those obtained in the harmonic approximation. If one can justify the neglect of the remaining diagrams, one can in this way obtain approximate solutions of the many-body problem. The main physical effect of anharmonicity is to give rise to interactions between the normal modes, which can be described as collision processes in which the number of phonons is not conserved. Thus Fig. 1.4 will contain processes in which two phonons are excited which interact to give one phonon. The effect of such processes is to shift the position of the poles of the single-particle

Green's function: the change in the real part leads to a frequency shift of the phonon modes, and the poles also acquire a finite imaginary part which gives a finite lifetime to the modes. This shows up experimentally as a broadening of the neutron scattering cross-section for one-phonon processes; the delta-function peaks acquire a finite width proportional to the inverse lifetime of the mode. The theory of the subject has been reviewed by Cowley (1963) and the experimental aspects by Martin (1965).

The Green's function approach also provides a powerful method for studying the lattice vibrations of crystals containing defects or impurities of various kinds. The main complication, as compared with the ideal lattice problem, is that the translational invariance property is lost, so that (in 3 dimensions) it is necessary to solve $3N$ equations of motion instead of only 3. Nevertheless, exact solutions can still be obtained in particular cases, and much work has been done on the modes of vibration of a defect lattice [for a review see Maradudin (1964)].

Chapter 2

Lattice Dynamics at Finite Temperatures

The formalism developed in Chap. 1 enables us to deal with ground state properties of many-body systems. It thus applies to systems which are, in the thermodynamic sense, at zero temperature. In this limit we obtain a theory which is free from the complicating effects of temperature motions and which can be applied to real systems provided that temperature effects are negligible. But there are many interesting properties which cannot be discussed in this way; for example, thermodynamic quantities such as specific heats are zero at $T = 0$, and in the case of other quantities (e.g., magnetic susceptibilities, scattering cross-sections) which are non-zero at $T = 0$ we may wish to study the variation with temperature and with other parameters at finite T. In this chapter we extend the techniques developed in Chap. 1 to deal with non-zero temperatures.

When $T \neq 0$ the ground state expectation values of Chap. 1 are replaced by thermal averages over an appropriate thermodynamic ensemble. [For a summary of relevant thermodynamic formulae, see, for example, Fetter and Walecka (1971), Chap. 2.] The Green's functions which enable us to calculate such averages have more complicated properties than the zero temperature functions (although we shall see that close analogies exist). In particular, we now have to consider two types of Green's function. We first study the so-called *temperature* Green's function, in which the time variable is replaced by a temperature variable. This has a perturbation theory analogous to that developed in Chap. 1, and it allows us to determine the equilibrium thermodynamic properties of the system. We shall use it to determine the free energy of the harmonic lattice at temperature T. For discussing excitations and the response of the system to an external perturbation we need a *time-dependent* Green's function at temperature T; we shall use such a function to obtain the neutron scattering cross-section at finite temperature. The relation between the two types of Green's function is discussed in Appendix 2.

We adopt a similar approach to that of Chap. 1 and begin by calculating the lattice free energy by a direct method, using the exact phonon eigenvalues and eigenstates. We then rederive the same result by calculating the temperature Green's function for the problem, and we finally show that the temperature Green's function needed for the free energy also allows us to determine the time-dependent Green's function needed for the neutron scattering cross-section.

2.1. THE FREE ENERGY IN THE HARMONIC APPROXIMATION

If we exclude fluctuation properties, the different ensembles of thermodynamics lead to equivalent results and we can use whichever is most convenient. We shall usually work with the grand canonical ensemble, in which neither the energy nor the number of particles in the system has to be precisely specified. This is the appropriate ensemble for use in a many-body theory expressed in the language of second quantization (see Appendix 1) which allows for the creation and destruction of particles. Furthermore, in some systems (e.g., the quantum condensates in superfluid helium and superconductors) the existence of a variable particle number is more than a mathematical convenience and corresponds to the physical reality of the situation.

In the grand canonical ensemble all the thermodynamic properties can be deduced from the *thermodynamic potential*

$$\Omega(T, V, \mu) = -k_B T \log Z_G, \tag{2.1.1}$$

which is a function of temperature T, volume V, and chemical potential μ. Z_G is the *grand partition function*, defined in terms of the many-body hamiltonian H as a trace

$$Z_G = \operatorname{Tr} e^{-\beta(H - \mu\hat{N})}, \tag{2.1.2}$$

where $\beta = 1/k_B T$ (k_B = Boltzmann's constant) and $\hat{N}$ is the (total) number operator (see Appendix 1). The entropy S, pressure P, and number of particles N in the system are obtained from Ω by differentiation:

$$S = -\left(\frac{\partial\Omega}{\partial T}\right)_{V,\mu}, \qquad P = -\left(\frac{\partial\Omega}{\partial V}\right)_{T,\mu}, \qquad N = -\left(\frac{\partial\Omega}{\partial \mu}\right)_{T,V}, \tag{2.1.3}$$

and from these quantities the specific heats, the equation of state, and other thermodynamic relations can be deduced.

In the canonical ensemble (where N is fixed) one deals instead of Ω

with the (Helmholtz) free energy

$$\mathscr{F}(T,\ V,\ N) = \Omega + \mu N, \tag{2.1.4}$$

in terms of which the chemical potential is given by

$$\mu = \left(\frac{\partial \mathscr{F}}{\partial N}\right)_{T,V}. \tag{2.1.5}$$

For the phonon system $\mu = 0$ [this is because there is no restriction on the number of phonons, so that the equilibrium state for fixed T and V is obtained by minimizing $\mathscr{F}$ with respect to N, leading to the condition $(\partial \mathscr{F}/\partial N)_{T,V} = 0$]; hence in this case the canonical and grand canonical ensembles are identical. We shall therefore work in this chapter with the free energy

$$\mathscr{F} = -k_B T \log \operatorname{Tr} e^{-\beta H}. \tag{2.1.6}$$

In the limit $T \to 0$ the trace (which is in general a sum over the eigenvalues of H) reduces to a single term $\exp(-\beta E_G)$, and $\mathscr{F}$ reduces to the ground state energy discussed before.

$\mathscr{F}$ may again be evaluated, when H is given in the harmonic approximation by Eqs. (1.1.3), (1.1.4), by diagonalizing H in terms of the normal modes. But it should be remembered that the linear terms in the expansion of the interatomic potential were zero since the expansion was about the equilibrium, *static*, configuration of the atoms. As the system is heated up, the lattice will in general expand and the equilibrium atomic spacing will change with temperature. This leads to a change in the harmonic frequencies of the normal modes. However, we will stick to an expansion about the static equilibrium configuration at $T = 0$ (neglecting zero-point vibrations), and remain in the idealized harmonic approximation for which, in the absence of anharmonic corrections, the thermal expansion does not occur.

The direct way to evaluate Eq. (2.1.6) is to perform the trace directly over the eigenstates of the set of normal mode oscillators whose frequencies are given by Eq. (1.1.11). To construct these eigenstates we need to generalize the creation and annihilation operators given in Eq. (1.5.3) for the single vibrating atom to the normal mode case. The main complication is that the normal mode displacement operators

$$Q_{\mathbf{k}}(t) = \sqrt{\left(\frac{M}{N}\right)} \sum_i e^{i\mathbf{k}\cdot\mathbf{R}_i} u_i \tag{2.1.7}$$

are now complex and, using the fact that the atomic displacements u_i are real, have to satisfy the reality condition

$$Q_{\mathbf{k}}^* = Q_{-\mathbf{k}} \tag{2.1.8}$$

obtained by taking the conjugate of (2.1.7). Eq. (2.1.8) means that the modes **k** and −**k** are not in fact independent dynamical variables. The definition (2.1.7) may be inverted by writing

$$u_i = \frac{1}{\sqrt{(NM)}} \sum_{\mathbf{k}} e^{-i\mathbf{k}\cdot\mathbf{R}_i} Q_{\mathbf{k}}, \tag{2.1.9}$$

which follows from (2.1.7) on using periodic boundary conditions.

Inserting (2.1.7), and the analogous definition of the normal mode momentum variable $P_{\mathbf{k}}$,

$$P_{\mathbf{k}} = \frac{1}{\sqrt{(NM)}} \sum_{i} e^{i\mathbf{k}\cdot\mathbf{R}_i} p_i, \tag{2.1.10}$$

into the hamiltonian (1.1.3), (1.1.6), and using the normal mode frequencies (1.1.11), the hamiltonian may be rewritten as

$$H = \tfrac{1}{2} \sum_{\mathbf{k}} (P_{\mathbf{k}} P_{-\mathbf{k}} + \Omega_{\mathbf{k}}^2 Q_{\mathbf{k}} Q_{-\mathbf{k}}). \tag{2.1.11}$$

From Eq. (2.1.7) the commutation rules are

$$[P_{\mathbf{k}}, Q_{\mathbf{k}'}] = -i\delta_{\mathbf{k},-\mathbf{k}'}, \tag{2.1.12}$$

and annihilation and creacion operators may be introduced as in Eq. (1.5.3) by the relations [using (2.1.8)]

$$b_{\mathbf{k}} = \frac{1}{\sqrt{(2\Omega_{\mathbf{k}})}} (Q_{\mathbf{k}} + iP_{\mathbf{k}}), \qquad b_{\mathbf{k}}^{\dagger} = \sqrt{\left(\frac{\Omega_{\mathbf{k}}}{2}\right)} (Q_{-\mathbf{k}} - iP_{-\mathbf{k}}), \tag{2.1.13}$$

from which

$$[b_{\mathbf{k}}, b_{\mathbf{k}'}^{\dagger}] = \delta_{\mathbf{k}\mathbf{k}'} \tag{2.1.14}$$

and H may be re-expressed as

$$H = \sum_{\mathbf{k}} \Omega_{\mathbf{k}} (b_{\mathbf{k}}^{\dagger} b_{\mathbf{k}} + \tfrac{1}{2}). \tag{2.1.15}$$

The normalized, n-phonon eigenstates may be written (see Appendix 1)

$$|n_{\mathbf{k}_1}, n_{\mathbf{k}_2}, \ldots\rangle = \prod_{\mathbf{k}} \frac{(b_{\mathbf{k}}^{\dagger})^{n_{\mathbf{k}}}}{\sqrt{(n_{\mathbf{k}}!)}} |0\rangle, \qquad (2.1.16)$$

where $|0\rangle$ is the ground state, defined by

$$b_{\mathbf{k}}|0\rangle = 0 \qquad (2.1.17)$$

for all $\mathbf{k}$. It may be shown then that $b_{\mathbf{k}}^{\dagger}b_{\mathbf{k}}$ is the phonon number operator for phonons with wave-vector $\mathbf{k}$

$$b_{\mathbf{k}}^{\dagger}b_{\mathbf{k}}|n_{\mathbf{k}_1}, n_{\mathbf{k}_2}, \ldots\rangle = n_{\mathbf{k}}|n_{\mathbf{k}_1}, n_{\mathbf{k}_2}, \ldots\rangle. \qquad (2.1.18)$$

Hence, using the fact that the $b_{\mathbf{k}}$'s commute for different $\mathbf{k}$,

$$\langle n_{\mathbf{k}_1}, n_{\mathbf{k}_2}, \ldots|e^{-\beta H}|n_{\mathbf{k}_1}, n_{\mathbf{k}_2}, \ldots\rangle = e^{-\beta \sum_{\mathbf{k}} \Omega_{\mathbf{k}}(n_{\mathbf{k}}+\frac{1}{2})}, \qquad (2.1.19)$$

and the free energy is given by

$$\mathscr{F} = -\frac{1}{\beta} \log \sum_{\{n_{\mathbf{k}}\}} e^{-\beta \sum_{\mathbf{k}} \Omega_{\mathbf{k}}(n_{\mathbf{k}}+\frac{1}{2})}$$

$$= -\frac{1}{\beta} \log \sum_{\{n_{\mathbf{k}}\}} \prod_{\mathbf{k}} e^{-\beta\Omega_{\mathbf{k}}(n_{\mathbf{k}}+\frac{1}{2})}. \qquad (2.1.20)$$

Here $\{n_{\mathbf{k}}\}$ indicates the set of choices of $n_{\mathbf{k}_1}, n_{\mathbf{k}_2}, \ldots$, where each $n_{\mathbf{k}}$ can take all integer values. Inverting sum and product we have, for fixed $\mathbf{k}$,

$$\sum_{n_{\mathbf{k}}=0,1,2,\ldots} e^{-\beta\Omega_{\mathbf{k}}(n_{\mathbf{k}}+\frac{1}{2})} = \frac{e^{-\frac{1}{2}\beta\Omega_{\mathbf{k}}}}{1-e^{-\beta\Omega_{\mathbf{k}}}} = \{2\sinh(\tfrac{1}{2}\beta\Omega_{\mathbf{k}})\}^{-1},$$

and hence we obtain the well-known result

$$\mathscr{F} = \frac{1}{\beta} \sum_{\mathbf{k}} \log\{2\sinh(\tfrac{1}{2}\beta\Omega_{\mathbf{k}})\}. \qquad (2.1.21)$$

At zero temperature ($\beta \to \infty$) this reduces to the ground state energy, Eq. (1.1.12),

$$\mathscr{F} \to \tfrac{1}{2} \sum_{\mathbf{k}} \Omega_{\mathbf{k}} \quad \text{as} \quad \beta \to \infty. \qquad (2.1.22)$$

2.2. THE PHONON TEMPERATURE GREEN'S FUNCTION

As in the previous chapter, another way of calculating the free energy is by integration of the average interaction energy over the coupling constant. Again, this is not really necessary for the harmonic lattice, for which the method of Sec. 2.1 provides the straightforward solution. However, as in the zero-temperature case the more general approach provides a powerful systematic method for dealing with more complicated many-body problems such as the anharmonic lattice. Here we apply the method to the harmonic lattice at non-zero temperatures in order to provide a simple illustration of the technique.

We begin by generalizing the formula (1.2.3), which expresses the ground state energy as an integral over coupling constant, to the free energy (2.1.6). Let $|\Psi_n(\lambda)\rangle$, $E_n(\lambda)$ be the exact eigenfunctions and eigenvalues of the hamiltonian $H(\lambda)$ of Eq. (1.1.5) with variable coupling constant. Then

$$-\beta\mathcal{F} = \log \sum_n e^{-\beta E_n(\lambda)},$$

and, since $\partial E_n(\lambda)/\partial\lambda = \langle\Psi_n(\lambda)|H_1|\Psi_n(\lambda)\rangle$ [compare Eq. (1.2.2)], we have

$$\frac{\partial\mathcal{F}}{\partial\lambda} = \frac{\sum_n e^{-\beta E_n}\langle\Psi_n(\lambda)|H_1|\Psi_n(\lambda)\rangle}{\sum_n e^{-\beta E_n}} = \langle H_1\rangle_\lambda, \tag{2.2.1}$$

where $\langle\cdot\cdot\cdot\rangle_\lambda$ now denotes the *thermal average*

$$\langle H_1\rangle_\lambda = \mathrm{Tr}\{e^{-\beta H(\lambda)}H_1\}/\mathrm{Tr}\, e^{-\beta H(\lambda)}. \tag{2.2.2}$$

Thus for the harmonic lattice case we have, instead of the expression (1.2.4) for ΔE_G,

$$\Delta\mathcal{F} = \tfrac{1}{2}M \sum_{i\neq j} D_{ij} \int_0^1 d\lambda\, \langle u_i u_j\rangle_\lambda, \tag{2.2.3}$$

where $\Delta\mathcal{F}$ is the change in free energy as λ is switched on.

To calculate thermal averages such as that in Eq. (2.2.3) (other examples will be encountered later) we define a *single-particle temperature Green's function* by

$$G_{ij}(\sigma - \sigma') = \langle T[\tilde{u}_i(\sigma)\tilde{u}_j(\sigma')]\rangle. \tag{2.2.4}$$

The brackets $\langle \cdots \rangle$ now denote the thermal average, defined as in Eq. (2.2.2), of the operator product at inverse temperature β, and the time variable in the argument of the operators has been replaced by a "temperature variable" σ. $\tilde{u}(\sigma)$ is a modified Heisenberg operator in the variable σ,

$$\tilde{u}(\sigma) = e^{\sigma H} u\, e^{-\sigma H}, \tag{2.2.5}$$

which satisfies the temperature analog of the Schrödinger equation, the *Bloch equation* [Bloch (1932)]

$$\frac{\partial \tilde{u}(\sigma)}{\partial \sigma} = [H, \tilde{u}]. \tag{2.2.6}$$

The T symbol orders the operators according to the value of σ with the larger on the left. The temperature Green's function is a function of the variables σ and σ' and, if H is independent of time, it depends only on the difference $\sigma - \sigma'$ (this follows, as at $T = 0$, from the analysis of Appendix 2).

It will be noted that Eqs. (2.2.5) and (2.2.6) can be obtained formally from the conventional time-dependent Heisenberg picture by replacing *it* by σ. σ may thus be called an *imaginary time* variable. By regarding σ as a complex variable which can take real and imaginary values, one can thus establish a correspondence between the Green's function methods at finite and at zero temperatures, and also between temperature and real-time Green's functions at finite T (see Appendix 2). This correspondence will become apparent as the analysis proceeds.

The thermal average required for the free energy may now be expressed in terms of G_{ij} by

$$\langle u_i u_j \rangle = \lim_{\sigma \to 0+} G_{ij}(\sigma). \tag{2.2.7}$$

It is understood here that H contains the coupling constant λ.

The calculation of G now proceeds in close analogy with the zero temperature case. The equation of motion with respect to the temperature variable σ follows, by the same analysis as led to Eq. (1.4.14), as

$$\left(\frac{d^2}{d\sigma^2} - \Omega_0^{\,2}\right) G_{ij}(\sigma) = -\frac{1}{M}\, \delta_{ij}\delta(\sigma) + \lambda \sum_l D_{il} G_{lj}(\sigma). \tag{2.2.8}$$

We shall again solve this by a Fourier transform method. First we need to examine the analytic form of the unperturbed Green's function

$G^0(\sigma)\delta_{ij}$. Introducing the creation operator for the atom with displacement u_i as in Eq. (1.5.3), we find

$$B_i(\sigma) = e^{-\Omega_0\sigma}B_i, \qquad B_i^\dagger(\sigma) = e^{+\Omega_0\sigma}B_i^\dagger, \tag{2.2.9}$$

so that

$$G^0(\sigma) = \frac{1}{2M\Omega_0}\langle T[(e^{-\Omega_0\sigma}B_i + e^{\Omega_0\sigma}B_i^\dagger)(B_i + B_i^\dagger)]\rangle. \tag{2.2.10}$$

[Note that, for real σ, the operator $B_i^\dagger(\sigma)$ is *not* the adjoint of $B_i(\sigma)$. The notation should not lead to any confusion.]

It is convenient to define

$$\left.\begin{aligned} G^{0>}(\sigma) &= G^0(\sigma) \qquad (\sigma > 0), \\ G^{0<}(\sigma) &= G^0(\sigma) \qquad (\sigma < 0). \end{aligned}\right\} \tag{2.2.11}$$

Using the fact that $B^\dagger B$ is a number operator for phonons of constant frequency Ω_0, we have

$$\langle B_i^\dagger B_i\rangle = n_0 = 1/(e^{\beta\Omega_0} - 1), \tag{2.2.12}$$

where n_0 is the Planck function (the distribution function for the phonons at inverse temperature β), and

$$\langle B_i B_i^\dagger\rangle = n_0 + 1 = 1/(1 - e^{-\beta\Omega_0}). \tag{2.2.13}$$

Hence

$$G^{0>}(\sigma) = \frac{1}{2M\Omega_0}\{(n_0 + 1)\, e^{-\Omega_0\sigma} + n_0 e^{\Omega_0\sigma}\},$$

and

$$G^{0<}(\sigma) = \frac{1}{2M\Omega_0}\{n_0 e^{-\Omega_0\sigma} + (n_0 + 1)\, e^{\Omega_0\sigma}\},$$

from which, for all σ,

$$G^0(\sigma) = \frac{1}{2M\Omega_0}\{(n_0 + 1)\, e^{-\Omega_0|\sigma|} + n_0\, e^{\Omega_0|\sigma|}\}. \tag{2.2.14}$$

It will be seen that $G^{0>}$ and $G^{0<}$ both contain the real exponentials $\exp(-\Omega_0\sigma)$ and $\exp(+\Omega_0\sigma)$. If we want to make a Fourier transform of G^0, the integrals involved will now diverge exponentially at the limits $\sigma \to \pm\infty$. However, if we restrict ourselves to the region $0 \leqslant \sigma \leqslant \beta$, it may be seen that the term $\exp(\beta\Omega_0)$ in the denominator of the Planck function will cancel the exponential growth of the exp $(\sigma\Omega_0)$ in the numerator, thus keeping the integrand finite for any value of β. This argument suggests that convergence problems will be overcome provided the Fourier transforms are made over this finite region of the σ axis. We therefore artificially repeat the function periodically in the regions $0 \leqslant \sigma \leqslant \beta$, $\beta \leqslant \sigma \leqslant 2\beta$ etc. Thus the function $G^0(\sigma)$ can be represented over the region $0 \leqslant \sigma \leqslant \beta$ by means of a *Fourier series* of the form [Abrikosov, Gor'kov, and Dzyaloshinskii (1959), Fradkin (1959), Martin and Schwinger (1959)]

$$G^0(\sigma) = \sum_{\mu} e^{2\pi i \mu\sigma/\beta}\, G^0(\mu), \tag{2.2.15}$$

where μ takes integer values $0, \pm 1, \pm 2, \ldots$. In fact one can prove directly (see Appendix 2) that the single-particle temperature Green's function $G(\sigma)$ for bosons is always periodic in σ with period β, for σ in the range $0 \leqslant \sigma \leqslant \beta$. [It is left to the reader to check this property for the function (2.2.14). For fermions the corresponding Green's function is antiperiodic with period β, and therefore periodic over the range 2β.]

By Fourier's theorem Eq. (2.2.15) may be inverted to give

$$G^0(\bar{\mu}) = \frac{1}{\beta}\int_0^\beta d\sigma\, e^{-i\bar{\mu}\sigma} G^0(\sigma), \tag{2.2.16}$$

where

$$\bar{\mu} = 2\mu\pi/\beta. \tag{2.2.17}$$

The Fourier coefficient $G^0(\bar{\mu})$, which replaces the Fourier transform $G^0(\omega)$ of the zero-temperature theory, is thus a function of a discrete frequency variable $\bar{\mu}$. For fermions there is a corresponding Fourier expansion with $\bar{\mu} = (2\mu + 1)\pi/\beta$. Using the fact that

$$\begin{aligned}\frac{1}{\beta}\int_0^\beta d\sigma\, e^{-i\bar{\mu}\sigma}\, e^{-\Omega_0\sigma}(n_0 + 1) &= \frac{1}{\beta}\,\frac{(e^{-\beta\Omega_0} - 1)}{-i\bar{\mu} - \Omega_0}\,(n_0 + 1)\\ &= \frac{1/\beta}{i\bar{\mu} + \Omega_0},\end{aligned} \tag{2.2.18}$$

we thus find, substituting (2.2.14) into (2.2.16),

$$G^0(\bar{\mu}) = \frac{1}{2\Omega_0 \beta M}\left\{\frac{1}{i\bar{\mu}+\Omega_0} - \frac{1}{i\bar{\mu}-\Omega_0}\right\}$$

$$= \frac{1}{\beta M}\frac{1}{\bar{\mu}^2+\Omega_0{}^2}. \tag{2.2.19}$$

The same result could also have been obtained from the equation of motion (2.2.8), by inserting the Fourier expansion (2.2.15) in the limit $\lambda = 0$ and using a Fourier series representation of the δ-function.

It is seen that $G^0(\bar{\mu})$ is a similar function to the zero-temperature Fourier transform $G^0(\omega)$, Eq. (1.6.9), with two poles corresponding to imaginary values of $\bar{\mu}$ at $i\bar{\mu} = \pm\Omega_0$. We obtained this simple result because we worked with a Green's function defined in terms of imaginary time variables. It is a consequence of the periodicity property of this Green's function, which permitted the Fourier series expansion (2.2.15). The *real-time* Green's function $G^0(t)$ at finite temperature T, defined as the thermal average of

$$-iT[\tilde{u}_i(t)\tilde{u}_i(0)]$$

for $\lambda = 0$, has evidently the same form as $G^0(\sigma)$, with $|\sigma|$ replaced by $i|t|$, but its Fourier transform $G^0(\omega)$ is a more complicated function than at $T = 0$ and still involves the Planck distribution functions n_0 and $(n_0 + 1)$. The real-time functions are discussed in Sec. 2.3 below.

The solution of the equation of motion (2.2.8) for $\lambda \neq 0$ now follows directly from the equation

$$G_{ij}(\bar{\mu}) = G^0(\bar{\mu})\delta_{ij} - \lambda\beta M G^0(\bar{\mu})\sum_l D_{il}G_{lj}(\bar{\mu}), \tag{2.2.20}$$

which corresponds to the Fourier-transformed integral equation (1.6.2). This is solved, as in the zero-temperature case of Eq. (1.6.6), in terms of k-space Green's functions as

$$G_{\mathbf{k}}(\bar{\mu}) = \frac{G^0(\bar{\mu})}{1 + \lambda\beta M G^0(\bar{\mu})D_{\mathbf{k}}}$$

$$= \frac{1/\beta M}{\bar{\mu}^2 + \Omega_{\mathbf{k}}{}^2}, \tag{2.2.21}$$

where $\Omega_{\mathbf{k}}{}^2 = \Omega_0{}^2 + \lambda D_{\mathbf{k}}$ as before. Thus the effect of the interaction is again to shift the poles of the Green's function from $\pm\Omega_0$ to $\pm\Omega_{\mathbf{k}}$.

Finally, the expression (2.2.3), (2.2.7) for the change in free energy may be expressed in terms of $G_{\mathbf{k}}(\bar{\mu})$:

$$\Delta \mathscr{F} = \frac{1}{2}\frac{M}{N}\sum_{i \neq j} D_{ij} \sum_{\mathbf{k}} e^{i\mathbf{k}\cdot(\mathbf{R}_i - \mathbf{R}_j)} \int_0^1 d\lambda \lim_{\sigma \to 0+} \sum_{\bar{\mu}} e^{i\bar{\mu}\sigma} G_{\mathbf{k}}(\bar{\mu}), \quad (2.2.22)$$

leading to

$$\Delta \mathscr{F} = \lim_{\sigma \to 0+} \tfrac{1}{2} M \int_0^1 d\lambda \sum_{\mathbf{k}} D_{\mathbf{k}} \sum_{\bar{\mu}} e^{i\bar{\mu}\sigma} G_{\mathbf{k}}(\bar{\mu}). \quad (2.2.23)$$

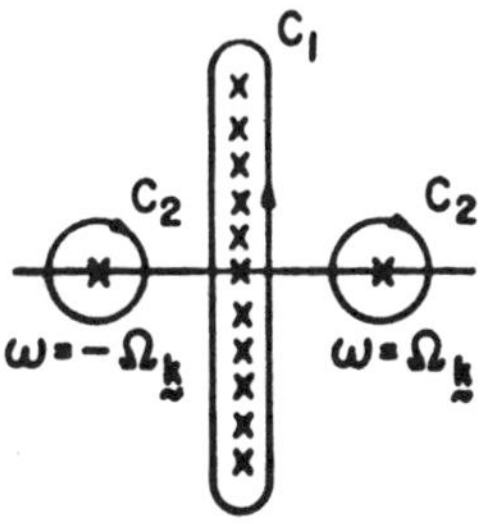

Fig. 2.1. Contours for the evaluation of $G_{\mathbf{k}}^{>}(\sigma)$.

For $\sigma > 0$ the function

$$\sum_{\bar{\mu}} e^{i\bar{\mu}\sigma} G_{\mathbf{k}}(\bar{\mu})$$

is $G_{\mathbf{k}}^{>}(\sigma)$, and in order to perform the frequency sum over $\bar{\mu}$ it is convenient to introduce a contour integral representation

$$G_{\mathbf{k}}^{>}(\sigma) = \frac{\beta}{2\pi i} \int_{C_1} d\omega \frac{e^{\sigma\omega}}{e^{\beta\omega} - 1} G_{\mathbf{k}}(i\omega), \quad (2.2.24)$$

where C_1 is the contour in the complex ω-plane encircling the poles of $1/(e^{\beta\omega} - 1)$ which lie on the imaginary axis at $\omega = 2\pi i\mu/\beta$ ($\mu = 0, \pm 1, \pm 2, \ldots$) (Fig. 2.1). Because of the fact that $0 \leqslant \sigma \leqslant \beta$, the $e^{\sigma\omega}$ in the integrand does not lead to any divergence, and if Eq. (2.2.24) is evaluated by residues, it is seen that the contour integral leads to the Fourier series for

$G_{\mathbf{k}}^{>}(\sigma)$. Now

$$G_{\mathbf{k}}(i\omega) = \frac{1/\beta M}{-\omega^2 + \Omega_{\mathbf{k}}^2},$$

and we can obtain (2.2.24) in closed form by deforming the contour C_1 to a contour C_2 encircling the poles $\omega = \pm\Omega_{\mathbf{k}}$ of $G_{\mathbf{k}}(i\omega)$ in the negative sense. (The integrals over the large arcs involved in the deformation vanish.) We now use the residue theorem again, evaluating the contributions of the two poles which lie inside C_2, to obtain in the limit $\sigma \to 0+$

$$G_{\mathbf{k}}(\sigma = 0+) = \frac{1}{2\Omega_{\mathbf{k}} M}\left\{\frac{1}{e^{\beta\Omega_{\mathbf{k}}} - 1} + \frac{1}{1 - e^{-\beta\Omega_{\mathbf{k}}}}\right\}$$

$$= \frac{1}{2\Omega_{\mathbf{k}} M} \coth\left(\tfrac{1}{2}\beta\Omega_{\mathbf{k}}\right). \qquad (2.2.25)$$

But by direct differentiation with respect to λ it may be seen that

$$\frac{\partial}{\partial\lambda} \log\left\{\sinh \frac{\beta\Omega_{\mathbf{k}}(\lambda)}{2}\right\} = \frac{\beta D_{\mathbf{k}}}{4\Omega_{\mathbf{k}}} \coth\left(\tfrac{1}{2}\beta\Omega_{\mathbf{k}}\right). \qquad (2.2.26)$$

Hence Eq. (2.2.23) reduces to

$$\Delta\mathcal{F} = \tfrac{1}{2}M \int_0^1 d\lambda \sum_{\mathbf{k}} D_{\mathbf{k}} G_{\mathbf{k}}(\sigma = 0+)$$

$$= \frac{1}{\beta} \sum_{\mathbf{k}} \log\left\{\frac{\sinh\left(\tfrac{1}{2}\beta\Omega_{\mathbf{k}}\right)}{\sinh\left(\tfrac{1}{2}\beta\Omega_0\right)}\right\}, \qquad (2.2.27)$$

which is in precise agreement with the result obtained by direct calculation for the change in the free energy on switching on λ.

From this example we see that there are close analogies between the time-dependent Green's function at $T = 0$ and the temperature Green's function at finite T. The main difference is that the imaginary time variable it has become transformed to a real temperature variable σ, and that σ is to be confined to a restricted range of values $0 \leqslant \sigma \leqslant \beta$. This analogy will be further exploited in the following chapter.

2.3. THE REAL-TIME GREEN'S FUNCTION AND NEUTRON SCATTERING AT FINITE TEMPERATURES

The temperature Green's function studied in the last section enabled us to calculate equilibrium thermodynamic properties which are independent of the time. To study excitations, however, we need to be able to calculate time-dependent correlation functions, and for this we require the finite-temperature generalization of the real-time Green's function studied in Chap. 1. As an example of such a quantity we now consider the generalization to non-zero temperatures of the neutron scattering cross-section evaluated for $T = 0$ in Sec. 1.7.

At $T \neq 0$ we cannot assume in calculating the cross-section that the target is in its ground state initially but have to consider a distribution of initial states $|\psi_I\rangle$ with energy E_I, occupied with probability p_I. The generalization of the formula (1.3.10) for the differential cross-section is thus

$$\frac{d^2\sigma}{d\Omega d\omega} = \frac{K'}{2\pi K}\left(\frac{M}{2\pi}\right)^2 |V_{\mathbf{q}}|^2 \sum_I \int_{-\infty}^{\infty} e^{i\omega t}\, dt \sum_{jl} p_I$$
$$\times \langle \psi_I | e^{-i\mathbf{q}\cdot\tilde{\mathbf{x}}_j(t)}\, e^{i\mathbf{q}\cdot\tilde{\mathbf{x}}_l(0)} | \psi_I \rangle. \tag{2.3.1}$$

For a canonical ensemble at inverse temperature β,

$$p_I = e^{-\beta E_I} \Big/ \sum_I e^{-\beta E_I}, \tag{2.3.2}$$

and thus the cross-section now involves a thermal average over the ensemble. We deduce that the correlation function (1.3.11), (1.3.12) is replaced by

$$F(\mathbf{q}, t) = \sum_{jl} e^{i\mathbf{q}\cdot(\mathbf{R}_l - \mathbf{R}_j)} \langle e^{-i\mathbf{q}\cdot\tilde{\mathbf{u}}_j(t)}\, e^{i\mathbf{q}\cdot\tilde{\mathbf{u}}_l(0)} \rangle$$
$$\simeq q^2 \sum_{jl} e^{i\mathbf{q}\cdot(\mathbf{R}_l - \mathbf{R}_j)} \langle \tilde{\mathbf{u}}_j(t)\tilde{\mathbf{u}}_l(0) \rangle, \tag{2.3.3}$$

where the brackets now denote a thermal average of the time-dependent quantities enclosed. (The Debye–Waller factor—now temperature-dependent—has again been omitted.)

Again, for the present problem this correlation function can be obtained directly from the exact phonon eigenstates. To obtain it by a

Green's function method, we define the single-particle real-time Green's function at temperature T as

$$G_{ij}(t - t') = -i\langle T[\tilde{u}_i(t)\tilde{u}_j(t')]\rangle, \tag{2.3.4}$$

which is an immediate generalization of the zero-temperature definition (1.4.6). The only difference is in the meaning of the brackets: the ground state expectation value has been replaced by a thermal average

$$\langle \mathcal{O} \rangle = \mathrm{Tr}\,(e^{-\beta H}\mathcal{O})/\mathrm{Tr}\, e^{-\beta H}. \tag{2.3.5}$$

$\tilde{u}(t)$is a time-dependent Heisenberg operator as in Chap. 1. The Green's function (2.3.4) is thus a function of both temperature and time variables.

To obtain $G_{ij}(t)$ we again appeal to the analogy between the time-dependent and the temperature formalism. It is shown in Appendix 2 that the time-dependent Green's function at temperature T can be obtained by an *analytic continuation* of the temperature Green's function. For this purpose it is in fact simpler to work not with the time-ordered but with the *retarded* Green's function at temperature T

$$G^R_{ij}(t - t') = -i\theta(t - t')\langle[\tilde{u}_i(t), \tilde{u}_j(t')]\rangle. \tag{2.3.6}$$

The essential formulae are Eqs. (A.2.8), (A.2.22) of Appendix 2:

$$G^R_{ij}(\omega) = \int_{-\infty}^{\infty} (1 - e^{-\beta\omega'}) \frac{J_{ij}(\omega')}{\omega - \omega' + i\eta} \frac{d\omega'}{2\pi} \qquad (\eta = 0+), \tag{2.3.7}$$

$$G_{ij}(\bar{\mu}) = -\frac{1}{\beta} \int_{-\infty}^{\infty} (1 - e^{-\beta\omega'}) \frac{J_{ij}(\omega')}{-i\bar{\mu} - \omega'} \frac{d\omega'}{2\pi}, \tag{2.3.8}$$

which express the Fourier transform of $G^R_{ij}(t)$ and the Fourier coefficient in the Fourier series for $G_{ij}(\sigma)$ in terms of the Fourier transform

$$J_{ij}(\omega) = \int_{-\infty}^{\infty} \langle \tilde{u}_i(t)\tilde{u}_j(0)\rangle\, e^{i\omega t}\, dt \tag{2.3.9}$$

of the time-correlation function $\langle \tilde{u}_i(t)\tilde{u}_j(0)\rangle$.

Eqs. (2.3.7), (2.3.8) apply equally to the **k**-space Green's functions, and we see that we obtain $G_{\mathbf{k}}{}^R(\omega)$ by replacing $-i\bar{\mu}$ in $-\beta G_{\mathbf{k}}(\bar{\mu})$,

Eq. (2.2.21), by $\omega + i\eta$, leading at once to

$$G_{\mathbf{k}}{}^{R}(\omega) = -\frac{1}{M}\frac{1}{-(\omega+i\eta)^2+\Omega_{\mathbf{k}}{}^2}$$
$$= \frac{1}{2M\Omega_{\mathbf{k}}}\left\{\frac{1}{\omega-\Omega_{\mathbf{k}}+i\eta} - \frac{1}{\omega+\Omega_{\mathbf{k}}+i\eta}\right\}. \tag{2.3.10}$$

We can invert the relation between $G_{\mathbf{k}}{}^{R}(\omega)$ and $J_{\mathbf{k}}(\omega)$ to obtain $J_{\mathbf{k}}(\omega)$ explicitly, using the formula [Messiah (1961, p. 469)] for the real and imaginary parts of the singular function $(x + i\eta)^{-1}$:

$$\frac{1}{x+i\eta} = \mathscr{P}\frac{1}{x} - i\pi\delta(x) \qquad (\eta = 0+), \tag{2.3.11}$$

where $\mathscr{P}$ denotes the principal part. For $J_{\mathbf{k}}(\omega)$ real, we thus have from Eqs. (2.3.7) and (2.3.10)

$$J_{\mathbf{k}}(\omega) = -\frac{2}{1-e^{-\beta\omega}}\,\mathrm{Im}\,G_{\mathbf{k}}{}^{R}(\omega)$$
$$= \frac{\pi}{M\Omega_{\mathbf{k}}(1-e^{-\beta\omega})}\{\delta(\omega-\Omega_{\mathbf{k}}) - \delta(\omega+\Omega_{\mathbf{k}})\}. \tag{2.3.12}$$

The δ-function peaks in the *spectral density function* $J_{\mathbf{k}}(\omega)$ correspond to the poles at $\pm\Omega_{\mathbf{k}}$ of the retarded Green's function.

The lattice function corresponding to $J_{\mathbf{k}}(\omega)$ is

$$J_{ij}(\omega) = \frac{1}{N}\sum_{\mathbf{k}} J_{\mathbf{k}}(\omega)\, e^{i\mathbf{k}\cdot(\mathbf{R}_i-\mathbf{R}_j)}, \tag{2.3.13}$$

and we finally obtain the time correlation function (2.3.3) by inverting the Fourier transform (2.3.9). Thus

$$F(\mathbf{q}, t) = q^2 \sum_{jl} e^{i\mathbf{q}\cdot(\mathbf{R}_l-\mathbf{R}_j)} \int_{-\infty}^{\infty} J_{jl}(\omega)\, e^{-i\omega t}\frac{d\omega}{2\pi},$$

and, substituting (2.3.13) and (2.3.12), we can perform the frequency integral to obtain the final result

$$F(\mathbf{q}, t) = \frac{q^2}{2NM}\sum_{jl}\sum_{\mathbf{k}}\frac{1}{\Omega_{\mathbf{k}}}\{(n_{\mathbf{k}}+1)\, e^{-i\Omega_{\mathbf{k}}t}\, e^{i(\mathbf{q}-\mathbf{k})\cdot(\mathbf{R}_l-\mathbf{R}_j)}$$
$$+ n_{\mathbf{k}}\, e^{i\Omega_{\mathbf{k}}t}\, e^{i(\mathbf{q}+\mathbf{k})\cdot(\mathbf{R}_l-\mathbf{R}_j)}\}. \tag{2.3.14}$$

Here

$$n_{\mathbf{k}} = \frac{1}{e^{\beta\Omega_{\mathbf{k}}} - 1} \tag{2.3.15}$$

is the distribution function for phonons of wave-vector **k**, and we have changed the sign of **k** in the second term.

Eq. (2.3.14) is the finite-temperature generalization of Eq. (1.7.3) and gives the temperature dependence of the neutron scattering cross-section. We see that $F(\mathbf{q}, t)$ now contains, in addition to the phonon emission processes described by the first term, phonon absorption processes described by the second term in which the target loses energy $\hbar\Omega_{\mathbf{k}}$ to the neutron. As the temperature is increased these absorption processes become more important until, when the temperature is large compared with the phonon frequencies, there is equipartition and emission and absorption processes are then equally frequent.

For completeness we finally use Eqs. (A.2.9) and (A.2.11) of Appendix 2 to evaluate the time-ordered real-time Green's function for the phonon problem at temperature T. (The details are left to the reader.) The Fourier transform $G_{\mathbf{k}}(\omega)$ is given by

$$\begin{aligned} G_{\mathbf{k}}(\omega) = &-\frac{n_{\mathbf{k}}+1}{2M\Omega_{\mathbf{k}}}\left\{\frac{1}{\omega+\Omega_{\mathbf{k}}-i\eta}-\frac{1}{\omega-\Omega_{\mathbf{k}}+i\eta}\right\} \\ &-\frac{n_{\mathbf{k}}}{2M\Omega_{\mathbf{k}}}\left\{\frac{1}{\omega-\Omega_{\mathbf{k}}-i\eta}-\frac{1}{\omega+\Omega_{\mathbf{k}}+i\eta}\right\}. \end{aligned} \tag{2.3.16}$$

The poles of this function now correspond to the energy differences between the excited states of the system. In fact, because in the harmonic approximation the matrix elements governing transitions between phonon states vanish unless only a single phonon is emitted or absorbed, the poles are at $\pm\Omega_{\mathbf{k}}$ as at $T = 0$ (but there are now two poles in the upper and two in the lower half-plane). In more general problems the function $G(\omega)$ contains both energies and lifetimes of excited states. Eq. (2.3.16) should be compared with the simpler expressions for $G_{\mathbf{k}}(\bar{\mu})$, Eq. (2.2.21), and $G_{\mathbf{k}}{}^{R}(\omega)$, Eq. (2.3.10); we note that, as at $T = 0$, the retarded Green's function is regular in the upper half of the ω-plane. The time-dependent Green's function corresponding to Eq. (2.3.16) is

$$G_{\mathbf{k}}(t) = -\frac{i}{2M\Omega_{\mathbf{k}}}(n_{\mathbf{k}}+1)\,e^{-i\Omega_{\mathbf{k}}|t|} - \frac{i}{2M\Omega_{\mathbf{k}}}\,n_{\mathbf{k}}\,e^{i\Omega_{\mathbf{k}}|t|}, \tag{2.3.17}$$

in close analogy to the form of the temperature Green's function [compare Eq. (2.2.14)].

Chapter 3

The Feynman-Dyson Expansion

In the first two chapters we calculated the phonon Green's function by setting up and solving its equation of motion. We obtained a closed set of equations with an exact closed solution; hence the perturbation series could be summed exactly. In more general many-body problems the situation is not so simple. For non-linear problems, when one attempts to set up an equation of motion for the Green's function by calculating its time derivatives, higher-order Green's functions appear and the system of equations does not close. One can in turn write down the equations of motion of the new Green's functions, thus obtaining Green's functions of still higher order, and in this way one generates an infinite hierarchy of differential equations. It is not usually easy to do infinite-order perturbation theory in this way, as the calculations quickly become very cumbersome. This does not mean that the formalism is useless, and in fact one can often extract relatively simple closed answers by use of a decoupling approximation in which, at some stage, the chain of equations is broken off and the higher order Green's functions are expressed approximately in terms of lower-order functions. We shall give examples of such an approach in later chapters. Decoupling approximations correspond, in some sense, to partial summations of the perturbation series, but it is not always easy to establish this correspondence or to assess the validity of the approximations involved.

We therefore now consider a powerful alternative method of developing the perturbation theory for the Green's function, which is based more directly on the methods of relativistic quantum field theory. This is the Feynman–Dyson expansion [Feynman (1949), Dyson (1949)]. In contrast to methods based on equations of motion, the idea is to expand the operators in the original form of the Green's function before one attempts to evaluate expectation values of operators. This leads to systematic rules for obtaining terms of arbitrary order in the perturbation series. The terms are in practice classified and displayed most conveniently by pictorial methods, in terms of the Feynman diagrams of quantum field theory. There are in fact two ways one can tackle the evaluation of the expectation values of operator products occurring in the expansion: the "time-labeled" form, which corresponds to the familiar Schrödinger perturbation theory for the one-body problem, and the "time-ordered" or Feynman–Dyson theory. The application of the time-labeled form to many-body problems has been developed by Hugenholtz (1957) and by Goldstone (1957). The two methods correspond to different ways of expressing and classifying the diagrams, but they are entirely equivalent and we confine ourselves here to the Feynman–Dyson form.

There are other methods, besides the equation-of-motion approach and the Feynman–Dyson perturbation theory, of generating approximate schemes for solving many-body problems. One approach, developed in recent years particularly in connection with magnetic problems, expresses the quantities of interest as functional integrals in which the integrand involves a simpler hamiltonian than appears in the original problem. One can then try to develop approximate methods for evaluating such integrals [for an example see Hamann (1970)]. We shall not, however, discuss this topic further.

In this chapter we illustrate the Feynman–Dyson formalism by using it to rederive the results of the last two chapters for the harmonic lattice. This enables us to display many of the techniques required for more general problems on this very simple example which also contains some of the physics characteristic of more general types of interacting systems. However, the details of the rules for writing down Feynman diagrams depend on the form of the hamiltonian and cannot be given once and for all for every problem; thus they will have to be reexamined for problems to be discussed in later chapters. The general quantum-theoretical formalism required for the analysis is presented in detail in the larger works on quantum field theory and many-body systems [see in particular Akhieser and Berestetskii (1965), Bogoliubov and Shirkov (1959), Abrikosov, Gor'kov and Dzyaloshinskii (1963), Fetter and Walecka (1971)]. We give a summary in Sec. 3.1, referring the reader to the larger works for detailed proofs. In Sec. 3.2 we use the formalism to evaluate the phonon Green's function at zero temperature, and in Secs. 3.3 and 3.4 we show how the theory can again be generalized to deal with the finite-temperature case.

3.1. ZERO-TEMPERATURE THEORY: GENERAL FORMALISM

It is convenient, for the type of perturbation theory to be developed, to transform state vectors and operators from the Heisenberg to the *interaction picture.* This is an intermediate form between the Schrödinger and Heisenberg pictures, in which both the state vectors and the operators are time-dependent but in which the operators have a particularly simple time dependence determined by the unperturbed hamiltonian. With $H = H_0 + H_1$, the state vector in the interaction picture is defined by

$$|\Psi_I(t)\rangle = e^{iH_0 t}|\Psi_S(t)\rangle, \tag{3.1.1}$$

where $|\Psi_S\rangle$is the Schrödinger state vector, and satisfies

$$i\frac{\partial|\Psi_I\rangle}{\partial t} = H_1(t)|\Psi_I\rangle, \tag{3.1.2}$$

where

$$H_1(t) = e^{iH_0 t}H_1 e^{-iH_0 t}. \tag{3.1.3}$$

(The origin of time has been chosen such that the Heisenberg, interaction and Schrödinger pictures all coincide at $t = 0$.) In general, operators in the interaction picture are related to Schrödinger operators by

$$\mathcal{O}_I(t) = e^{iH_0 t}\mathcal{O}_S e^{-iH_0 t}. \tag{3.1.4}$$

We now define the *time development operator* $U(t, t')$ by

$$|\Psi_I(t)\rangle = U(t, t')|\Psi_I(t')\rangle, \tag{3.1.5}$$

and it follows from the time variation (3.1.1) and (1.4.2) of the state vectors that

$$U(t, t') = e^{iH_0 t} e^{-iH(t-t')} e^{-iH_0 t'}. \tag{3.1.6}$$

This expression cannot in general be simplified further since H and H_0 do not commute. The operator U obviously satisfies the Schrödinger equation

$$i\frac{\partial U(t, t')}{\partial t} = H_1(t)U(t, t'), \tag{3.1.7}$$

and possesses the *group property*

$$U(t, t')U(t', t'') = U(t, t'') \tag{3.1.8}$$

and the properties of *hermiticity*

$$U(t, t') = U^{\dagger}(t', t) \tag{3.1.9}$$

and *unitarity*

$$U(t, t')U^{\dagger}(t, t') = 1. \tag{3.1.10}$$

State vectors $|\Psi_H\rangle$ and operators $\mathcal{O}_H(t)$ in the Heisenberg picture are related to those in the interaction picture by

$$|\Psi_H\rangle = e^{iHt}|\Psi_S(t)\rangle = e^{iHt}\, e^{-iH_0t}|\Psi_I(t)\rangle = U^{\dagger}(t, 0)|\Psi_I(t)\rangle, \tag{3.1.11}$$

and

$$\mathcal{O}_H(t) = U^{\dagger}(t, 0)\mathcal{O}_I(t)U(t, 0). \tag{3.1.12}$$

We now introduce the *adiabatic hypothesis*. We wish to express the Green's function in terms of expectation values with respect to the ground state $|\Phi_0\rangle$ of the unperturbed hamiltonian H_0. We assume that at time $t = -\infty$ the system is in the state $|\Phi_0\rangle$, that the interaction is then "switched on" to increase infinitely slowly so as to reach its full value H_1 at time $t = 0$, and is then switched off again to vanish adiabatically at $t = +\infty$. To formulate this time variation mathematically we multiply H_1 by a factor $\exp(-\epsilon|t|)$ and allow ϵ to tend to zero at the end of the calculations. This somewhat artificial procedure ensures that all the integrals appearing in the formalism have well-defined values, and one can show [Gell–Mann and Low (1951)] that the quantity

$$\frac{U(0, -\infty)|\Phi_0\rangle}{\langle\Phi_0|U(0, -\infty)|\Phi_0\rangle} \tag{3.1.13}$$

obtained in this way at time $t = 0$ exists in the limit $\epsilon = 0$ and is an eigenstate of the Hamiltonian $H_0 + H_1$. [The argument has to be formulated with care, because the numerator and denominator of (3.1.13) are separately infinite in the limit $\epsilon = 0$.] The theorem defines the state that develops from the non-interacting hamiltonian by adiabatic switching on of the interaction, but it does not require this state to be the ground state of the interacting system. To obtain the new ground state by this procedure it is necessary to assume in addition that the eigenstates of the perturbed system evolve from the unperturbed eigenstates without any crossing of states. This excludes the possibility of instabilities such as superconductivity which lead to a new kind of ground state with a different symmetry and a lower energy than the state obtained adiabatically from $|\Phi_0\rangle$. In fact the ground state energy in this case does not have a perturbation series in the coupling constant. In Chap. 10 we shall discuss how the Green's function approach can be adapted to deal with such phenomena.

We now use this formalism to re-express the Green's function in a form in which the expectation values refer to the unperturbed ground state $|\Phi_0\rangle$. The state vector used in the original definition (1.4.6) of G is $U(0, -\infty)|\Phi_0\rangle$, and we have

$$U(0,-\infty)|\Phi_0\rangle = U(0,\infty)U(\infty,-\infty)|\Phi_0\rangle = U(0,\infty)S|\Phi_0\rangle, \qquad (3.1.14)$$

where we have introduced the *S-matrix* $S = U(\infty, -\infty)$. Since the interaction is again zero at $t = +\infty$, the state $S|\Phi_0\rangle$ can differ from $|\Phi_0\rangle$ only by a phase factor of unit modulus, *provided* that the state which develops from $|\Phi_0\rangle$ is non-degenerate. We make this assumption and thus have

$$S|\Phi_0\rangle = e^{i\alpha}|\Phi_0\rangle. \qquad (3.1.15)$$

Transforming the Heisenberg operators in the Green's function (1.4.6) to the interaction picture by means of Eq. (3.1.12) and using the properties (3.1.8) and (3.1.9) of the operator U we have, for $t > 0$,[1]

$$\begin{aligned} G_{ij}(t) &= -i\langle\Phi_0|S^\dagger U^\dagger(0,\infty)U^\dagger(t,0)u_i(t)U(t,0)u_j(0)U(0,-\infty)|\Phi_0\rangle \\ &= -i\,e^{-i\alpha}\langle\Phi_0|U(\infty,t)u_i(t)U(t,0)u_j(0)U(0,-\infty)|\Phi_0\rangle \\ &= -i\,e^{-i\alpha}\langle\Phi_0|T[u_i(t)u_j(0)U(\infty,t)U(t,0)U(0,-\infty)]|\Phi_0\rangle \\ &= -i\,e^{-i\alpha}\langle\Phi_0|T[u_i(t)u_j(0)S]|\Phi_0\rangle. \end{aligned}$$

Also, using Eq. (3.1.15),

$$\langle\Phi_0|S|\Phi_0\rangle = e^{i\alpha}\langle\Phi_0|\Phi_0\rangle = e^{i\alpha}.$$

The same expression is obtained for G_{ij} when $t < 0$, and we thus have the result

$$G_{ij}(t) = -i\,\frac{\langle\Phi_0|T[u_i(t)u_j(0)S]|\Phi_0\rangle}{\langle\Phi_0|S|\Phi_0\rangle}. \qquad (3.1.16)$$

Thus, by use of the time-ordering symbol, we have obtained a compact form for the Green's function in the interaction picture. Note that the T symbol applies to the operators in S as well as to the product $u_i(t)u_j(0)$.

[1] This transformation can again be justified by an $\epsilon \to 0$ limiting procedure; see Fetter and Walecka (1971), p. 83. In the rest of this section all operators are in the interaction picture, and we omit the suffix I.

The use of the T operator here is a generalization of its use in Chap. 1. It is to be understood to act in terms of the time labels of the $u(t)$ operators implicit in the expansion of S [see Eq. (3.1.23)]. The idea is that, in any given term in the expansion, the T operator automatically orders all operators $u(t_1)u(t_2)\ldots u(t_n)$ in such a term in order of occurrence of the t_i labels. Thus, if $t_{i_1} > t_{i_2} > \ldots > t_{in}$, then

$$T\{u(t_1)u(t_2)\ldots u(t_n)\} = u(t_{i_1})u(t_{i_2})\ldots u(t_{in}). \tag{3.1.17}$$

The form (3.1.16) has the advantage that the interaction H_1 is contained entirely in S, so that we can confine attention to the perturbation series for this quantity. To derive this, standard field-theoretic methods are available.

To obtain the required series we note that the differential equation (3.1.7) together with the initial condition $U(t, t) = 1$ is equivalent to the integral equation

$$U(t, t') = 1 - i\lambda \int_{t'}^{t} dt_1 H_1(t_1)U(t_1, t'). \tag{3.1.18}$$

Here we have again introduced a coupling constant λ into the interaction. We solve Eq. (3.1.18) by direct iteration, and obtain

$$\begin{aligned} U(t, t') = 1 - i\lambda \int_{t'}^{t} dt_1 H_1(t_1) + (i\lambda)^2 \int_{t'}^{t} dt_1 \int_{t'}^{t_1} dt_2 H_1(t_1)H_1(t_2) \\ + \cdots + (-i\lambda)^n \int_{t'}^{t} dt_1 \int_{t'}^{t_1} dt_2 \ldots \int_{t'}^{t_{n-1}} dt_n H_1(t_1)\ldots H_1(t_n) + \cdots. \end{aligned} \tag{3.1.19}$$

The order of the factors is important in these integrals, since $H_1(t_1)$ and $H_1(t_2)$ do not commute if $t_1 \neq t_2$. The essential step in the Feynman–Dyson evaluation of the terms in this series now comes in a trick, explicitly formulated by Dyson, in which the "time-labeling" of the operators is removed in favor of the effect of the time-ordering operator T. Consider, in second order,

$$\int_{t'}^{t} dt_1 \int_{t'}^{t_1} dt_2 H_1(t_1)H_1(t_2) + \int_{t'}^{t} dt_1 \int_{t_1}^{t} dt_2 H_1(t_2)H_1(t_1). \tag{3.1.20}$$

Interchange of the order of integration in the second term shows that the two terms of (3.1.20) are equal. In the first term $t_1 \geqslant t_2$; in the second term $t_2 \geqslant t_1$. Hence, using the T symbol, we can write the first term alone as

$$\tfrac{1}{2} \int_{t'}^{t} dt_1 \int_{t'}^{t} dt_2 T[H_1(t_1)H_1(t_2)].$$

This transformation can be extended to the general case of n arguments, when there are $n!$ equal terms of the type (3.1.20), corresponding to the $n!$ possible orderings of the time labels $t_1, \ldots, t_n$. The new form of the series (3.1.19) is thus

$$U(t, t') = 1 - i\lambda \int_{t'}^{t} dt_1 H_1(t_1) + \frac{(i\lambda)^2}{2!} \int_{t'}^{t} dt_1 \int_{t'}^{t} dt_2 T[H_1(t_1)H_1(t_2)] + \cdots + \frac{(-i\lambda)^n}{n!} \int_{t'}^{t} dt_1 \int_{t'}^{t} dt_2 \ldots \int_{t'}^{t} dt_n T[H_1(t_1) \ldots H_1(t_n)] + \cdots. \tag{3.1.21}$$

This can be summed symbolically as an exponential series to give

$$U(t, t') = T \exp \left\{ -i\lambda \int_{t'}^{t} dt_1 H_1(t_1) \right\}; \tag{3.1.22}$$

in practice, however, this expression must always be interpreted as the series (3.1.21). Thus S has the expansion

$$S = 1 - i\lambda \int_{-\infty}^{\infty} dt_1 H_1(t_1) + \frac{(i\lambda)^2}{2!} \int_{-\infty}^{\infty} dt_1 \int_{-\infty}^{\infty} dt_2 T[H_1(t_1)H_1(t_2)] + \cdots + \frac{(-i\lambda)^n}{n!} \int_{-\infty}^{\infty} dt_1 \ldots \int_{-\infty}^{\infty} dt_n T[H_1(t_1) \ldots H_1(t_n)] + \cdots, \tag{3.1.23}$$

and this gives for the Green's function (3.1.16) the series

$$G_{ij}(t) = G^0(t)\delta_{ij} - i \sum_{n=1}^{\infty} \frac{(-i\lambda)^n}{n!} \int_{-\infty}^{\infty} dt_1 \ldots \int_{-\infty}^{\infty} dt_n \times \langle \Phi_0 | T[u_i(t)u_j(0)H_1(t_1) \ldots H_1(t_n)] | \Phi_0 \rangle. \tag{3.1.24}$$

Here, however, we have ignored the terms coming from the expansion of the denominator $\langle\Phi_0|S|\Phi_0\rangle$; we return later to the reason for this omission.

To evaluate the terms in this series we have to express the operators $u_i(t)$, $H_1(t)$ in terms of the creation and annihilation operators $B_i^\dagger(t)$ and $B_i(t)$. The mathematical problem is then that of evaluating the expectation value of a time-ordered product of an arbitrary number of such non-commuting operators. There is no difficulty in principle in carrying out such a calculation, since in the interaction picture the expectation values are expressed with respect to the unperturbed ground state $|\Phi_0\rangle$, and the time dependence of the operators is determined by the unperturbed frequencies according to

$$B_i(t) = B_i\, e^{-i\Omega_0 t}, \qquad B_i^\dagger(t) = B_i^\dagger e^{i\Omega_0 t}. \tag{3.1.25}$$

The time factors can thus be removed from the bracket, and the calculation is similar to the determination of the unperturbed Green's function in Sec. 1.5. But in practice a direct evaluation quickly becomes very unwieldy because of the rapidly increasing number of terms as the order n increases, and it is essential to proceed systematically. The mathematical result which allows us to classify the terms in a simple way is *Wick's theorem* [Wick (1950)]. The idea is to rearrange the order of factors in the operator products (using the commutation rules) so that the annihilation operators stand on the right. The expectation values of the rearranged terms with respect to the unperturbed ground state then vanish, since, by definition of $|\Phi_0\rangle$,

$$B_i|\Phi_0\rangle = \langle\Phi_0|B_i^\dagger = 0 \qquad \text{for all } i. \tag{3.1.26}$$

But of course the process introduces additional terms, proportional to the commutators of the operators involved in the rearrangement, which survive to give the final result. In its most general form Wick's theorem is a relation between operators, but we shall only state the restricted version which applies to ground state averages, since this is what is required in practice.

The theorem can then be stated very simply. Define the *contraction* (or *pairing*) $\underbrace{A(t)B(t')}$ of any two creation or annihilation operators $A(t)$, $B(t')$ as the time-ordered ground state expectation value

$$\langle\Phi_0|T[A(t)B(t')]|\Phi_0\rangle,$$

which we write as $\langle T[A(t)B(t')]\rangle_0$. This is a c-number (not an operator) and, if non-zero, is in fact just a free propagator or Green's function. Then, from any arbitrary time-ordered product of creation and annihilation operators, construct the product of the contractions obtained by pairing off all the terms in some particular way, thus forming a "fully paired" product. Repeat the process by pairing off the terms in a different way to form another fully paired product, and continue until the operators have been paired off in all possible ways. The required form of Wick's theorem states that the ground state expectation value of the original operator product is the sum of *all the fully paired products* obtained in this way. (Clearly the result is non-zero only if the number of terms in the original product is even.) In the form stated here the theorem applies to operators satisfying the boson commutation rules (1.5.4). For fermion operators with anticommuting properties there is an analogous theorem incorporating certain changes in sign (see Chap. 6).

Thus Wick's theorem leads to an expansion of the exact Green's function in terms of the known unperturbed single-particle Green's functions. This expansion will be obtained explicitly for the phonon problem in the next section. Here we illustrate the theorem by evaluating as a simple example the expectation value

$$\mu = \langle T[B(t_1)B^\dagger(t_2)B(t_3)B^\dagger(t_3)]\rangle_0 . \qquad (3.1.27)$$

There are three ways of pairing off the operators, and hence by Wick's theorem we can write down that

$$\begin{aligned}\mu = &\langle T[B(t_1)B^\dagger(t_2)]\rangle_0\langle T[B(t_3)B^\dagger(t_3)]\rangle_0\\ &+\langle T[B(t_1)B(t_3)]\rangle_0\langle T[B^\dagger(t_2)B^\dagger(t_3)]\rangle_0\\ &+\langle T[B(t_1)B^\dagger(t_3)]\rangle_0\langle T[B^\dagger(t_2)B(t_3)]\rangle_0 .\end{aligned} \qquad (3.1.28)$$

Each term varies with time according to the factor $\exp\{i\Omega_0(t_2 - t_1)\}$, and the second term is evidently identically zero. Suppose $t_1 > t_2 > t_3$; then the third term also vanishes, and (omitting the time factor) we have

$$\mu = \langle BB^\dagger\rangle_0\langle BB^\dagger\rangle_0 = 1, \qquad (3.1.29)$$

since $BB^\dagger = B^\dagger B + [B, B^\dagger]$, $[B, B^\dagger] = 1$ and $\langle B^\dagger B\rangle_0 = 0$. If, on the other hand, $t_1 > t_3 > t_2$, the first and third terms are non-zero, and

$$\mu = \langle BB^\dagger\rangle_0\langle BB^\dagger\rangle_0 + \langle BB^\dagger\rangle_0\langle BB^\dagger\rangle_0 = 2. \qquad (3.1.30)$$

As a check we find by direct evaluation of (3.1.27) that, in the first case

$$\langle BB^\dagger BB^\dagger\rangle_0 = \langle BB^\dagger B^\dagger B\rangle_0 + \langle BB^\dagger[B, B^\dagger]\rangle_0 = \langle BB^\dagger\rangle_0 = 1,$$

and in the second case

$$\langle BBB^\dagger B^\dagger\rangle_0 = \langle BB^\dagger BB^\dagger\rangle_0 + \langle B[B, B^\dagger]B^\dagger\rangle_0 = \langle BB^\dagger\rangle_0 + \langle BB^\dagger\rangle_0 = 2.$$

There are altogether six possible time orderings, and consideration of the remaining four cases is left to the reader.

It is seen that Wick's theorem, by including all possible contractions, adds in additional terms which are in fact zero but which allow a simple general result to be stated. A general proof of the theorem can be given by the method of induction and will be found in the works cited earlier. The applicability of the theorem is seen to depend on the use of time ordering in the original definition of the Green's function.

3.2. EVALUATION OF THE PHONON GREEN'S FUNCTION AT $T = 0$ BY FEYNMAN-DYSON PERTURBATION THEORY

For our oscillator problem the interaction hamiltonian is obtained by expressing Eq. (1.1.6) in the interaction picture. Thus

$$H_1(t) = \tfrac{1}{2}M \sum_{i \neq j} D_{ij}u_i(t)u_j(t), \tag{3.2.1}$$

where u_i is given in terms of the annihilation and creation operators B_i, $B_i^\dagger$ by Eq. (1.5.8) and the time dependence of the operators is determined by Eq. (3.1.25). In applying Wick's theorem to the evaluation of the expectation value $\langle T[u_i(t)u_j(0)H_1(t_1) \ldots H_1(t_n)]\rangle_0$, a considerable simplification is possible since only the operators $u_i(t)$ occur in H_1. The B_i, $B_i^\dagger$ operators thus occur always in the linear combinations of Eq. (1.5.8); this means that we need not introduce these operators explicitly, but can apply Wick's theorem directly to the u_i's. The results are thus expressed in terms of the contraction of two u's, and the essential point is that this is just the Green's function for the unperturbed oscillator problem which has already been evaluated in Chap. 1. In fact [see Eq. (1.5.2)]

$$\langle T[u_i(t)u_j(t')]\rangle_0 = iG^0_{ij}(t - t') = iG^0(t - t')\delta_{ij}, \tag{3.2.2}$$

where $G^0(t)$ is given by (1.5.11).

Thus, for $n = 1$, we have to evaluate

$$\langle T[u_i(t)u_j(0)H_1(t_1)]\rangle_0$$

$$= \tfrac{1}{2}M \sum_{i_1 \neq j_1} D_{i_1 j_1} \langle T[u_i(t)u_j(0)u_{i_1}(t_1)u_{j_1}(t_1)]\rangle_0. \qquad (3.2.3)$$

It is clear that, for the present problem, contractions at a *single time* vanish (this is because we have only interactions between atoms with *different* labels i, j). Hence, in pairing the operators, there are only two non-zero possibilities as indicated in Eq. (3.2.3), and we obtain

$$\tfrac{1}{2}M \sum_{i_1 \neq j_1} D_{i_1 j_1} \{iG^0(t - t_1)\delta_{ii_1} iG^0(0 - t_1)\delta_{jj_1}$$

$$+ iG^0(t - t_1)\delta_{ij_1} iG^0(0 - t_1)\delta_{ji_1}\}$$

$$= \tfrac{1}{2}M\{iG^0(t - t_1)D_{ij}iG^0(t_1) + iG^0(t - t_1)D_{ji}iG^0(t_1)\} \qquad (3.2.4)$$

[note that $G^0(t - t') = G^0(t' - t)$]. Since $D_{ij} = D_{ji}$ the two terms in Eq. (3.2.4) are equal.

We again draw Feynman diagrams to represent the terms of the perturbation series. The rules for constructing the diagrams are as follows. As in Sec. 1.5 we imagine the time to be increasing from left to right. A factor $iG^0(t^\alpha - t^\beta)$ is represented by a straight line running from time t^α to t^β, with an arrow pointing from t^α towards t^β. Each such line is labeled with an atomic index i. At each intermediate time t^α between 0 and t at which a G^0 line begins and another ends, there is an interaction factor D_{ij} or D_{ji} represented by a dotted line and carrying the labels of the two G^0 lines. The convention regarding the order of the suffixes of D_{ij} is that Fig. 3.1(a) corresponds to a factor $G^0(t - t_1)D_{ji}G^0(t_1 - t_2)$, whereas Fig. 3.1(b) corresponds to $G^0(t - t_1)D_{ij}G^0(t_1 - t_2)$.

Fig. 3.1. First-order contributions to G_{ij}.

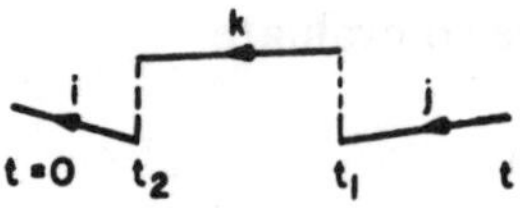

Fig. 3.2. A second-order diagram.

The two equal first-order terms in Eq. (3.2.4) are thus represented by the two diagrams of Fig. 3.1, with t_2 equal to zero. For $n = 2$ a typical diagram is shown in Fig. 3.2 and contributes a term

$$(\tfrac{1}{2}M)^2 \sum_k iG^0(t - t_1)D_{jk}iG^0(t_1 - t_2)D_{ik}iG^0(t_2).$$

To this we have to add all possible permutations of Fig. 3.2 which all give the same contribution to the Green's function (3.1.24) after integration over t_1 and t_2. Since each G^0 line can be connected to either the bottom or the top of a dotted interaction line, and since the intermediate times t_1, t_2 can be inserted in any order, there are altogether $4 \times 2 = 8$ diagrams equivalent to Fig. 3.2 (for general n, the number is clearly $2^n n!$).

The diagrams are essentially the same as those representing the perturbation series of Sec. 1.5 obtained from the equation of motion for G_{ij}, except that in the equation-of-motion approach all the separate permutations obtained here in each order are already summed into a single diagram.

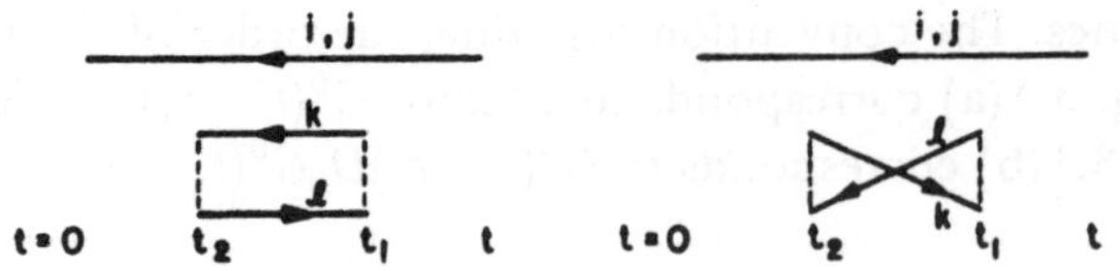

Fig. 3.3. Unlinked second-order diagrams.

In the Feynman–Dyson series there are, however, additional diagrams in second (and higher) order, different from those shown in Fig. 3.2. For

$n = 2$ these are the two diagrams shown in Fig. 3.3. They are *disconnected* or *unlinked* diagrams, containing parts which are not connected to the rest of the diagram by any lines. Such diagrams arise from contractions of the $H_1(t)$ terms in $\langle T[u_i(t)u_j(0)H_1(t_1)\ldots H_1(t_n)]\rangle_0$ which are not linked to $u_i(t)u_j(0)$; for example,

$$u_i(t)u_j(0)u_k(t_1)u_l(t_1)u_m(t_2)u_n(t_2).$$

For $n = 3$ we obtain linked diagrams of the type of Fig. 3.4 and unlinked diagrams of the types of Fig. 3.5; and so on.

Fig. 3.4. A third-order linked diagram.

Fig. 3.5. Third-order unlinked diagrams.

How do we deal with the contribution of the unlinked diagrams? At this stage we have to consider the effect of the denominator $\langle\Phi_0|S|\Phi_0\rangle$ in the expression (3.1.16) for the Green's function which has been ignored until now. The perturbation series for $\langle\Phi_0|S|\Phi_0\rangle$, which can of course also be obtained by means of Wick's theorem, gives the so-called vacuum fluctuation diagrams, representing the vacuum expectation value of the S-matrix. Now one can show generally that the contributions in the numerator of Eq. (3.1.16) from the unlinked diagrams are exactly canceled to all orders by corresponding contributions coming from the denominator, so that the unlinked diagrams in fact disappear from the final expression for $G_{ij}(t)$. This "*linked-cluster theorem*" was first derived (for the ground state energy) by Brueckner (1955) and Goldstone (1957); for a formal proof see Abrikosov, Gor'kov, and Dzyaloshinskii (1963),

Sec. 8. The idea is simply that each *linked* Green's function diagram has associated with it other diagrams obtained by adding to it all possible vacuum fluctuation diagrams. Therefore if, for any particular linked diagram, we take the sum over all the associated vacuum fluctuation diagrams, we obtain always a factor $\langle\Phi_0|S|\Phi_0\rangle$ which is therefore a constant in the numerator of Eq. (3.1.16) multiplying each linked diagram. This factorization of the numerator is represented diagrammatically by Fig. 3.6, where the second bracket is the expansion

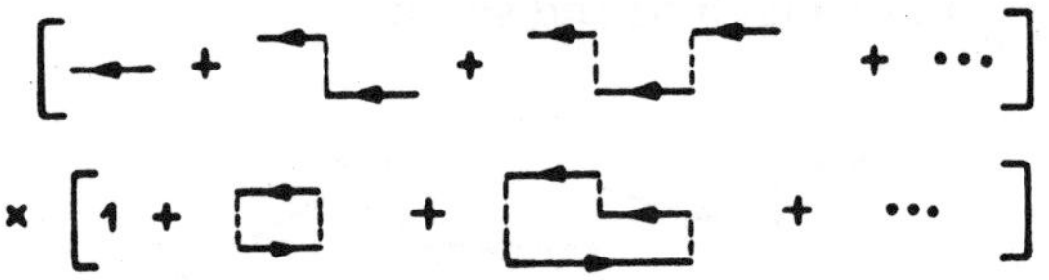

Fig. 3.6. The numerator of the Green's function.

of $\langle\Phi_0|S|\Phi_0\rangle$. Canceling this with the denominator, we can replace Eq. (3.1.16) by

$$G_{ij}(t) = -i\langle\Phi_0|T[u_i(t)u_j(0)S]|\Phi_0\rangle_{\text{linked}}, \tag{3.2.5}$$

and we accordingly reinterpret the perturbation series (3.1.24) for $G_{ij}(t)$ as containing only linked diagrams.

For the present problem we can easily write down the general nth order contribution of the linked diagrams. A particular diagram in nth order has n dotted interaction lines corresponding to intermediate times $t_1, t_2, \ldots, t_n$, and a summation over $(n-1)$ intermediate atoms between i and j. The contribution is thus

$$(\tfrac{1}{2}M)^n \sum_{i_1,\ldots,i_{n-1}} iG^0(t-t_1)D_{ii_1}iG^0(t_1-t_2)D_{i_1i_2}iG^0(t_2-t_3)$$
$$\ldots D_{i_{n-1}j}iG^0(t_n). \tag{3.2.6}$$

There are $2^n n!$ equivalent linked diagrams in nth order, all contributing the same amount. Hence, multiplying the term (3.2.6) by $2^n n!$ and putting the resulting expression for

$$\langle\Phi_0|T[u_i(t)u_j(0)H_1(t_1)\ldots H_1(t_n)]|\Phi_0\rangle_{\text{linked}}$$

into Eq. (3.1.24), we obtain for the Green's function the series

$$G_{ij}(t) = G^0(t)\delta_{ij} + \sum_{n=1}^{\infty} (\lambda M)^n \sum_{i_1,\ldots,i_{n-1}} \int_{-\infty}^{\infty} dt_1 \ldots \int_{-\infty}^{\infty} dt_n G^0(t - t_1)$$

$$\times D_{ii_1} G^0(t_1 - t_2) D_{i_1 i_2} G^0(t_2 - t_3) \ldots D_{i_{n-1} j} G^0(t_n). \qquad (3.2.7)$$

At this point we establish contact with Chap. 1, since this is precisely the iteration solution (1.5.14) of the equation of motion for G_{ij}. Because of the convolution form of the integrals the series (3.2.7) can now be summed in closed form as in Sec. 1.5 by going over to the Fourier transform of $G_{ij}(t)$, and we regain all the previous results.

In particular, we can evaluate the ground state energy from $G(t)$, as in Sec. 1.7, by an integration over the coupling constant. In fact however, if one is only interested in the ground state energy, one can proceed more directly by expressing E_G as an expectation value of the hamiltonian, using the exact ground state eigenfunction (3.1.13), and then developing a linked-cluster expansion for this quantity analogous to the expansion of the Green's function [Goldstone (1957)]. For the oscillator problem this leads to a logarithmic series which can again be summed exactly. We shall develop such an expansion for the free energy at non-zero temperature in Sec. 3.3 below.

In the present case, because of the simple chain form of the diagrams, it was easy to write down and evaluate the general term in the expansion of the time-dependent Green's function. In general, however, the evaluation of the time integrals in the higher terms of the perturbation series is complicated because each non-interacting Green's function $G^0(t - t')$ has a different analytic form for positive and for negative values of the time argument. It is therefore in general preferable to work with the Fourier transform $G^0(\omega)$ which has a simpler analytic form, and to use the corresponding Feynman diagrams in ω-space. In the present problem the diagrams corresponding to the Fourier-transformed perturbation series (1.6.3) are in fact the same as the diagrams for the time-dependent Green's function, except that time labels are absent. In Chap. 6, when we apply the diagram technique to the interacting electron gas, we shall work with Feynman diagrams in (four-dimensional) momentum space.

3.3. THE FEYNMAN-DYSON EXPANSION AT FINITE TEMPERATURES

The operator perturbation expansion of the Green's function studied in the previous section may also be generalized to apply to the thermodynamic Green's functions introduced in Chap. 2. That this should be possible seems likely from the observation that the thermodynamic Heisenberg representation arises from the continuation $it \to \sigma$ of the time-dependent form. In this section we show how this continuation is carried out. As in Chap. 2 we use the canonical ensemble for the phonon problem, and we are therefore interested in

$$G_{ij}(\sigma - \sigma') = \mathrm{Tr}\,[e^{-\beta H} T[\tilde{u}_i(\sigma)\tilde{u}_j(\sigma')]]/\mathrm{Tr}\, e^{-\beta H}. \tag{3.3.1}$$

To evaluate this by perturbation theory we introduce the interaction picture development operator for the Bloch equation, analogous to the time-dependent definition (3.1.6):

$$U(\sigma, \sigma') = e^{\sigma H_0}\, e^{-(\sigma - \sigma')(H_0 + H_1)}\, e^{-\sigma' H_0}. \tag{3.3.2}$$

This satisfies

$$-\frac{\partial U}{\partial \sigma} = H_1(\sigma) U(\sigma, \sigma'), \tag{3.3.3}$$

where

$$H_1(\sigma) = e^{\sigma H_0} H_1\, e^{-\sigma H_0}. \tag{3.3.4}$$

The operator U is no longer unitary, but it still has the group property

$$U(\sigma, \sigma')U(\sigma', \sigma'') = U(\sigma, \sigma''). \tag{3.3.5}$$

We have

$$e^{-\beta H} = e^{-\beta H_0} U(\beta, 0), \tag{3.3.6}$$

and we may therefore rewrite $G(\sigma - \sigma')$ in the form, for $\sigma > \sigma'$,

$$\begin{aligned} G_{ij}(\sigma - \sigma') &= \frac{1}{Z}\, \mathrm{Tr}\,\{e^{-\beta H_0} U(\beta, 0) U(0, \sigma) u_i(\sigma) U(\sigma, \sigma') u_j(\sigma') U(\sigma', 0)\} \\ &= \frac{1}{Z}\, \mathrm{Tr}\,\{e^{-\beta H_0} U(\beta, \sigma) u_i(\sigma) U(\sigma, \sigma') u_j(\sigma') U(\sigma', 0)\}, \end{aligned}$$

where $Z = \mathrm{Tr}\, e^{-\beta H}$ is the partition function and $u_i(\sigma)$, $u_j(\sigma')$ are operators in the interaction picture. Generalizing the T operator to σ-ordering as in the t-ordering case, we thus have

$$G_{ij}(\sigma - \sigma') = \frac{\mathrm{Tr}\{e^{-\beta H_0} T[U(\beta, 0) u_i(\sigma) u_j(\sigma')]\}}{\mathrm{Tr}\{e^{-\beta H_0} U(\beta, 0)\}} \tag{3.3.7}$$

as the form which corresponds to Eq. (3.1.16) (the same result is obtained for $\sigma < \sigma'$).

$U(\beta, 0)$ possesses the analogous perturbation expansion to the time-dependent case: by iterating the Dyson integral equation

$$U(\sigma, 0) = 1 - \int_0^{\sigma} d\sigma' H_1(\sigma') U(\sigma', 0) \tag{3.3.8}$$

one has

$$U(\beta, 0) = 1 - \int_0^{\beta} H_1(\sigma)\, d\sigma + \int_0^{\beta} d\sigma_1 \int_0^{\sigma_1} d\sigma_2 H_1(\sigma_1) H_1(\sigma_2) \ldots,$$

which may be rewritten in terms of T-products as

$$U(\beta, 0) = 1 - \int_0^{\beta} H_1(\sigma_1)\, d\sigma_1 + \frac{1}{2!} \int_0^{\beta} d\sigma_1 \int_0^{\beta} d\sigma_2 T[H_1(\sigma_1) H_1(\sigma_2)]$$
$$- \cdots + \frac{(-1)^n}{n!} \int_0^{\beta} d\sigma_1 \ldots \int_0^{\beta} d\sigma_n T[H_1(\sigma_1) \ldots H_1(\sigma_n)] + \cdots. \tag{3.3.9}$$

In this way the numerator and denominator of the Green's function (3.3.7) can be expressed as series in increasing powers of H_1 which are similar in structure to the series for the zero-temperature Green's function. To evaluate the terms in these series, we now need a thermodynamic generalization of Wick's theorem. At first sight it would seem that the idea which works at zero temperature cannot apply, since, whereas the ground state expectation value of an operator product with an annihilation operator on the right is zero, this is not true for the ensemble average at $T \neq 0$. Nevertheless, as first shown by Matsubara (1955), a generalized Wick's theorem is valid when $T \neq 0$, applicable to the ensemble average of an operator product and similar in structure to the zero-temperature theorem for ground state expectation values. The contractions of pairs of creation and annihilation operators must now be defined as thermal averages with respect to exp $(-\beta H_0)$:

$$\underbrace{AB} = \langle AB \rangle_0 = \mathrm{Tr}\{e^{-\beta H_0} T(AB)\}/\mathrm{Tr}\, e^{-\beta H_0}. \tag{3.3.10}$$

Wick's theorem for thermodynamic averages then simply states that the thermodynamic average [with respect to exp $(-\beta H_0)$] of any product of operators is equal to the sum of products obtained by *all possible pairwise contractions* of the operators in the product [for a proof see Gaudin (1960)].

Finally, the dependence of each operator $B(\sigma)$, $B^\dagger(\sigma)$ in the product on the temperature variables σ, σ' may be evaluated directly using the commutation rules. For the hamiltonian

$$H_0 = \sum_i \Omega_0(B_i^\dagger B_i + \tfrac{1}{2}) \tag{3.3.11}$$

we have [see Eq. (2.2.9)]

$$B_i(\sigma) = e^{-\Omega_0 \sigma} B_i, \qquad B_i^\dagger(\sigma) = e^{+\Omega_0 \sigma} B_i^\dagger. \tag{3.3.12}$$

Hence the terms in the perturbation series for G may be evaluated directly in terms of products of contractions of the form

$$G^0(\sigma - \sigma')\delta_{ij} = \langle T[u_i(\sigma) u_j(\sigma')] \rangle_0. \tag{3.3.13}$$

Thus the algebraic structure of the finite-T expansion is identical with that of the $T = 0$ expansion, and the temperature Green's function therefore has the same set of Feynman diagrams as the zero-temperature Green's function. There are again both linked and unlinked diagrams, and as in the time-dependent case it may be shown that the contributions from unlinked diagrams cancel out exactly with the partition function in the denominator of G.

These are general results, and we are now in a position to apply them to our lattice dynamics example with

$$H_1 = \tfrac{1}{2}\lambda M \sum_{i \neq j} D_{ij} u_i u_j. \tag{3.3.14}$$

Just as in the previous section the nth term in the series becomes

$$G_{ij}^{(n)}(\sigma - \sigma') = (-\lambda M)^n \sum_{i_1,\ldots,i_{n-1}} \int_0^\beta d\sigma_1 \int_0^\beta d\sigma_2 \ldots \int_0^\beta d\sigma_n$$

$$\times\, G^0(\sigma - \sigma_1) D_{ii_1} G^0(\sigma_1 - \sigma_2) \ldots D_{i_{n-1}j} G^0(\sigma_n - \sigma'), \tag{3.3.15}$$

where the factor $2^n n!$ has been inserted to count the possible number of ways of performing the above contraction. $G^0(\sigma)$ is given by Eq. (2.2.14) and, because of the different form of $G^0(\sigma)$ for $\sigma > 0$ and $\sigma < 0$, direct evaluation of (3.3.15) leads to large numbers of terms and is very cumbersome. Here again the Fourier series representation (2.2.15) of $G^0(\sigma)$ over the interval $0 \leqslant \sigma \leqslant \beta$ provides an essential simplification, analogous to the use of the convolution theorem in the zero-temperature formalism. As in Chap. 2 we define

$$G^0(\sigma) = \sum_{\bar{\mu}} e^{i\bar{\mu}\sigma} G^0(\bar{\mu}), \tag{3.3.16}$$

($\bar{\mu} = 2\mu\pi/\beta$, μ = integer), with the inverse

$$G^0(\bar{\mu}) = \frac{1}{\beta} \int_0^\beta d\sigma\, e^{-i\bar{\mu}\sigma} G^0(\sigma). \tag{3.3.17}$$

The transform $G_{ij}^{(n)}(\bar{\mu})$ of $G_{ij}^{(n)}(\sigma)$ can then be evaluated explicitly by carrying out the σ-integrations successively, using

$$\int_0^\beta e^{i(\bar{\mu}'-\bar{\mu})\sigma}\, d\sigma = \beta \delta_{\bar{\mu},\bar{\mu}'}. \tag{3.3.18}$$

The result is

$$G_{ij}^{(n)}(\bar{\mu}) = (-\lambda M)^n \beta^n \sum_{i_1,\ldots,i_{n-1}} D_{ii_1} \ldots D_{i_{n-1}j} [G^0(\bar{\mu})]^{n+1}. \tag{3.3.19}$$

This is precisely the result obtained by iteration of the equation of motion (2.2.20) for $G_{ij}(\bar{\mu})$. Then, introducing **k**-space Green's functions as before, we may sum the perturbation series for $G_{\mathbf{k}}(\bar{\mu})$ in closed form to regain the earlier expression, Eq. (2.2.21).

Thus we see that the application of the Feynman–Dyson technique to the evaluation of temperature Green's functions is no more complicated than the zero-temperature formalism, and (with the appropriate formal modifications) the technique is essentially the same in both cases.

3.4. DIRECT EVALUATION OF THE FREE ENERGY BY FEYNMAN-DYSON PERTURBATION THEORY

If we are only interested in calculating the free energy of the system, we can work directly with a perturbation expansion of the partition function $Z = \text{Tr}\, e^{-\beta H}$, and it is then not necessary to proceed via the calculation of the full Green's function followed by an integration over the coupling constant. We now verify that this simpler treatment leads to the same final expression for $\mathscr{F}$.

$e^{-\beta H}$ is given in terms of $U(\beta, 0)$ by Eq. (3.3.6), and therefore

$$\frac{Z}{Z_0} = \frac{\text{Tr}\, e^{-\beta H}}{\text{Tr}\, e^{-\beta H_0}} = \frac{\text{Tr}\{e^{-\beta H_0} U(\beta, 0)\}}{\text{Tr}\, e^{-\beta H_0}} = \langle U(\beta, 0)\rangle_0, \tag{3.4.1}$$

where Z_0 is the unperturbed partition function. $\langle U(\beta, 0)\rangle_0$ is obtained, using the perturbation expansion (3.3.9), as a series containing both linked and unlinked "vacuum-fluctuation" diagrams. The sum of all these diagrams can be expressed in an elegant way in terms of the linked diagrams only. Suppose L is the sum of *all linked diagrams* that contribute to $\langle U(\beta, 0)\rangle_0$:

$$L = \sum_{i=1}^{\infty} L_i, \tag{3.4.2}$$

where L_i is the linked diagram of order i. Consider an arbitrary unlinked diagram which is made up of n_1 identical linked diagrams L_1, n_2 identical linked diagrams L_2, and so on. Its contribution to $\langle U(\beta, 0)\rangle_0$ is

$$\frac{L_1^{n_1}}{n_1!} \cdot \frac{L_2^{n_2}}{n_2!} \cdots,$$

where the denominators $n_i!$ allow for the arbitrary ordering of the n_i identical factors in the product. If we let $n_1, n_2, \ldots$ run independently over the values $0, 1, 2, \ldots$, we obtain all the diagrams contributing to

$\langle U(\beta, 0)\rangle_0$, and thus

$$\langle U(\beta, 0)\rangle_0 = \sum_{n_1=0}^{\infty} \sum_{n_2=0}^{\infty} \cdots \frac{L_1^{n_1}}{n_1!}\frac{L_2^{n_2}}{n_2!} \cdots$$

$$= e^{L_1} e^{L_2} \ldots = e^{\sum_i L_i} = e^L. \tag{3.4.3}$$

The change in free energy can thus be expressed in terms of the linked diagrams only, and we have

$$\Delta \mathscr{F} = -\frac{1}{\beta} \log \frac{Z}{Z_0} = -\frac{L}{\beta}. \tag{3.4.4}$$

We now have to evaluate the sum of all linked vacuum fluctuation diagrams, i.e., linked diagrams which begin and end at the same value σ_1 of the temperature variable (Fig. 3.7):

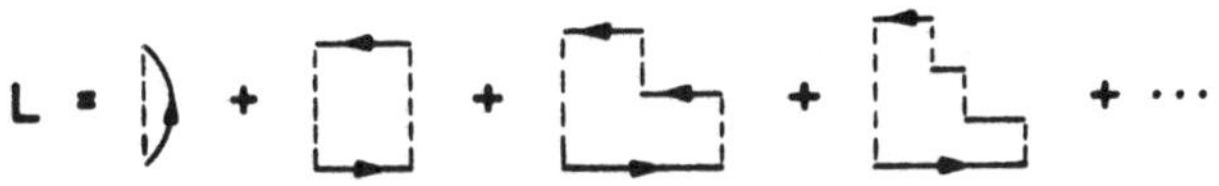

Fig. 3.7. Linked diagrams contributing to $\langle U(\beta, 0)\rangle_0$.

The nth-order contribution to L contains one G^0 line less than the nth-order contribution to the Green's function $G(\sigma - \sigma')$, so that there are now only $2^{n-1}(n-1)!$ identical diagrams in nth order; also all σ variables and atomic suffixes are now internal and must be summed. Hence, with

$$H_1 = \tfrac{1}{2}M \sum_{ij} D_{ij}u_i u_j,$$

we can write down that

$$\Delta \mathscr{F} = -\frac{1}{\beta} \sum_{n=1}^{\infty} (-1)^n \frac{(\frac{1}{2}M)^n}{n!} 2^{n-1}(n-1)! \sum_{i_1,\ldots,i_n} \int_0^\beta d\sigma_1 \ldots \int_0^\beta d\sigma_n$$
$$\times\, G^0(\sigma_1 - \sigma_2)D_{i_1 i_2} G^0(\sigma_2 - \sigma_3)D_{i_2 i_3} \ldots G^0(\sigma_n - \sigma_1)D_{i_n i_1}. \tag{3.4.5}$$

We simplify this, as in Sec. 3.3, by expanding $G^0(\sigma)$ as the Fourier series (3.3.16), which enables the σ integrals to be evaluated, and then transforming to k-space and performing the lattice sums. We obtain

$$\Delta \mathscr{F} = \frac{1}{2\beta} \sum_{\mathbf{k}} \sum_{\bar{\mu}} \sum_{n=1}^{\infty} \frac{(-1)^{n-1}}{n} \{\beta M D_{\mathbf{k}} G^0(\bar{\mu})\}^n, \tag{3.4.6}$$

where $\bar{\mu} = 2\mu\pi/\beta$ ($\mu = 0, \pm 1, \pm 2, \ldots$). The nth term in the sum over n now has denominator n, so that we have a logarithmic series instead of the geometric series obtained for the Green's function. Summing the series and using the expression (2.2.19) for $G^0(\bar{\mu})$, we have

$$\begin{aligned} \Delta \mathscr{F} &= \frac{1}{2\beta} \sum_{\mathbf{k}} \sum_{\bar{\mu}} \log \{1 + \beta M D_{\mathbf{k}} G^0(\bar{\mu})\} \\ &= \frac{1}{2\beta} \sum_{\mathbf{k}} \sum_{\bar{\mu}} \log \left(\frac{\bar{\mu}^2 + \Omega_{\mathbf{k}}^2}{\bar{\mu}^2 + \Omega_0^2} \right), \end{aligned} \tag{3.4.7}$$

where $\Omega_{\mathbf{k}}^2 = \Omega_0^2 + D_{\mathbf{k}}$.

At $T = 0$, one obtains, instead of (3.4.7),

$$\Delta E_G = \frac{i}{2} \sum_{\mathbf{k}} \int_{-\infty}^{\infty} \frac{d\omega}{2\pi} \log \left(\frac{1}{1 - M D_{\mathbf{k}} G^0(\omega)} \right), \tag{3.4.8}$$

which can be evaluated by contour integration and leads directly to the ground state energy (1.1.12).

The frequency sum at finite T is most easily evaluated by using the infinite product expansion for the hyperbolic sine:

$$\sinh x = x \left(1 + \frac{x^2}{\pi^2} \right) \left(1 + \frac{x^2}{2^2 \pi^2} \right) \cdots . \tag{3.4.9}$$

With this we obtain

$$\sum_{\bar{\mu}} \log \left(\frac{\bar{\mu}^2 + \Omega_{\mathbf{k}}^2}{\bar{\mu}^2 + \Omega_0^2} \right) = \log \prod_{\bar{\mu}} \left(\frac{\bar{\mu}^2 + \Omega_{\mathbf{k}}^2}{\bar{\mu}^2 + \Omega_0^2} \right) = \log \left\{ \frac{\sinh \frac{\beta \Omega_{\mathbf{k}}}{2}}{\sinh \frac{\beta \Omega_0}{2}} \right\}^2, \tag{3.4.10}$$

and hence, finally,

$$\Delta \mathscr{F} = \frac{1}{\beta} \sum_{\mathbf{k}} \log \left(\frac{\sinh \frac{\beta \Omega_{\mathbf{k}}}{2}}{\sinh \frac{\beta \Omega_0}{2}} \right), \tag{3.4.11}$$

in exact agreement with the result (2.2.27) obtained previously. In the limit $T \to 0$ this reduces correctly to the ground state energy formula

$$\Delta E_G = \tfrac{1}{2} \sum_{\mathbf{k}} (\Omega_{\mathbf{k}} - \Omega_0). \tag{3.4.12}$$

Thus in this example one can obtain the zero-temperature result either by working directly with the zero-temperature theory, or as the $T \to 0$ limit of the finite-temperature theory. It seems self-evident that these two approaches should give the same result for the ground state energy, but in fact, for interacting particles obeying *fermi* statistics, it unexpectedly turns out that this is not always true. There is in this case a rather subtle distinction between the two formalisms which arises from the singular nature of fermi distribution functions in the limit $T \to 0$. The problem has been analyzed by Kohn and Luttinger (1960) and Luttinger and Ward (1960), who give conditions under which the two approaches lead to the same result.

We have seen that, in our special soluble example of the harmonic lattice, every term in the perturbation series can be rigorously evaluated and the result summed exactly. In a non-linear problem such as that of the anharmonic lattice exact results cannot be obtained, but the terms in the perturbation series can be analyzed and classified in a similar way in terms of the topology of the associated Feynman diagrams, and approximate solutions can be obtained by summing appropriate subsets of diagrams. In this way the Feynman–Dyson expansion provides a very powerful approach to the perturbation-theoretic analysis of the Green's function for many-body systems.

Chapter 4

The Scattering of Fermions by a Localized Perturbation

The effect of impurity atoms on the properties of a metal is of considerable physical interest, both because of the possibility of deliberately introducing impurities so as to study the physics of electron–atom interactions in metals, and because most real materials contain impurities which affect their physical properties.

In this chapter we consider the effect of a single impurity atom on an electron gas. The electron gas is taken to be non-interacting, but is perturbed by the potential due to the impurity atom. (In real metals this will be a screened potential of the Hartree–Fock type.) The problem can immediately be seen to be expressible in terms of the quantum mechanics of scattering of a single electron from the impurity potential. One can imagine that one has solved the Schrödinger equation for the problem and produced a complete set of scattering eigenstates (and bound states if the potential is right). The statistical mechanics of the non-interacting many-electron system then follows by filling up the new set of energy levels according to fermi statistics. Thus in principle we do not need the methods of many-body theory to discuss the effect of impurities in metals, provided the effects of interactions between electrons are neglected. Indeed, historically, this problem was well understood before the techniques of many-body theory were developed.

However, as soon as interactions between electrons are included, we obtain a true many-body problem. It is then no longer possible to treat the statistical mechanics separately from the determination of the energy levels, and we do need to use many-body theory to understand the physics of the situation. It is therefore helpful to formulate the initial problem of a potential acting on a non-interacting fermi gas in terms of Green's functions, so that one is later in a position to generalize to the case of interactions. It turns out that, like the phonon problem, this problem is also a rather nice illustration of many of the properties of

Green's functions which one later requires in the theory of interacting systems.

The end product of the calculation of this chapter will be the effect of the impurity potential on the measurable physical properties of metals. In contrast to the situation in elementary particle problems, where scattering cross-sections can be measured directly as a function of energy and scattering angle, in a metal one has to rely on the effect of the scattering on the bulk thermodynamic properties of the material, such as the low-temperature electronic specific heat and the Pauli susceptibility. We will show, both by the traditional method and by a Green's function calculation, how the effect of impurities on these quantities can be expressed directly in terms of the phase shift of scattering on the impurity potential of the electrons at the fermi level.

4.1. SCATTERING OF A SINGLE ELECTRON

We start by formulating the problem of scattering of a single electron from a potential $U(\mathbf{x})$, in terms of scattering eigenstates $|\Psi\rangle$ satisfying the Schrödinger equation

$$(H_0 + U)|\Psi\rangle = \epsilon|\Psi\rangle \tag{4.1.1}$$

[for a fuller account of scattering theory, see for example Messiah (1961), Chaps. X and XIX]. In order to prescribe the scattering boundary conditions, which we take to be of outgoing wave form, $|\Psi\rangle$ is reexpressed as the solution of the scattering integral equation

$$|\Psi\rangle = |\mathbf{p}\rangle + \frac{1}{\epsilon - H_0 + i\eta} U|\Psi\rangle, \tag{4.1.2}$$

where $|\mathbf{p}\rangle$ is the incident wave. To ensure that $|\Psi\rangle - |\mathbf{p}\rangle$ represents an outgoing scattered wave, η must be chosen real and positive, with $\eta \to 0+$. The asymptotic behavior of the scattering solutions $|\Psi\rangle$ may be expressed in terms of the transfer or "T" matrix, defined by its matrix elements between states $|\mathbf{p}\rangle$ and $|\mathbf{p}'\rangle$, which are related to the matrix elements of U by means of

$$T(\mathbf{p}', \mathbf{p}) \equiv \langle\mathbf{p}'|T|\mathbf{p}\rangle = \langle\mathbf{p}'|U|\Psi\rangle. \tag{4.1.3}$$

The operator $(\epsilon - H_0 + i\eta)^{-1}$ is diagonal in the $\mathbf{p}$-representation, and its matrix elements form the outgoing wave Green's function of the free particle Schrödinger equation:

$$\left\langle \mathbf{p} \left| \frac{1}{\epsilon - H_0 + i\eta} \right| \mathbf{p}' \right\rangle = \frac{1}{\epsilon - \epsilon_{\mathbf{p}} + i\eta} \delta_{\mathbf{pp}'} = G^0(\mathbf{p}, \epsilon)\delta_{\mathbf{pp}'}. \tag{4.1.4}$$

In the x-representation [writing $\langle \mathbf{x}|\Psi\rangle = \Psi(\mathbf{x})$] it is then found that the integral equation (4.1.2) has the well-known form

$$\Psi(\mathbf{x}) = e^{i\mathbf{p}\cdot\mathbf{x}} - \frac{m}{2\pi}\int \frac{e^{ip|\mathbf{x}-\mathbf{x}'|}}{|\mathbf{x}-\mathbf{x}'|} U(\mathbf{x}')\Psi(\mathbf{x}')\, d^3x',$$

which has the required asymptotic behavior. By studying the asymptotic form of $\Psi(\mathbf{x})$, $T(\mathbf{p}', \mathbf{p})$ can be related to the phase shifts in a partial wave analysis of $\Psi(\mathbf{x})$:

$$\Psi(\mathbf{x}) \underset{x\to\infty}{\simeq} (px)^{-1} \sum_{l=0}^{\infty} (2l+1) i^l e^{i\delta_l} P_l(\cos\theta) \sin(px - \tfrac{1}{2}l\pi + \delta_l), \tag{4.1.5}$$

leading to

$$\frac{m}{2\pi} T(\mathbf{p}', \mathbf{p}) = \sum_{l=0}^{\infty} P_l(\cos\theta_{\mathbf{p}'\mathbf{p}}) \frac{e^{2i\delta_l} - 1}{2ip}. \tag{4.1.6}$$

The differential scattering cross-section is given in terms of $T(\mathbf{p}', \mathbf{p})$ by

$$\frac{d\sigma}{d\Omega} = \left| \frac{m}{2\pi} T(\mathbf{p}', \mathbf{p}) \right|^2.$$

We shall be particularly interested in potentials of very short range. Under conditions such that the wavelength of the incident electron is much greater than the range of the potential (assuming this is not a Coulomb potential), only the *s*-wave phase shift δ_0 is significant and all the formulas simplify; in particular

$$\frac{m}{2\pi} T = \frac{e^{2i\delta_0} - 1}{2ip}, \tag{4.1.7}$$

from which

$$\frac{d\sigma}{d\Omega} = \frac{\sin^2\delta_0}{p^2} = \left| \frac{m}{2\pi} T \right|^2, \tag{4.1.8}$$

and the total cross-section is

$$\sigma = \frac{4\pi \sin^2 \delta_0}{p^2}. \tag{4.1.9}$$

4.2. FORMULATION OF THE MANY-ELECTRON SCATTERING PROBLEM IN TERMS OF FERMION CREATION AND ANNIHILATION OPERATORS

We now consider a set of N electrons, all scattering from the same potential, with hamiltonian

$$H = \sum_i \frac{\mathbf{p}_i^2}{2m} + \sum_i U(\mathbf{x}_i). \tag{4.2.1}$$

The N-electron wave functions for the unperturbed problem ($U = 0$) can be written as Slater determinants of plane wave solutions

$$u_{\mathbf{p}}(\mathbf{x}) = V^{-1/2}\, e^{i\mathbf{p}\cdot\mathbf{x}}, \tag{4.2.2}$$

which we take to be normalized in a box of volume V, with periodic boundary conditions.

It is now convenient to go over to the formalism of second quantization (see Appendix 1) and to introduce the creation and annihilation operators $a_{\mathbf{p}}^\dagger$, $a_{\mathbf{p}}$ which for fermions satisfy the anticommutation rules

$$\{a_{\mathbf{p}}, a_{\mathbf{q}}^\dagger\} = \delta_{\mathbf{pq}}, \qquad \{a_{\mathbf{p}}, a_{\mathbf{q}}\} = \{a_{\mathbf{p}}^\dagger, a_{\mathbf{q}}^\dagger\} = 0, \tag{4.2.3}$$

where $\{A, B\}$ denotes $AB + BA$.

For present purposes it is not necessary to refer explicitly to the spin states of the particles, but, if required, we can regard the suffix **p** as referring to both momentum and spin states, and assume that a spin function is included in (4.2.2). The number operator associated with the state **p** is

$$n_{\mathbf{p}} = a_{\mathbf{p}}^\dagger a_{\mathbf{p}}; \tag{4.2.4}$$

in accordance with the Pauli principle it has only the two eigenvalues 0 and 1, corresponding to the unoccupied state $|0\rangle$ and the singly occupied state $a_{\mathbf{p}}^\dagger|0\rangle = |\mathbf{p}\rangle$. The many-particle states are of the form

$$|\mathbf{p}_1\mathbf{p}_2 \ldots \mathbf{p}_r\rangle = a_{\mathbf{p}_1}^\dagger a_{\mathbf{p}_2}^\dagger \ldots a_{\mathbf{p}_r}^\dagger |0\rangle; \tag{4.2.5}$$

in order to make the sign unique it is necessary, because of the anticommutation rules (4.2.3), to postulate a definite order for the one-particle states in the state vector.

The hamiltonian (4.2.1) is a sum $\sum_i h_i$ of one-body operators, and its second-quantized form is (see Appendix 1)

$$H = \sum_{\mathbf{pp}'} \langle \mathbf{p}'|h|\mathbf{p}\rangle a_{\mathbf{p}'}{}^{\dagger} a_{\mathbf{p}}, \tag{4.2.6}$$

where $\langle \mathbf{p}'|h|\mathbf{p}\rangle$ is the matrix element

$$\int_V u_{\mathbf{p}'}^*(\mathbf{x}) \left\{ \frac{\mathbf{p}^2}{2m} + U(\mathbf{x}) \right\} u_{\mathbf{p}}(\mathbf{x})\, d^3x. \tag{4.2.7}$$

With (4.2.2) we obtain at once

$$\langle \mathbf{p}'|h|\mathbf{p}\rangle = \epsilon_{\mathbf{p}} \delta_{\mathbf{p}'\mathbf{p}} + U(\mathbf{p}' - \mathbf{p}), \tag{4.2.8}$$

where $\epsilon_{\mathbf{p}} = \mathbf{p}^2/2m$, and $U(\mathbf{q})$ is the Fourier coefficient

$$U(\mathbf{q}) = \frac{1}{V} \int_V U(\mathbf{x})\, e^{-i\mathbf{q}\cdot\mathbf{x}}\, d^3x \tag{4.2.9}$$

in the Fourier decomposition of the scattering potential

$$U(\mathbf{x}) = \sum_{\mathbf{q}} U(\mathbf{q})\, e^{i\mathbf{q}\cdot\mathbf{x}}. \tag{4.2.10}$$

Since $U(\mathbf{x})$ is real we must have $U(\mathbf{q})^* = U(-\mathbf{q})$.

The hamiltonian is now

$$H = \sum_{\mathbf{p}} \epsilon_{\mathbf{p}} a_{\mathbf{p}}{}^{\dagger} a_{\mathbf{p}} + \sum_{\mathbf{q}} U(\mathbf{q}) \sum_{\mathbf{p}} a_{\mathbf{p}+\mathbf{q}}^{\dagger} a_{\mathbf{p}}. \tag{4.2.11}$$

In order to investigate the effect of the potential on physical properties of the metal we now need to calculate various expectation values of products of the $a_{\mathbf{p}}{}^{\dagger}$ and $a_{\mathbf{p}}$ operators with respect either to the perturbed N-particle ground state (at $T = 0$) or to a grand canonical average at finite temperatures.

Examples of such expectation values are the *ground state energy*

$$E_G = \langle \Psi_G|H|\Psi_G\rangle$$

$$= \sum_{\mathbf{pp}'} \{\delta_{\mathbf{pp}'}\epsilon_{\mathbf{p}} + U(\mathbf{p} - \mathbf{p}')\} \langle \Psi_G|a_{\mathbf{p}}{}^{\dagger} a_{\mathbf{p}'}|\Psi_G\rangle, \tag{4.2.12}$$

and the *thermodynamic potential* at $T \neq 0$, which may be obtained as in Chap. 2 by inserting a parameter λ in the strength of the potential U and differentiating:

$$\Omega - \Omega_0 = \int_0^1 d\lambda \sum_{\mathbf{pp}'} U(\mathbf{p} - \mathbf{p}') \frac{\mathrm{Tr}\{e^{-\beta[H(\lambda)-\mu\hat{N}]} a_{\mathbf{p}}{}^\dagger a_{\mathbf{p}'}\}}{\mathrm{Tr}\, e^{-\beta[H(\lambda)-\mu\hat{N}]}} \tag{4.2.13}$$

(here we have used a grand canonical average in which the equilibrium electron density is determined by the chemical potential μ). We shall also be interested in the *charge density* in the ground state, given by

$$\langle \rho(\mathbf{x}) \rangle_G = \frac{1}{V} \sum_{\mathbf{pp}'} e^{i(\mathbf{p}-\mathbf{p}')\cdot\mathbf{x}} \langle \Psi_G | a_{\mathbf{p}'}^\dagger a_{\mathbf{p}} | \Psi_G \rangle \tag{4.2.14}$$

(with units such that $e = 1$). All these quantities require a knowledge of expectation values of the form $\langle \Psi_G | a_{\mathbf{p}}{}^\dagger a_{\mathbf{q}} | \Psi_G \rangle$ (or a thermal average at $T \neq 0$), and this leads us again to study a one-particle Green's function for the many-electron system which we define (for $T = 0$) by

$$G^{(N)}(\mathbf{p}, t) = -i \langle \Psi_G{}^{(N)} | T[\tilde{a}_{\mathbf{p}}(t) \tilde{a}_{\mathbf{p}}{}^\dagger(0)] | \Psi_G{}^{(N)} \rangle, \tag{4.2.15}$$

where $|\Psi_G{}^{(N)}\rangle$ is the exact ground state of the N-electron system and $\tilde{a}_{\mathbf{p}}(t)$ is the Heisenberg operator

$$\tilde{a}_{\mathbf{p}}(t) = e^{iHt} a_{\mathbf{p}}\, a^{-iHt}. \tag{4.2.16}$$

This definition of G is analogous to Eq. (1.4.6), but for fermions obeying the rules (4.2.3) it is convenient to define the T operator with a change of sign such that

$$\begin{aligned} T[a(t) a^\dagger(t')] &= a(t) a^\dagger(t') \qquad (t > t'), \\ &= -a^\dagger(t') a(t) \qquad (t < t'). \end{aligned} \tag{4.2.17}$$

4.3. SINGLE-ELECTRON GREEN'S FUNCTION

We will show below that, because of the absence of interactions between electrons, the properties defined above in terms of the N-particle states of the system can actually be calculated in terms of a Green's function representing a single electron injected into an *empty* system. This is given by the value of (4.2.15) with N set equal to zero, i.e., by

$$G(\mathbf{p}, t) = -i \langle 0 | T[\tilde{a}_{\mathbf{p}}(t) \tilde{a}_{\mathbf{p}}{}^\dagger(0)] | 0 \rangle, \tag{4.3.1}$$

where $|0\rangle$ is the vacuum state with no particles present, for which we have $a_{\mathbf{p}}|0\rangle = 0$ for all $\mathbf{p}$. Hence the hamiltonian (4.2.11) is such that $H|0\rangle = 0$. Therefore $e^{iHt}|0\rangle = |0\rangle$, and for $t > 0$ we have

$$\begin{aligned} G(\mathbf{p}, t) &= -i\langle 0|e^{iHt}a_{\mathbf{p}}\, e^{-iHt}a_{\mathbf{p}}^{\dagger}|0\rangle \\ &= -i\langle 0|a_{\mathbf{p}}\, e^{-iHt}a_{\mathbf{p}}^{\dagger}|0\rangle = -i\langle \mathbf{p}|e^{-iHt}|\mathbf{p}\rangle, \end{aligned} \tag{4.3.2}$$

while, for $t < 0$,

$$\begin{aligned} G(\mathbf{p}, t) &= i\langle 0|a_{\mathbf{p}}^{\dagger}\, e^{iHt}a_{\mathbf{p}}\, e^{-iHt}|0\rangle \\ &= i\langle 0|a_{\mathbf{p}}^{\dagger}\, e^{iHt}a_{\mathbf{p}}|0\rangle = 0. \end{aligned}$$

Note that, for this particular example, the time-ordered Green's function defined by Eq. (4.3.1) is identical with the *retarded* function

$$G^R(\mathbf{p}, t) = -i\theta(t)\langle 0|\{\tilde{a}_{\mathbf{p}}(t), \tilde{a}_{\mathbf{p}}^{\dagger}(0)\}|0\rangle, \tag{4.3.3}$$

where $\{\cdot\cdot\cdot\}$ denotes an anticommutator.

The quantity which will be used below to express the properties of the N-electron system in terms of those of the one-electron system is the *density of one-electron states*, denoted by $\rho(\epsilon)$. If we confine the system in a box of finite volume V, then the scattering eigenstates of the Schrödinger equation (4.1.1) form a discrete set which we can denote by $|\Psi_m\rangle$ with eigenvalues ϵ_m. $\rho(\epsilon)$ is then defined by

$$\rho(\epsilon) = \sum_m \delta(\epsilon - \epsilon_m). \tag{4.3.4}$$

We now show that $\rho(\epsilon)$ can be obtained from the one-particle Green's function (4.3.1). This follows from the fact that the sum $\sum_{\mathbf{p}} \langle \mathbf{p}|e^{-iHt}|\mathbf{p}\rangle$ is the trace of the operator e^{-iHt}, and this is invariant under a change of basis from $|\mathbf{p}\rangle$ to $|\Psi_m\rangle$. Therefore, for $t > 0$,

$$\begin{aligned} \sum_{\mathbf{p}} G(\mathbf{p}, t) &= -i \operatorname{Tr} e^{-iHt} = -i \sum_m \langle \Psi_m|e^{-iHt}|\Psi_m\rangle \\ &= -i \sum_m e^{-iEmt}, \end{aligned} \tag{4.3.5}$$

assuming the $|\Psi_m\rangle$ to be normalized; (4.3.5) relates the exact eigenvalues of the problem to the properties of $G(\mathbf{p}, t)$. As in Sec. 1.6 we introduce the Fourier transform of $G(\mathbf{p}, t)$,

$$G(\mathbf{p}, \epsilon) = \int_{-\infty}^{\infty} G(\mathbf{p}, t)\, e^{i\epsilon t}\, dt. \tag{4.3.6}$$

Since $G(\mathbf{p}, t) = 0$ for $t < 0$ we then obtain, replacing ϵ by $\epsilon + i\eta$ where η is a positive infinitesimal,

$$\sum_{\mathbf{p}} G(\mathbf{p}, \epsilon) = -i \sum_m \int_0^\infty e^{i(\epsilon - \epsilon_m + i\eta)t}\, dt$$

$$= \sum_m \frac{1}{\epsilon - \epsilon_m + i\eta}. \tag{4.3.7}$$

The imaginary part of this is [see Eq. (2.3.11)]

$$-\pi \sum_m \delta(\epsilon - \epsilon_m),$$

and we thus obtain from Eq. (4.3.7) the desired result

$$\rho(\epsilon) = -\frac{1}{\pi} \operatorname{Im} \sum_{\mathbf{p}} G(\mathbf{p}, \epsilon). \tag{4.3.8}$$

To set up an equation of motion from which $G(\mathbf{p}, t)$ can be calculated, we note that $G(\mathbf{p}, t)$ is the diagonal part $F(\mathbf{p}, \mathbf{p}; t)$ of the function

$$F(\mathbf{p}, \mathbf{p}'; t) = -i\langle 0 | T[\tilde{a}_{\mathbf{p}}(t)\tilde{a}^\dagger_{\mathbf{p}'}(0)] | 0\rangle. \tag{4.3.9}$$

We have

$$i\frac{\partial}{\partial t} F(\mathbf{p}, \mathbf{p}'; t) = i\delta(t)\{F(\mathbf{p}, \mathbf{p}'; 0+) - F(\mathbf{p}, \mathbf{p}'; 0-)\}$$
$$-i\langle 0 | T[[\tilde{a}_{\mathbf{p}}(t), H]\tilde{a}^\dagger_{\mathbf{p}'}(0)] | 0\rangle,$$

and, with the form (4.2.11) of H and the anticommutation rules (4.2.3), we easily find that

$$[\tilde{a}_{\mathbf{p}}, H] = \epsilon_{\mathbf{p}}\tilde{a}_{\mathbf{p}} + \sum_{\mathbf{q}} U(\mathbf{q})\tilde{a}_{\mathbf{p}+\mathbf{q}};$$

also

$$F(\mathbf{p}, \mathbf{p}'; 0+) - F(\mathbf{p}, \mathbf{p}'; 0-) = -i(a_{\mathbf{p}}a^\dagger_{\mathbf{p}'} + a^\dagger_{\mathbf{p}'}a_{\mathbf{p}}) = -i\delta_{\mathbf{p}\mathbf{p}'}.$$

The equation of motion for $F(\mathbf{p}, \mathbf{p}'; t)$ is thus

$$\left(i\frac{\partial}{\partial t} - \epsilon_{\mathbf{p}}\right) F(\mathbf{p}, \mathbf{p}'; t) = \delta_{\mathbf{p}\mathbf{p}'}\delta(t) + \sum_{\mathbf{q}} U(\mathbf{q})F(\mathbf{p} + \mathbf{q}, \mathbf{p}'; t) \tag{4.3.10}$$

[i.e., an inhomogeneous form of the original Schrödinger equation (4.1.1)]. The zero-order approximation is of the diagonal form

$$F^0(\mathbf{p}, \mathbf{p}'; t) = \delta_{\mathbf{pp}'} G^0(\mathbf{p}, t), \tag{4.3.11}$$

and we can again calculate $G^0(\mathbf{p}, t)$ directly: keeping only the first term H_0 in H, we have

$$i \frac{\partial}{\partial t} \tilde{a}_{\mathbf{p}}(t) = [\tilde{a}_{\mathbf{p}}, H_0] = \epsilon_{\mathbf{p}} \tilde{a}_{\mathbf{p}}(t), \quad \text{so that} \quad \tilde{a}_{\mathbf{p}}(t) = a_{\mathbf{p}}\, e^{-i\epsilon_{\mathbf{p}} t},$$

and, for $t > 0$,

$$\begin{aligned} G^0(\mathbf{p}, t) &= -i\, e^{-i\epsilon_{\mathbf{p}} t} \langle 0 | a_{\mathbf{p}} a_{\mathbf{p}}^{\dagger} | 0 \rangle \\ &= -i\, e^{-i\epsilon_{\mathbf{p}} t} \langle 0 | (1 - a_{\mathbf{p}}^{\dagger} a_{\mathbf{p}}) | 0 \rangle = -i\, e^{-i\epsilon_{\mathbf{p}} t}, \end{aligned} \tag{4.3.12}$$

while, for $t < 0$, $G^0(\mathbf{p}, t) = 0$.

As in the phonon problem (Sec. 1.5), the differential equation (4.3.10) and the boundary condition (4.3.11) are together equivalent to an integral equation:

$$\begin{aligned} F(\mathbf{p}, \mathbf{p}'; t) = \delta_{\mathbf{pp}'} G^0(\mathbf{p}, t) + \int_{-\infty}^{\infty} dt' G^0(\mathbf{p}, t - t') \sum_{\mathbf{q}} U(\mathbf{q}) \\ \times F(\mathbf{p} + \mathbf{q}, \mathbf{p}'; t'). \end{aligned} \tag{4.3.13}$$

To discuss this equation it is again convenient to work with the Fourier transforms of the time-dependent functions, which are defined as in (4.3.6). The transformed equation is

$$\begin{aligned} F(\mathbf{p}, \mathbf{p}'; \epsilon) &= \delta_{\mathbf{pp}'} G^0(\mathbf{p}, \epsilon) + G^0(\mathbf{p}, \epsilon) \sum_{\mathbf{q}} U(\mathbf{q}) F(\mathbf{p} + \mathbf{q}, \mathbf{p}'; \epsilon) \\ &= \delta_{\mathbf{pp}'} G^0(\mathbf{p}, \epsilon) + G^0(\mathbf{p}, \epsilon) \sum_{\mathbf{q}} U(\mathbf{q} - \mathbf{p}) F(\mathbf{q}, \mathbf{p}'; \epsilon). \end{aligned} \tag{4.3.14}$$

Using the Fourier transform of $G^0(\mathbf{p}, t)$,

$$G^0(\mathbf{p}, \epsilon) = \frac{1}{\epsilon - \epsilon_{\mathbf{p}} + i\eta}, \tag{4.3.15}$$

it may now be seen that Eq. (4.3.14) is equivalent to the original scattering integral equation (4.1.2).

We complete this section by discussing the iterative solution of Eq.

(4.3.14) in powers of U (the Born series). This series solution is defined by

$$F(\mathbf{p},\mathbf{p}') = \sum_{n=0}^{\infty} F^{(n)}(\mathbf{p},\mathbf{p}'), \qquad F^0(\mathbf{p},\mathbf{p}') = \delta_{\mathbf{pp}'}G^0(\mathbf{p}),$$

$$F^{(n)}(\mathbf{p},\mathbf{p}') = G^0(\mathbf{p}) \sum_{\mathbf{q}} U(\mathbf{q}-\mathbf{p})F^{(n-1)}(\mathbf{q},\mathbf{p}') \qquad (n = 1, 2, 3, \ldots),$$

and we obtain for the diagonal part $G(\mathbf{p})$

$$\begin{aligned} G(\mathbf{p}) = F(\mathbf{p},\mathbf{p}) = G^0(\mathbf{p}) &+ G^0(\mathbf{p})U(\mathbf{q}=0)G^0(\mathbf{p}) \\ &+ G^0(\mathbf{p}) \sum_{\mathbf{q}} U(\mathbf{q})G^0(\mathbf{p}+\mathbf{q})U(-\mathbf{q})G^0(\mathbf{p}) \\ &+ G^0(\mathbf{p}) \sum_{\mathbf{qq}'} U(\mathbf{q})G^0(\mathbf{p}+\mathbf{q})U(\mathbf{q}')G^0(\mathbf{p}+\mathbf{q}+\mathbf{q}') \\ &\times U(-\mathbf{q}-\mathbf{q}')G^0(\mathbf{p}) + \cdots. \end{aligned} \qquad (4.3.16)$$

The terms in this series can be represented by diagrams, describing multiple scattering processes of successively higher orders, of the form shown in Fig. 4.1. An equivalent expansion may be made of the original equation (4.1.2), and hence of the T matrix: we obtain for the diagonal

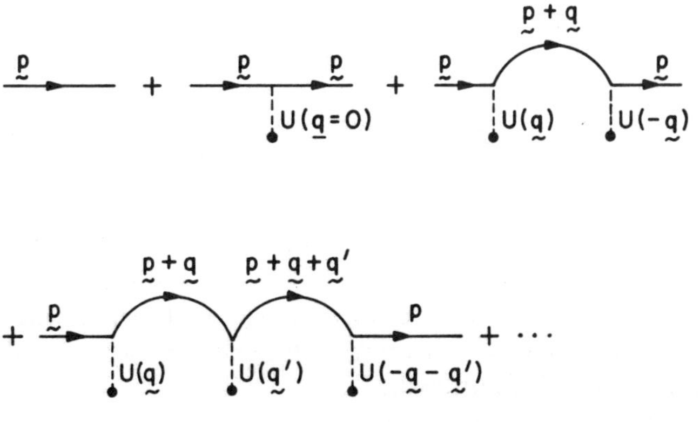

Fig. 4.1. The iteration series for $G(\mathbf{p}, \epsilon)$.

part of this the series

$$\langle \mathbf{p}|T|\mathbf{p}\rangle = \langle \mathbf{p}|U|\mathbf{p}\rangle + \left\langle \mathbf{p}|U \frac{1}{\epsilon - H_0 + i\eta} U|\mathbf{p}\right\rangle$$

$$+ \left\langle \mathbf{p}|U \frac{1}{\epsilon - H_0 + i\eta} U \frac{1}{\epsilon - H_0 + i\eta} U|\mathbf{p}\right\rangle + \cdots. \quad (4.3.17)$$

Writing out these matrix products explicitly, we have

$$\langle \mathbf{p}|T|\mathbf{p}\rangle = U(\mathbf{q}=0) + \sum_{\mathbf{q}} U(\mathbf{q})G^0(\mathbf{p}+\mathbf{q})U(-\mathbf{q})$$

$$+ \sum_{\mathbf{q}\mathbf{q}'} U(\mathbf{q})G^0(\mathbf{p}+\mathbf{q})U(\mathbf{q}')G^0(\mathbf{p}+\mathbf{q}+\mathbf{q}')$$

$$\times U(-\mathbf{q}-\mathbf{q}') + \cdots. \quad (4.3.18)$$

Comparison with (4.3.16) shows that the series for $G(\mathbf{p}) - G^0(\mathbf{p})$ and $\langle \mathbf{p}|T|\mathbf{p}\rangle$ differ only through extra factors $G^0(\mathbf{p})$ on the left and right of each term in (4.3.16). In fact G and T are connected by the general operator relation

$$G = G^0 + G^0 T G^0 \quad (4.3.19)$$

which can be derived directly in closed form by formal manipulation of the operators [see Messiah (1961), Chap. XIX, Secs. 13, 14].

4.4. CLOSED SOLUTION FOR SHORT-RANGE POTENTIAL

The series for G and T can be summed in closed form for the special case of a potential $U(\mathbf{x})$ of zero range, for which

$$U(\mathbf{x}) = U\delta(\mathbf{x}), \qquad U(\mathbf{q}) = U/V = \text{constant}. \quad (4.4.1)$$

This model, although artificial, allows us to derive simple closed expressions for various physical quantities of interest. It has proved very successful in the discussion of qualitative and semiquantitive effects of the short-range potentials due to impurities in real metals where the calculation of the realistic scattering of Bloch electrons by real atomic potentials is quite a difficult task.

When $U(\mathbf{q})$ is independent of $\mathbf{q}$, the repeated summations over $\mathbf{q}$ in each order of (4.3.16) are independent of each other. For example, the third-order term gives

$$G^0(\mathbf{p})\left(\frac{U}{V}\right)^3\left\{\sum_{\mathbf{p}} G^0(\mathbf{p})\right\}^2 G^0(\mathbf{p}) = G^0(\mathbf{p})\left(\frac{U}{V}\right)^3 \{V\overline{G^0(\epsilon)}\}^2 G^0(\mathbf{p}), \tag{4.4.2}$$

where

$$\overline{G^0(\epsilon)} = \frac{1}{V}\sum_{\mathbf{p}} G^0(\mathbf{p}, \epsilon). \tag{4.4.3}$$

For free electrons for which $\epsilon_{\mathbf{p}} = \mathbf{p}^2/2m$, this sum diverges. However, in real metals we can treat (4.4.1) as only acting on electrons within a given atomic cell, as defined by Wannier wave functions of the electronic bands in the metal. This leads to a model (the Slater–Koster model) in which the sum over p-states in (4.4.3) is confined to states lying within a given band, $\epsilon_0 \leqslant \epsilon_{\mathbf{p}} \leqslant \epsilon_B$ say, for which the sum converges.

Introducing a density of states function per unit volume for the unperturbed hamiltonian of the metal

$$\rho^0(\epsilon) = \frac{1}{V}\sum_{\mathbf{p}\in \text{band}} \delta(\epsilon - \epsilon_{\mathbf{p}}), \tag{4.4.4}$$

we can separate $\overline{G^0(\epsilon)}$ into its real and imaginary parts, and write

$$\overline{G^0(\epsilon)} = F(\epsilon) - i\pi\rho^0(\epsilon), \tag{4.4.5}$$

where

$$F(\epsilon) = \mathscr{P}\int \frac{\rho^0(\epsilon')\,d\epsilon'}{\epsilon - \epsilon'}. \tag{4.4.6}$$

In the regions of the band where an effective mass approximation holds, so that $\epsilon_{\mathbf{p}} = \mathbf{p}^2/2m^*$, $\rho^0(\epsilon)$ may be evaluated as

$$\rho^0(\epsilon) = \frac{m^* p(\epsilon)}{2\pi^2}. \tag{4.4.7}$$

The nth-order term in the series for G may now be evaluated explicitly to give

$$G^0(\mathbf{p})\left(\frac{U}{V}\right)^n \{V\overline{G^0(\epsilon)}\}^{n-1} G^0(\mathbf{p}) = G^0(\mathbf{p})\,\frac{U}{V}\,(\overline{G^0}U)^{n-1} G^0(\mathbf{p}).$$

Hence the series (4.3.16) for $G(\mathbf{p})$ becomes a geometric series:

$$G(\mathbf{p}) = G^0(\mathbf{p}) + G^0(\mathbf{p}) \frac{U}{V} \left\{ \sum_{n=1}^{\infty} (\overline{G^0} U)^{n-1} \right\} G^0(\mathbf{p})$$

$$= G^0(\mathbf{p}) + G^0(\mathbf{p}) \frac{U/V}{1 - \overline{G^0} U} G^0(\mathbf{p}). \tag{4.4.8}$$

The T matrix [see Eq. (4.3.19)] is therefore given by

$$T = \frac{U/V}{1 - \overline{G^0(\epsilon)} U}. \tag{4.4.9}$$

This is isotropic, showing that the short-range potential model corresponds to neglecting the higher angular momentum components in the scattering amplitude. Using Eqs. (4.1.7) and (4.4.7) we can express the s-wave phase shift in terms of the real and imaginary parts of $\overline{G^0(\epsilon)}$:

$$e^{2i\delta_0} - 1 = \frac{2\pi i \rho^0(\epsilon)}{F(\epsilon) - \dfrac{1}{U} - i\pi\rho^0(\epsilon)}, \tag{4.4.10}$$

which gives the simple formula [Friedel (1958), Clogston (1962)]

$$\tan \delta_0 = \frac{\pi\rho^0(\epsilon)}{F(\epsilon) - \dfrac{1}{U}}. \tag{4.4.11}$$

The cross-section σ is

$$\sigma = \frac{4\pi}{p^2} \frac{\pi^2(\rho^0)^2}{\left(F - \dfrac{1}{U}\right)^2 + \pi^2(\rho^0)^2}. \tag{4.4.12}$$

The dependence of the electron-impurity scattering on electron energy ϵ is thus expressed directly in terms of the density of states $\rho^0(\epsilon)$ of the electron in the original crystal, its *Hilbert transform* $F(\epsilon)$ defined by Eq. (4.4.6), and the potential strength parameter U. In practice $\rho^0(\epsilon)$ is a complicated function of ϵ, but a qualitative idea of the scattering may be obtained by using a model in which $\rho^0(\epsilon)$ is parabolic within a band of width $2\epsilon_0$ and is zero outside. $F(\epsilon)$ is then easily evaluated, and the form of the functions $\rho^0(\epsilon)$ and $F(\epsilon)$ is shown in Fig. 4.2. Suppose that the potential is attractive, so that $U < 0$, and that $|U|$ is so large that the equation $F(\epsilon) - U^{-1} = 0$ has two real roots, ϵ_1 and ϵ_2, as indicated in

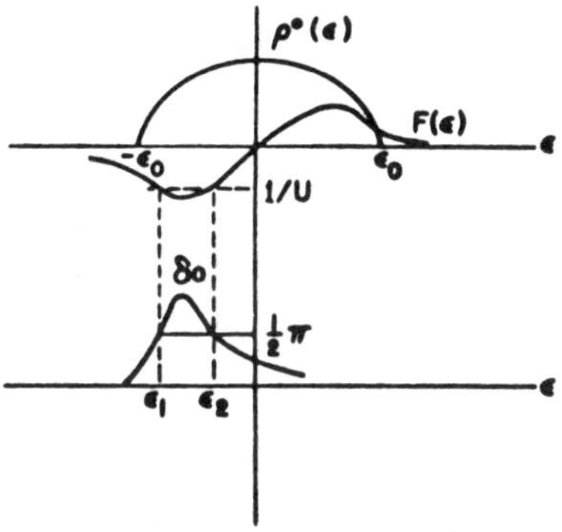

Fig. 4.2. $\rho^{\circ}(\epsilon)$, $F(\epsilon)$ and *s*-wave phase shift for parabolic model. [After Clogston (1962).]

Fig. 4.2. Using (4.4.11) $\delta^0(\epsilon)$ can be evaluated, and its dependence on ϵ is also shown in Fig. 4.2. As may be seen in the figure, δ^0 reaches $\frac{1}{2}\pi$ at the energy ϵ_1. At this point the cross-section reaches a maximum and the scattering goes through a "resonance." If U is increased further, ϵ_1 moves below the bottom of the band at $-\epsilon_0$. In this region $G^0(\epsilon)$ is real, and the T matrix (4.4.9) now has a *pole* at the energy ϵ_1. This corresponds to a negative-energy, or bound-state, solution of the Schrödinger equation lying below the bottom of the band.

The closed result (4.4.9) can also be obtained without use of a perturbation expansion, by noting that, when $U(\mathbf{q})$ is constant, the kernel of the integral equation (4.3.14) is "separable." Under these circumstances a solution in closed form can be obtained by elementary methods, and this leads directly to Eq. (4.4.9).

4.5. THE FRIEDEL SUM RULE

In order to apply fermi statistics to the one-particle scattering solutions of the Schrödinger equation discussed in Secs. 4.3 and 4.4, we require some way of calculating how the number of energy levels per unit energy interval is affected by the impurity. This will be evaluated explicitly in Sec. 4.6 using Green's functions. Here we give the more intuitive argument due to Friedel (1952). We consider the impurity to be placed at the center of a large sphere of radius R and confine ourselves to a short-range potential at which only s-wave phase shifts need be considered, as in Sec. 4.4.

We enumerate the energy levels by imposing the requirement that all wave functions vanish at $r = R$. (It can be shown that, for sufficiently large volumes, the number of states with energy less than a given value is independent of the shape of the region considered.) From the asymptotic form (4.1.5) of the wave function, it may be seen that the eigenvalues are given by the condition.

$$pR + \delta_0(p) = m\pi \qquad (m \text{ an integer}), \tag{4.5.1}$$

and the change Δp in p between two successive wave functions is given by

$$\left(R + \frac{d\delta_0}{dp}\right) \Delta p = \pi. \tag{4.5.2}$$

The number of states per unit change of p is thus

$$\frac{1}{\Delta p} = \frac{1}{\pi}\left(R + \frac{d\delta_0}{dp}\right), \tag{4.5.3}$$

and the *change* in the number of states, due to the perturbation, per unit change of p is $\pi^{-1}\, d\delta_0/dp$. Integrating this, and assuming that $\delta_0(0) = 0$, we see that the total change in the number of states up to some momentum p (or energy ϵ) is just $\pi^{-1}\, \delta_0(\epsilon)$. When the perturbation is a point charge Ze at $r = 0$, it will be neutralized by an appropriate change in the number of electrons in the neighborhood of the origin. For neutrality at infinity we require that $\frac{1}{2}Z$ new levels appear below the fermi energy ϵ_f, allowing for the two spin states of the electron and assuming that the position of the fermi level is not affected by the impurity. This gives us the required result

$$\frac{1}{\pi}\, \delta_0(\epsilon_f) = \tfrac{1}{2}Z, \tag{4.5.4}$$

which is a special case of the *Friedel sum rule*. Since δ_0 is related to U by Eq. (4.4.11), the relation (4.5.4) is a condition for the self-consistency of the perturbation.

Thus, as a result of the screening of the point charge by the scattered electrons, the self-consistent potential is in fact short-ranged, and the use of an s-wave phase shift approximation is not too bad. Note that Z represents the valence difference between the impurity and the solvent metal, and may be either positive or negative. Referring back to the solution (4.4.11) for the s-wave model, we thus see that, for an impurity whose valence differs from that of the host metal by exactly unit charge, the strength of the self-consistent potential U must be such as to cause the resonance to sit exactly at the fermi surface:

$$\delta_0(\epsilon_f) = \tfrac{1}{2}\pi, \tag{4.5.5}$$

so that (counting both spins) the impurity potential is exactly neutralized. This screening charge is thus effectively bound at the impurity site, although it does not form a true bound state of the potential (for which ϵ_1 would have to be below $-\epsilon_0$). We refer to it as forming a "virtual bound state." Because of the rapid change of phase shift with energy near the virtual level (see Fig. 4.2), the properties of an alloy with its fermi level in this energy region are very sensitive to the electron concentration.

It will be noted that, with this model, the maximum possible value of δ_0 is π, so that not more than one state can be displaced below the fermi level. Hence it is only possible to satisfy the self-consistency condition (4.5.4) if Z is not too large. For larger values of Z it becomes necessary to include the higher angular momentum states (p, d, etc.) which then become significant. This leads to a set of degenerate orbital states which must (neglecting correlation effects) be filled up independently. Such considerations become important for transition metal impurities where 5 d-states are present.

The above theory can be used to discuss the effect of a small concentration of charged impurities on the Pauli magnetic susceptibility χ_p and the coefficient γ of the electronic specific heat of a metal [Clogston (1962)]. According to the elementary electron theory of metals, χ_p and γ are proportional to the density of states at the fermi surface, provided electron–electron interactions may be neglected. From the arguments leading to (4.5.4) the change in the density of states at the fermi level per added impurity atom is

$$\delta\rho(\epsilon_f) = \frac{1}{\pi}\left.\frac{d\delta_0(\epsilon)}{d\epsilon}\right|_{\epsilon_f}. \tag{4.5.6}$$

Using (4.5.4) and

$$\frac{d\delta_0}{d\epsilon} = \frac{1}{2\rho^0}\frac{d\rho^0}{d\epsilon}\sin 2\delta_0 - \frac{1}{\pi\rho^0}\frac{dF}{d\epsilon}\sin^2\delta_0, \tag{4.5.7}$$

obtained by differentiating the expression (4.4.11) for the *s*-wave phase shift, we thus obtain

$$\delta\rho(\epsilon_f) = \frac{1}{2\pi}\left(\frac{1}{\rho^0}\frac{d\rho^0}{d\epsilon}\right)_{\epsilon f} \sin \pi Z - \frac{1}{\pi^2}\left(\frac{1}{\rho^0}\frac{dF}{d\epsilon}\right)_{\epsilon f} \sin^2 (\tfrac{1}{2}\pi Z). \quad (4.5.8)$$

This result may be compared with the prediction of the "rigid-band" model, which postulates that in the presence of the perturbation the band is displaced by an energy increment Δ, such that the number of levels displaced is $\rho^0(\epsilon_f)\Delta = \frac{1}{2}Z$. The change in the density of states at ϵ_f is then

$$\delta\rho(\epsilon_f) = \Delta\left(\frac{d\rho^0}{d\epsilon}\right)_{\epsilon f} = \tfrac{1}{2}Z\left(\frac{1}{\rho^0}\frac{d\rho^0}{d\epsilon}\right)_{\epsilon f}. \quad (4.5.9)$$

We note that this is positive or negative, depending upon the sign of Z, and that it represents a first approximation to (4.5.8) in the limit of very small Z.

For $Z = \pm 1$, however, (4.5.8) shows that the predicted change in χ_p or γ is proportional to

$$-\left(\frac{1}{\rho^0}\frac{dF}{d\epsilon}\right)_{\epsilon f},$$

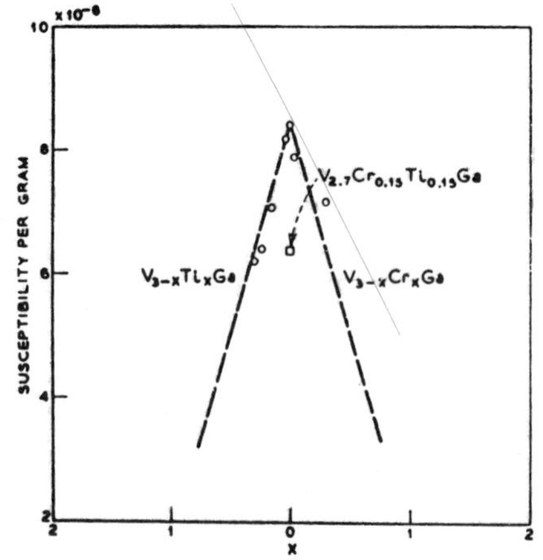

Fig. 4.3. Susceptibility versus fractional concentration of impurities in doped V_3Ga [After Clogston (1962).]

which is independent of the sign of Z. If ϵ_f is near the bottom of the band, ρ^0 is small and $dF/d\epsilon$ is negative; the theory, therefore, predicts a large increase in χ_p or γ, due to the virtual state at the fermi surface associated with the impurity. Near the center of the band $dF/d\epsilon$ is positive and χ_p and γ are expected to decrease. The theory explains the results of measurements on alloys of V_3Ga with Ti and Cr; the susceptibility decreases for both types of impurity, although the signs of Z are different in the two cases (Fig. 4.3).

4.6. MANY-ELECTRON FORMULATION OF THE FRIEDEL SUM RULE

We now make the connection between the above theory of a *single* electron scattering from a potential and the many-electron Green's function formulation introduced in Sec. 4.2. We start by rederiving the Friedel sum rule for the short-range potential model of Sec. 4.4. If we consider the equation of motion of the Green's function defined in (4.2.15) [with a corresponding redefinition of $F(\mathbf{p}, \mathbf{p}'; t)$ which replaces (4.3.9)], we find that the differential equation (4.3.10) is unchanged in form, but that the solution requires different boundary conditions. The unperturbed ground state is now the N-particle fermi sea, in which all single-particle states with momenta p less than the fermi momentum p_f are occupied, and all states with momenta greater than p_f are unoccupied; thus

$$\left.\begin{aligned} &a_{\mathbf{p}}|\Psi_G^{(N)}\rangle = 0 \qquad \text{for } p > p_f, \\ &\text{but} \\ &a_{\mathbf{p}}^\dagger|\Psi_G^{(N)}\rangle = 0 \qquad \text{for } p < p_f. \end{aligned}\right\} \tag{4.6.1}$$

The zero-order Green's function $G^0(\mathbf{p}, t)$ is obtained by evaluating Eq. (4.2.15) with respect to this state. As before, we have the time dependence $\tilde{a}_{\mathbf{p}}(t) = a_{\mathbf{p}}\, e^{-i\epsilon_{\mathbf{p}} t}$, and

$$\begin{aligned} G^0(\mathbf{p}, t) &= -i\, e^{-i\epsilon_{\mathbf{p}} t}\langle a_{\mathbf{p}} a_{\mathbf{p}}^\dagger\rangle \qquad \text{for } t > 0, \\ &= i\, e^{-i\epsilon_{\mathbf{p}} t}\langle a_{\mathbf{p}}^\dagger a_{\mathbf{p}}\rangle \qquad \text{for } t < 0, \end{aligned} \tag{4.6.2}$$

but because of (4.6.1) the second expectation value is no longer zero for all $\mathbf{p}$. We write

$$\left.\begin{aligned} &\langle a_{\mathbf{p}}^\dagger a_{\mathbf{p}}\rangle = \langle n_{\mathbf{p}}\rangle = f_{\mathbf{p}}^-, \\ &\text{and} \\ &\langle a_{\mathbf{p}} a_{\mathbf{p}}^\dagger\rangle = 1 - \langle n_{\mathbf{p}}\rangle = 1 - f_{\mathbf{p}}^- = f_{\mathbf{p}}^+; \end{aligned}\right\} \tag{4.6.3}$$

then

$$\left.\begin{aligned} &f_{\mathbf{p}}^{-}=1, \qquad f_{\mathbf{p}}^{+}=0 \quad \text{for} \quad p<p_f, \\ &\text{and} \\ &f_{\mathbf{p}}^{-}=0, \qquad f_{\mathbf{p}}^{+}=1 \quad \text{for} \quad p>p_f. \end{aligned}\right\} \tag{4.6.4}$$

$f_{\mathbf{p}}^{-}$ is an *electron* distribution function and $f_{\mathbf{p}}^{+}$ is a *hole* distribution function. In terms of these, we can now write

$$\left.\begin{aligned} G^0(\mathbf{p}, t) &= -i\, e^{-i\epsilon_{\mathbf{p}} t} f_{\mathbf{p}}^{+} \quad \text{for} \quad t>0, \\ &= i\, e^{-i\epsilon_{\mathbf{p}} t} f_{\mathbf{p}}^{-} \quad \text{for} \quad t<0, \end{aligned}\right\} \tag{4.6.5}$$

and we see that $G^0(\mathbf{p})$ now describes both electron and hole propagation. The Fourier transform is (introducing appropriate convergence factors)

$$G^0(\mathbf{p}, \epsilon) = -i f_{\mathbf{p}}^{+} \int_0^{\infty} dt\, e^{i\epsilon t}\, e^{-i\epsilon_{\mathbf{p}} t}\, e^{-\eta t} + i f_{\mathbf{p}}^{-} \int_{-\infty}^{0} dt\, e^{i\epsilon t}\, e^{-i\epsilon_{\mathbf{p}} t}\, e^{\eta t}$$

$$= \frac{f_{\mathbf{p}}^{+}}{\epsilon - \epsilon_{\mathbf{p}} + i\eta} + \frac{f_{\mathbf{p}}^{-}}{\epsilon - \epsilon_{\mathbf{p}} - i\eta} \quad (\eta > 0). \tag{4.6.6}$$

Thus $G^0(\mathbf{p}, \epsilon)$ has a pole in the complex ϵ plane at the single-particle energy $\epsilon_{\mathbf{p}}$ which lies *below* the real axis for $p>p_f$, and *above* the real axis for $p<p_f$; the imaginary part of G changes sign at the fermi level. This description of electrons and holes is similar to Feynman's formulation of the theory of electrons and positrons [Feynman (1949)], and we can regard a positive hole, like a positron, as an electron propagating "backwards in time." Going over to the short-range potential model of Sec. 4.4 we can write the solution of the equation of motion for the Green's function in the identical form (4.4.8), except that now $G^0(\mathbf{p})$ has changed from the form (4.3.15)

$$G^0(\mathbf{p}, \epsilon) = \frac{1}{\epsilon - \epsilon_{\mathbf{p}} + i\eta} \qquad (N=0)$$

to the above form (4.6.6).[1]

[1] We can also (compare Sec. 1.4) define *retarded* and *advanced* Green's functions for the N-particle system which have simpler analytic properties than the time-ordered function. In fact the zero-order retarded function, obtained by evaluating (4.3.3) with respect to the N-particle ground state, is identical with the Green's function (4.3.15) for a single electron in an empty system.

We can now use this result to establish the Friedel sum rule. Consider the total charge in the metal. This is obtained from the charge density (4.2.14) in the form

$$\langle Q_{\text{tot}}\rangle = \sum_{\mathbf{p}} \langle \Psi_G{}^{(N)}| a_{\mathbf{p}}{}^{\dagger} a_{\mathbf{p}} | \Psi_G{}^{(N)}\rangle$$
$$= \lim_{t\to 0-} \left\{ -i \sum_{\mathbf{p}} G^{(N)}(\mathbf{p}, t)\right\}. \qquad (4.6.7)$$

We now calculate the *change* in charge resulting from the effect of the impurity. We assume the position of the fermi level is unaltered by the single impurity. Strictly speaking the use of an N-particle ground state would imply that the number of electrons in the box cannot change! However, the form (4.6.6) really implies that we are using a grand canonical ensemble with variable N and fixed chemical potential $\mu = \epsilon_f$ (at $T = 0$), so that electrons can flow in as a result of the presence of the impurity potential.

Then we have

$$\langle \Delta Q_{\text{tot}}\rangle = \lim_{t\to 0-} \sum_{\mathbf{p}} (-i)[G^{(N)}(\mathbf{p}, t) - G^{(N, U=0)}(\mathbf{p}, t)], \qquad (4.6.8)$$

where $G(\mathbf{p})$ in the presence of the potential U is just the diagonal element $F(\mathbf{p}, \mathbf{p})$ of the function $F(\mathbf{p}, \mathbf{p}')$ [Eq. (4.3.9)], now defined for the N-particle system.

Using our solution (4.4.8) giving $G(\mathbf{p}, \epsilon)$ for the short-range problem we thus see that

$$\langle \Delta Q_{\text{tot}}\rangle = -\frac{i}{2\pi} \int_{-\infty}^{\infty} d\epsilon\, e^{-i0^{-}\epsilon} \sum_{\mathbf{p}} [G^0(\mathbf{p}, \epsilon)]^2 \frac{U/V}{1 - \overline{G^0(\epsilon)}U}, \qquad (4.6.9)$$

where we have inverted the Fourier transform (4.3.6). How do we evaluat this? We notice that, at $T = 0$,

$$\frac{1}{V}\sum_{\mathbf{p}} [G^0(\mathbf{p}, \epsilon)]^2 = -\frac{1}{V}\sum_{p<p_f} \frac{\partial}{\partial\epsilon} G^0(\mathbf{p}, \epsilon) - \frac{1}{V}\sum_{p>p_f} \frac{\partial}{\partial\epsilon} G^0(\mathbf{p}, \epsilon)$$
$$= -\frac{\partial}{\partial\epsilon} \overline{G^0(\epsilon)}, \qquad (4.6.10)$$

where we have used the fact that $f_{\mathbf{p}}{}^{+} f_{\mathbf{p}}{}^{-} = 0$, so that cross terms in $(G^0)^2$

disappear. Eq. (4.6.9) may thus be rewritten

$$\langle \Delta Q_{\text{tot}} \rangle = -\frac{i}{2\pi} \int_{-\infty}^{\infty} d\epsilon \, e^{-i0^-\epsilon} \frac{\partial}{\partial \epsilon} \log\{1 - \overline{G^0(\epsilon)}U\}. \tag{4.6.11}$$

Now to perform the integral, which we do by contour integration, we examine the singularities of $\overline{G^0(\epsilon)}$ in the complex ϵ plane.

From the form

$$\overline{G^0(\epsilon)} = \frac{1}{V} \sum_{p<p_f} \frac{1}{\epsilon - \epsilon_{\mathbf{p}} - i\eta} + \frac{1}{V} \sum_{p>p_f} \frac{1}{\epsilon - \epsilon_{\mathbf{p}} + i\eta}, \tag{4.6.12}$$

we see that $\overline{G^0(\epsilon)}$ has a cut (dense series of poles) *above* the real axis for $\epsilon < \epsilon_f$ and *below* the axis for $\epsilon > \epsilon_f$ (Fig. 4.4). Now the integral (4.6.11)

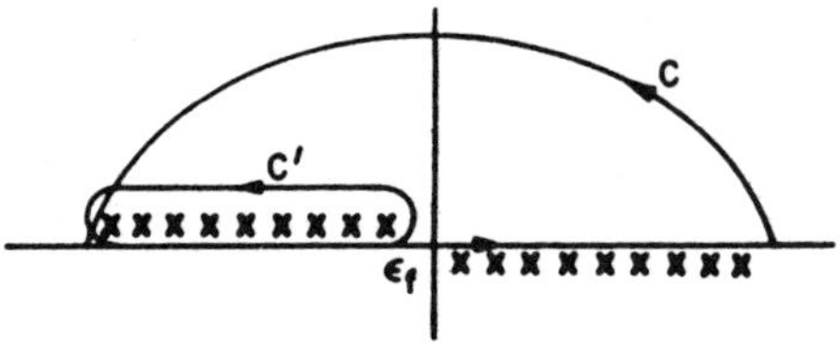

Fig. 4.4. Singularities of $\overline{G^0(\epsilon)}$.

can be replaced by a contour integral over the closed contour C consisting of the real axis from $-\infty$ to $+\infty$ and a large semicircle in the *upper* half-plane; the contribution from the semicircular arc is zero because of the Fourier factor $e^{-i0^-\epsilon}$. As the only singularities inside the contour C occur just above the real axis, C can be deformed into a contour C' running from $-\infty$ to ϵ_f below the real axis and back above the axis. Thus

$$\langle \Delta Q_{\text{tot}} \rangle = -\frac{i}{2\pi} \int_{-\epsilon_0}^{\epsilon_f} d\epsilon \frac{\partial}{\partial \epsilon} [\log\{1 - \overline{G^0(\epsilon - i\eta)}U\} - \log\{1 - \overline{G^0(\epsilon + i\eta)}U\}], \tag{4.6.13}$$

where $-\epsilon_0$ is the lowest value of $\epsilon_{\mathbf{p}}$ in (4.6.12), i.e., the bottom of the band.

Using Eq. (4.4.9), the logarithms in (4.6.13) can now be reexpressed in terms of the T matrix at energies $\epsilon \pm i\eta$ and hence in terms of the phase shift $\delta_0(\epsilon)$:

$$\langle \Delta Q_{\text{tot}} \rangle = -\frac{i}{2\pi} \int_{-\epsilon_0}^{\epsilon_f} d\epsilon \frac{\partial}{\partial \epsilon} \log \frac{T(\epsilon + i\eta)}{T(\epsilon - i\eta)}. \tag{4.6.14}$$

From Eq. (4.1.7) and its complex conjugate we have

$$\frac{T(\epsilon + i\eta)}{T(\epsilon - i\eta)} = -\frac{e^{2i\delta_0} - 1}{e^{-2i\delta_0} - 1} = e^{2i\delta_0}; \tag{4.6.15}$$

hence (4.6.14) finally becomes

$$\langle \Delta Q_{\text{tot}} \rangle = \frac{1}{\pi} \int_{-\epsilon_0}^{\epsilon_f} d\epsilon \frac{\partial \delta_0(\epsilon)}{\partial \epsilon} = \frac{1}{\pi} \delta_0(\epsilon_f), \tag{4.6.16}$$

[assuming $\delta_0(-\epsilon_0) = 0$], and this is precisely the Friedel sum rule (4.5.4).

This result has been obtained by calculating the total electronic charge in an infinite crystal. The Friedel condition $\langle \Delta Q_{\text{tot}} \rangle = \frac{1}{2}Z$ ensures the charge neutrality of the entire crystal in the presence of the charged impurity. It does not tell us where the extra electrons are located. If one makes a calculation of the screening charge inside a large finite sphere of radius R [Kittel (1963), Chap. 18], one finds an additional term which is an oscillatory function of R. This oscillatory term makes a negligible contribution to $\langle \Delta Q_{\text{tot}} \rangle$, but it leads to long-range oscillations in the *charge density* around an impurity atom. These "Friedel oscillations" show up as line shape effects in nuclear magnetic resonance experiments on dilute alloys [Bloembergen and Rowland (1953), Kohn and Vosko (1960)].

We see that the cunning thing about the many-electron Green's function is that, once the fermion commutation relations have been postulated for the creation and annihilation operators, the effects of the Pauli principle (filling all states up to ϵ_f) are *automatically* built into the calculations. In the next section we see that this works equally well at non-zero temperature for the thermodynamic properties of the electrons.

4.7. EFFECT OF IMPURITIES ON THERMODYNAMIC PROPERTIES OF THE ELECTRON GAS

We will use the temperature Green's functions introduced in Chap. 2 to give a calculation from first principles of the effect of impurities on the low-temperature thermodynamic properties (in particular the linear coefficient γ of the electronic specific heat) which were discussed in an *ad hoc* way in Sec. 4.5.

As explained in Chap. 2, the appropriate statistical mechanical formulation for a gas of fermions is in terms of the grand canonical ensemble. We work in terms of a hamiltonian adjusted to the chemical potential:

$$H(T) = H - \mu\hat{N}, \tag{4.7.1}$$

where H is the original hamiltonian of the system, $\mu(T)$ is the chemical potential at temperature T, and $\hat{N} = \sum_{\mathbf{p}} a_{\mathbf{p}}^{\dagger} a_{\mathbf{p}}$ is the number operator. In what follows we take it as understood that the symbol H refers to $H(T)$. The appropriate thermodynamic average of a many-electron operator X now becomes

$$\langle X \rangle = \mathrm{Tr}\,\{e^{-\beta H} X\}/Z_G, \tag{4.7.2}$$

where Z_G is the grand partition function, Eq. (2.1.2).

Now consider the one-fermion temperature Green's function defined by

$$\mathscr{G}(\mathbf{p}, \sigma - \sigma') = \langle T[\tilde{a}_{\mathbf{p}}(\sigma)\tilde{a}_{\mathbf{p}}^{\dagger}(\sigma')]\rangle, \tag{4.7.3}$$

and the temperature version of the non-diagonal function (4.3.9) given by

$$F(\mathbf{p}, \mathbf{p}'; \sigma - \sigma') = \langle T[\tilde{a}_{\mathbf{p}}(\sigma)\tilde{a}_{\mathbf{p}'}^{\dagger}(\sigma')]\rangle. \tag{4.7.4}$$

The change of the thermodynamic potential (2.1.1) appropriate to the grand canonical ensemble, due to the introduction of an impurity, is given, as indicated in Eq. (4.2.13), by

$$\Omega_U - \Omega_0 = -\int_0^1 d\lambda \sum_{\mathbf{pp}'} U(\mathbf{p} - \mathbf{p}') \lim_{\sigma \to 0^-} F(\mathbf{p}, \mathbf{p}'; \sigma). \tag{4.7.5}$$

The change of electronic specific heat can then be obtained via the entropy [Eq. (2.1.3)] by differentiating with respect to temperature.

Since we will be dealing with very few impurities (which we treat in terms of the effect of a single impurity) we can neglect the change of chemical potential.

To calculate (4.7.5) we can use the same equation-of-motion method (now in terms of the Bloch rather than the Schrödinger equation) as in the zero-temperature case. However, as in Chap. 2, the resulting equation of motion is solved by making a Fourier series expansion. As we are now dealing with fermions, the temperature variable σ ranges over the interval $-\beta \leqslant \sigma \leqslant \beta$.

We start by examining the *unperturbed* Green's function

$$G^0(\mathbf{p}, \sigma) = \langle T[\tilde{a}_{\mathbf{p}}(\sigma)\tilde{a}_{\mathbf{p}}^{\dagger}(0)]\rangle_0. \tag{4.7.6}$$

Using the anticommutation rules of Sec. 4.2 it is easy to show that

$$\left.\begin{aligned} \tilde{a}_{\mathbf{p}}^{\dagger}(\sigma) &= e^{\sigma H_0} a_{\mathbf{p}}^{\dagger}\, e^{-\sigma H_0} = e^{\epsilon_{\mathbf{p}}\sigma} a_{\mathbf{p}}^{\dagger}, \\ \tilde{a}_{\mathbf{p}}(\sigma) &= e^{-\epsilon_{\mathbf{p}}\sigma} a_{\mathbf{p}}, \end{aligned}\right\} \tag{4.7.7}$$

from which

$$\left.\begin{aligned} G^0(\mathbf{p}, \sigma) &= e^{-\epsilon_{\mathbf{p}}\sigma}\langle a_{\mathbf{p}} a_{\mathbf{p}}^{\dagger}\rangle_0 = e^{-\epsilon_{\mathbf{p}}\sigma} f_{\mathbf{p}}^{+} && \text{for } \sigma > 0, \\ &= -\, e^{-\epsilon_{\mathbf{p}}\sigma}\langle a_{\mathbf{p}}^{\dagger} a_{\mathbf{p}}\rangle_0 = -e^{-\epsilon_{\mathbf{p}}\sigma} f_{\mathbf{p}}^{-} && \text{for } \sigma < 0, \end{aligned}\right\} \tag{4.7.8}$$

as in (4.6.5), where $f_{\mathbf{p}}^{-}$ is now the fermi function for electrons at inverse temperature β

$$f_{\mathbf{p}}^{-} = \frac{1}{e^{\beta\epsilon_{\mathbf{p}}} + 1} \tag{4.7.9}$$

and $\epsilon_{\mathbf{p}}$ is given by $(p^2/2m - \mu)$, i.e., is zero at the fermi level, while

$$f_{\mathbf{p}}^{+} = 1 - f_{\mathbf{p}}^{-} = \frac{1}{1 + e^{-\beta\epsilon_{\mathbf{p}}}} \tag{4.7.10}$$

is the fermi function for holes.

Let us examine the Fourier coefficients of the expansion

$$G^0(\mathbf{p}, \sigma) = \sum_{\bar{\nu}} e^{i\bar{\nu}\sigma} G^0(\mathbf{p}, \bar{\nu}), \tag{4.7.11}$$

where σ is in the interval $-\beta < \sigma < \beta$. In general $\bar{\nu}$ will take the values

$2\pi\nu/2\beta$, where $\nu = 0, \pm 1, \pm 2, \ldots$. The coefficients are

$$G^0(\mathbf{p}, \bar{\nu}) = \frac{1}{2\beta} \int_{-\beta}^{\beta} e^{-i\bar{\nu}\sigma} G^0(\mathbf{p}, \sigma)\, d\sigma$$

$$= \frac{1}{2\beta}\left[\int_0^{\beta} e^{-i\bar{\nu}\sigma} f_{\mathbf{p}}^{+} e^{-\epsilon_{\mathbf{p}}\sigma}\, d\sigma - \int_{-\beta}^{0} e^{-i\bar{\nu}\sigma} f_{\mathbf{p}}^{-} e^{-\epsilon_{\mathbf{p}}\sigma}\, d\sigma\right]$$

$$= \frac{1}{2\beta}\left[\frac{(e^{-i\bar{\nu}\beta} e^{-\beta\epsilon_{\mathbf{p}}} - 1)}{(-i\bar{\nu} - \epsilon_{\mathbf{p}})(1 + e^{-\beta\epsilon_{\mathbf{p}}})} - \frac{(1 - e^{i\bar{\nu}\beta} e^{\beta\epsilon_{\mathbf{p}}})}{(-i\bar{\nu} - \epsilon_{\mathbf{p}})(e^{\beta\epsilon_{\mathbf{p}}} + 1)}\right].$$

We thus see that, if ν is an *even* integer, the two terms exactly cancel, so that the only non-zero Fourier coefficients are those for which

$$\nu = \pm 1, \pm 3, \ldots,$$

and we then have

$$G^0(\mathbf{p}, \bar{\nu}) = \frac{1/\beta}{i\bar{\nu} + \epsilon_{\mathbf{p}}}. \tag{4.7.12}$$

(The restriction to odd values of ν is a general result for fermions. It comes from the fact that the temperature Green's function is antiperiodic in σ with period β.)

We can now use these results to evaluate (4.7.5) for which we again restrict ourselves to the short-range potential model of Sec. 4.4. The series for $F(\mathbf{p}, \mathbf{p}'; \bar{\nu})$, the Fourier coefficient of $F(\mathbf{p}, \mathbf{p}'; \sigma)$, becomes

$$F(\mathbf{p}, \mathbf{p}'; \bar{\nu}) = \delta_{\mathbf{p}\mathbf{p}'} G^0(\mathbf{p}, \bar{\nu}) - G^0(\mathbf{p}, \bar{\nu})\beta \frac{U}{V} G^0(\mathbf{p}', \bar{\nu}) + \cdots$$
$$- G^0(\mathbf{p}, \bar{\nu})\beta \frac{U}{V}\left[-\sum_{\mathbf{p}_1} \frac{U}{V} \beta G^0(\mathbf{p}_1, \bar{\nu})\right]^n G^0(\mathbf{p}', \bar{\nu}) + \cdots \tag{4.7.13}$$

[this corresponds to the zero-temperature series (4.4.8)]. Thus, by writing

$$\overline{G^0(\bar{\nu})} = \frac{1}{V}\sum_{\mathbf{p}} G^0(\mathbf{p}, \bar{\nu}), \tag{4.7.14}$$

we can evaluate $F(\mathbf{p}, \mathbf{p}'; \bar{\nu})$ in closed form as

$$F(\mathbf{p}, \mathbf{p}'; \bar{\nu}) = \delta_{\mathbf{pp}'} G^0(\mathbf{p}, \bar{\nu}) - G^0(\mathbf{p}, \bar{\nu}) \frac{\beta U/V}{1 + \beta U \overline{G^0(\bar{\nu})}} G^0(\mathbf{p}', \bar{\nu}), \tag{4.7.15}$$

and the change in thermodynamic potential (4.7.5) becomes [replacing U by λU in (4.7.15)]

$$\begin{aligned} \Omega_U - \Omega_0 &= -U \sum_{\bar{\nu}} e^{i\bar{\nu}0^-} \overline{G^0(\bar{\nu})} \left\{ 1 - \overline{G^0(\bar{\nu})} \int_0^1 d\lambda \frac{\beta\lambda U}{1 + \beta U \overline{G^0(\bar{\nu})}} \right\} \\ &= -\frac{1}{\beta} \sum_{\bar{\nu}} e^{i\bar{\nu}0^-} \log\{1 + \beta U \overline{G^0(\bar{\nu})}\}, \end{aligned} \tag{4.7.16}$$

closely analogous to the Friedel sum rule formula (4.6.11).

To evaluate the Fourier sum in (4.7.16), as in the analogous sum (2.2.23), we convert to a contour integral

$$\sum_{\bar{\nu}} e^{i\bar{\nu}0^-} A(-i\bar{\nu}) = -\frac{\beta}{2\pi i} \int_{C_1} dz \frac{e^{z0^-}}{e^{\beta z} + 1} A(z), \tag{4.7.17}$$

where C_1 surrounds the set of imaginary poles $z = \pi i\nu/\beta$ ($\nu = \pm 1, \pm 3, \ldots$) of $1/(e^{\beta z} + 1)$. In the present case the function $A(z)$ has the real axis as a branch line, its imaginary part being discontinuous across this line. We deform the contour from C_1 to C' together with the large arcs Γ (Fig. 4.5). Because of the factor e^{z0^-} the integrand tends to zero exponentially

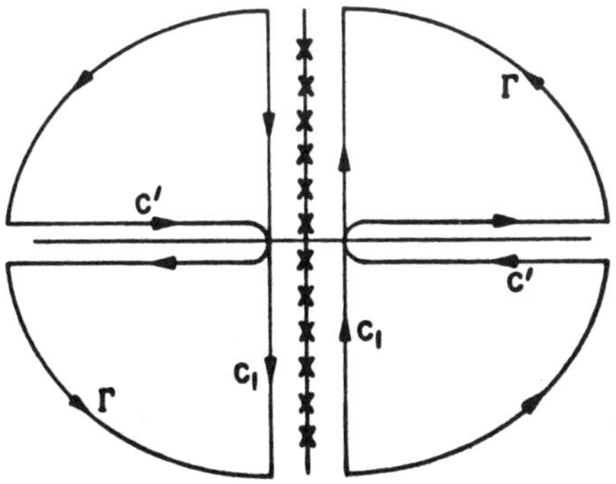

Fig. 4.5. Contour for the evaluation of the frequency sum (4.7.17).

as $|z| \to \infty$ both for Re $(z) > 0$ and for Re $(z) < 0$. We are therefore left with an integral going from $-\infty$ to $+\infty$ just above the real axis and returning to $-\infty$ below the real axis. Eq. (4.7.17) becomes

$$\sum_{\bar{\nu}} e^{i\bar{\nu}0^-} A(-i\bar{\nu}) = -\frac{\beta}{2\pi i} \int_{-\infty}^{\infty} d\epsilon \, \frac{1}{e^{\beta\epsilon}+1} \{A(\epsilon + i\eta) - A(\epsilon - i\eta)\}. \tag{4.7.18}$$

Thus we see that the change in thermodynamic potential at finite T can be evaluated in terms of the real and imaginary parts of the zero-temperature Green's function $\overline{G^0(\epsilon)}$ [Eq. (4.4.5)], and we can then proceed to introduce the T matrix and the phase shift δ_0 exactly as in the derivation of the Friedel sum rule at the end of Sec. 4.6.

We obtain, combining Eqs. (4.7.16) and (4.7.18),

$$\begin{aligned} \Omega_U - \Omega_0 &= \frac{1}{2\pi i} \int_{-\infty}^{\infty} d\epsilon \, f(\epsilon) \log \left\{ \frac{1 - U\overline{G^0(\epsilon + i\eta)}}{1 - U\overline{G^0(\epsilon - i\eta)}} \right\} \\ &= \frac{1}{2\pi i} \int_{-\infty}^{\infty} d\epsilon \, f(\epsilon) \log \frac{T(\epsilon - i\eta)}{T(\epsilon + i\eta)} \\ &= -\frac{1}{\pi} \int_{-\infty}^{\infty} d\epsilon \, f(\epsilon) \, \delta_0(\epsilon), \end{aligned} \tag{4.7.19}$$

where $f(\epsilon)$ is the fermi function. At $T = 0$ the upper limit of the integral is $\mu = \epsilon_f$, and we regain the Friedel rule (4.6.17) via the thermodynamic formula $N = -\partial\Omega/\partial\mu$.

To find the specific heat from this thermodynamic potential we use the fact that the change of entropy due to the impurity is

$$\Delta S = -\frac{d(\Omega_U - \Omega_0)}{dT},$$

and hence

$$\frac{\Delta C_v}{T} = -\frac{d^2}{dT^2} (\Omega_U - \Omega_0). \tag{4.7.20}$$

Differentiating (4.7.19) we have

$$\frac{\Delta C_v}{T} = \frac{1}{\pi} \int_{-\infty}^{\infty} d\epsilon \frac{\partial^2 f}{\partial T^2} \delta_0(\epsilon)$$

$$= \frac{1}{\pi} \int_{-\infty}^{\infty} d\epsilon\, \delta_0(\epsilon) \left\{ \frac{\epsilon^2}{T^2} f''(\epsilon) + \frac{2\epsilon}{T^2} f'(\epsilon) \right\}$$

$$= -\frac{1}{\pi} \int_{-\infty}^{\infty} d\epsilon\, \delta_0'(\epsilon) \frac{\epsilon^2}{T^2} f'(\epsilon). \qquad (4.7.21)$$

Since the derivatives of $f(\epsilon)$ are only large in the region of ϵ_f we expand $\delta_0'(\epsilon)$ about ϵ_f as

$$\delta_0'(\epsilon) = \delta_0'(\epsilon_f) + \epsilon\delta_0''(\epsilon_f) + \cdots,$$

where ϵ is measured relative to the fermi level. So finally (putting $\epsilon/k_B T = x$)

$$\frac{\Delta C_v}{T} = k_B{}^2 \frac{\delta_0'(\epsilon_f)}{\pi} \left\{ -\int_{-\infty}^{\infty} dx\, x^2 f'(x) \right\} + O(T^2), \qquad (4.7.22)$$

and this is just what one gets from the usual density of states argument since, from (4.5.6), $\pi^{-1}\partial\delta_0(\epsilon)/\partial\epsilon$ measures the change in the density of states due to the presence of the impurity.

Chapter 5

Electrons in the Presence of Many Impurities—the Theory of Electrical Resistance in Metals

5.1. THE PHYSICS OF IRREVERSIBLE BEHAVIOR

In the previous chapter we discussed the thermodynamic properties of a non-interacting electron gas in the presence of (essentially) a single impurity. These effects will be proportional to the concentration of impurities at sufficiently low concentrations, so that in real systems we can simply multiply the change of free energy calculated for a single impurity by the concentration of impurities in the system to get the change per unit volume.

However, when we come to consider time-dependent properties of the electron gas, the physics of the effect of impurities differs in an essential way: the response of the system to a single impurity is *no longer* an indication of how the many-impurity system behaves.

To be explicit we will develop, below, the theory of how a particle density disturbance in the electron gas formed at time $t = 0$ spreads out as a function of time t. If there were only one impurity center in the gas, the whole enclosed in a box of side L (periodic boundary conditions), then we could resolve the density disturbance $\rho(\mathbf{x})$ at $t = 0$ in terms of eigenstates of the gas plus impurity. Suppose these eigenstates had wave functions $\varphi_m(\mathbf{x})$, then we could set up a corresponding set of creation operators $b_m{}^\dagger$ and write [see Appendix 1, Eq. (A.1.12)]

$$\rho(\mathbf{x}) = \sum_{mm'} \langle \varphi_m | \rho(\mathbf{x}) | \varphi_{m'} \rangle b_m{}^\dagger b_{m'}. \tag{5.1.1}$$

Since the φ_m are exact eigenstates, energy ϵ_m, the resulting time dependence is completely determined:

$$\rho(\mathbf{x}, t) = \sum_{mm'} e^{i(\epsilon_{m'} - \epsilon_m)t} \langle \varphi_m | \rho | \varphi_{m'} \rangle b_m{}^\dagger b_{m'}. \tag{5.1.2}$$

Now, provided the box is finite, the level spacing between the states ϵ_m

will be finite (apart from accidental degeneracies), so that each term in (5.1.2) is periodic in time. If the numbers ϵ_m are commensurable (i.e., if their ratios are rational numbers), the sum (5.1.2) is also strictly periodic. More generally, it follows from the theory of almost periodic functions that, *if the energy levels are discrete*, $\rho(\mathbf{x}, t)$ is periodic in the sense that, given any positive δ, a time T exists such that $|\rho(\mathbf{x}, T) - \rho(\mathbf{x}, 0)| < \delta$ [Bocchieri and Loinger (1957), Percival (1961)]. For a large box the level spacing is very small and the cycle time T is correspondingly very long, but nevertheless it is finite. (In classical mechanics this is the so-called Poincaré cycle time.) This means the charge disturbance will always return to its starting value if one waits long enough, and the system is "reversible".

Now let us put in a macroscopic number of impurities placed at random positions in the box. Again we can, in principle, form exact eigenfunctions and for a finite system we will again in principle find a Poincaré time. However, at this point we inject an assertion about the realization of such a measurement in the laboratory. What we assert is that because of the complex and *extensive* nature of the distortion of the wave function by the macroscopic number of impurity scatterers, the "repeat" or Poincaré time is now effectively infinite! We will not try and elucidate further the epistemological problems implied by this assertion [for a review and references see Chester (1963)], but instead we go on to explain how this effect, namely, irreversibility of the response of the system, is built into the mathematics for the particular case of the electron-impurity system.

There are two components to the process of accounting for irreversibility. The most important is the following limiting procedure: in all calculations of the time dependence of the system we will take the limit in which the volume tends to infinity (the particle density remaining constant) *before* considering the asymptotic ($t \to \infty$) time dependence. The other component in the impurity case is a trick to take care of the randomness of the impurity locations. We will perform an "ensemble" average over the impurity positions. This is not the standard thermodynamic ensemble, but a process whereby we envisage making up a huge number of ingots of our given metal, each with the same impurity concentration, but each with different (random) locations for the impurities. The calculated response will be averaged over members of the ensemble by simply allowing unrestricted averages over the impurity positions. (In the case of liquid metals one would want to include some correlations between positions of the scattering centers—the metallic ions themselves.) This averaging introduces the phase incoherence of the electronic motion which is a second ingredient of the irreversibility.

The procedure outlined above is sufficient to produce irreversible effects provided the nature of the disturbance in the system (impurities, electron–phonon scattering) extends throughout space. It then transpires that there are two different types of response which can be measured within our model. The theoretically simpler one is the time dependence of the one-electron Green's function. It is hard to think of a measurable physical effect directly related to this Green's function (in fact some tunneling measurements are a way of getting at averages over it), but it is conceptually the easiest to calculate. What comes out is a one-electron lifetime for decay of particle number from a given momentum state $\mathbf{p}$. A *different* response is that of the particle number density perturbation mentioned above. This is the response measured in an electrical conductivity experiment, as will be shown below, and it has a characteristic decay time which is in general different from the one-particle lifetime. The physical reason is that a conservation law is involved. As charge density is scattered out of a given momentum state, other particles are scattered back in. This is the process classically treated by setting up a Boltzmann equation. What we will show below is that these results, in the simple impurity case, can actually be derived in a systematic way from first principles by examining the Green's functions of the electron gas in the presence of the impurities.

5.2. ONE-ELECTRON GREEN'S FUNCTION IN A MANY-IMPURITY SYSTEM

We start by considering the motion of a *single* electron in an array of potentials $U(\mathbf{x} - \mathbf{X}_i)$ located at positions $\mathbf{X}_i$ which are randomly placed. The hamiltonian reads

$$H = \frac{\mathbf{p}^2}{2m} + \sum_i U(\mathbf{x} - \mathbf{X}_i). \tag{5.2.1}$$

It is convenient to rewrite this in second-quantized form in which we allow the possibility of many-electron states. However, it is important to emphasize that the electrons do not interact with each other (except for the constraints on the wave functions imposed by the antisymmetry condition, or Pauli principle). Following the single-impurity problem we now have

$$H = \sum_{\mathbf{p}} \epsilon_{\mathbf{p}} a_{\mathbf{p}}^{\dagger} a_{\mathbf{p}} + \sum_{\mathbf{q}} U(\mathbf{q}) \rho_{\mathbf{q}} \sum_{\mathbf{p}} a_{\mathbf{p}+\mathbf{q}}^{\dagger} a_{\mathbf{p}}. \tag{5.2.2}$$

The only difference between this and the single-impurity hamiltonian (4.2.11) appears in the factor

$$\rho_{\mathbf{q}} = \sum_{j} e^{-i\mathbf{q}\cdot\mathbf{X}_j}, \tag{5.2.3}$$

which results on making a Fourier representation of the sum over impurity potentials in Eq. (5.2.1). $\rho_{\mathbf{q}}$ is the Fourier transform of the density function $\sum_j \delta(\mathbf{x} - \mathbf{X}_j)$ for the scattering centers, and $U(\mathbf{q})$ is as before the Fourier transform of the potential function $U(\mathbf{x})$.

Just as in Sec. 4.3 we can study the equation of motion of the single-electron Green's function $G(\mathbf{p}, t)$ by considering the off-diagonal function $F(\mathbf{p}, \mathbf{p}'; t)$, defined as in (4.3.9). The equation of motion now reads

$$\left(i\frac{\partial}{\partial t} - \epsilon_{\mathbf{p}}\right) F(\mathbf{p}, \mathbf{p}'; t) = \delta_{\mathbf{p}\mathbf{p}'}\delta(t) + \sum_{\mathbf{q}} U(\mathbf{q})\rho_{\mathbf{q}} F(\mathbf{p} + \mathbf{q}, \mathbf{p}'; t). \tag{5.2.4}$$

The difference from the earlier equation of motion (4.3.10) is that the potential term appearing on the right-hand side now has a *randomly varying* factor $\rho_{\mathbf{q}}$. There is no longer a scattering solution in closed form as in the single-impurity case, but we can proceed as in Sec. 4.3 by taking the Fourier transform of (5.2.4) and iterating to produce a series solution to the problem:

$$\begin{aligned} F(\mathbf{p}, \mathbf{p}') = {} & G^0(\mathbf{p})\delta_{\mathbf{p}\mathbf{p}'} + G^0(\mathbf{p})U(\mathbf{p} - \mathbf{p}')\rho_{\mathbf{p}-\mathbf{p}'} G^0(\mathbf{p}') \\ & + \sum_{\mathbf{q}} G^0(\mathbf{p})U(\mathbf{q})\rho_{\mathbf{q}} G^0(\mathbf{p} + \mathbf{q})U(\mathbf{p} - \mathbf{q} - \mathbf{p}') \\ & \times \rho_{\mathbf{p}-\mathbf{q}-\mathbf{p}'} G^0(\mathbf{p}') + \cdots . \end{aligned}$$

The corresponding series for the one-electron Green's function $G(\mathbf{p}) = F(\mathbf{p}, \mathbf{p})$ is

$$\begin{aligned} G(\mathbf{p}) = {} & G^0(\mathbf{p}) + G^0(\mathbf{p})[U(\mathbf{q})\rho_{\mathbf{q}}]_{\mathbf{q}=0} G^0(\mathbf{p}) \\ & + \sum_{\mathbf{q}} G^0(\mathbf{p})U(\mathbf{q})\rho_{\mathbf{q}} G^0(\mathbf{p} + \mathbf{q})U(-\mathbf{q})\rho_{-\mathbf{q}} G^0(\mathbf{p}) + \cdots . \end{aligned} \tag{5.2.5}$$

In order to examine the terms in this series it is convenient to represent a given term by a series of lines representing G^0 connecting at vertices with dashed lines representing $U(\mathbf{q})\rho_{\mathbf{q}}$ (Fig. 5.1):

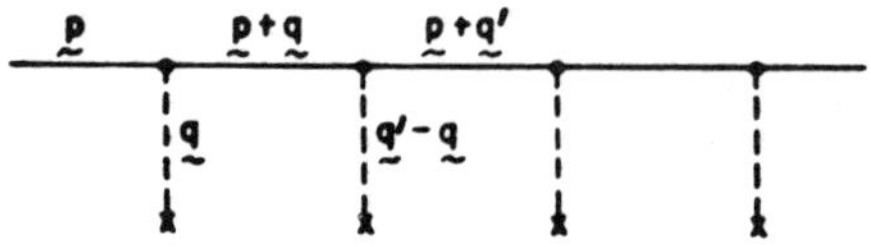

Fig. 5.1. Diagrammatic representation of the Green's function.

At this point we introduce the averaging discussed qualitatively in Sec. 5.1. We assert that the quantity of interest is the *ensemble average* of the Green's function. This is not in itself a very obvious statement–later in this chapter we will show that a measured quantity (the electrical conductivity) can in fact be written as an ensemble average of a response function which is a generalization of the one-particle Green's function. So the present discussion should be taken as introducing a mathematical construct which will be a step in the calculation of a measured quantity (the conductivity).

If there are N impurities in the system at positions $\mathbf{X}_1, \mathbf{X}_2, \ldots, \mathbf{X}_N$, then G is a functional of this set of position vectors

$$G \equiv G[\mathbf{X}_1, \ldots, \mathbf{X}_N], \tag{5.2.6}$$

and the ensemble average is defined as

$$\bar{G} = \prod_{i=1}^{N} \frac{1}{V} \int d^3X_i G[\mathbf{X}_1, \ldots, \mathbf{X}_N]. \tag{5.2.7}$$

(V is the volume of the system.) In this ensemble we assume that each $\mathbf{X}_i$ is uncorrelated with any other. More generally, a distribution function of relative atomic positions could be introduced into the definition to take care of liquids, etc., where correlations between atomic positions have to be taken into account.

This averaging procedure may now be applied, term by term, to the expansion (5.2.5). We then need to consider averages of the form

$\overline{\rho_{\mathbf{q}_1}\rho_{\mathbf{q}_2}\cdots\rho_{\mathbf{q}n}}$. On applying (5.2.7) one finds

$$\overline{\rho_{\mathbf{q}}}=\frac{N}{V}\int d^3X\, e^{-i\mathbf{q}\cdot\mathbf{X}}=N\delta_{\mathbf{q},0}, \tag{5.2.8}$$

where $\delta_{\mathbf{q},0}$ is a Kronecker delta (for the $\mathbf{q}$-vectors quantized in a large box). For the second-order term we have

$$\overline{\rho_{\mathbf{q}_1}\rho_{\mathbf{q}_2}}=\sum_{ij}\overline{e^{-i\mathbf{q}_1\cdot\mathbf{X}_i}\,e^{-i\mathbf{q}_2\cdot\mathbf{X}_j}}\ .$$

There are two contributions to this average—one for $i \neq j$ and the other from the terms in the sum for which $i = j$:

$$\begin{aligned}\overline{\rho_{\mathbf{q}_1}\rho_{\mathbf{q}_2}} &= \sum_{i\neq j} e^{-i\mathbf{q}_1\cdot\mathbf{X}_i}\,e^{-i\mathbf{q}_2\cdot\mathbf{X}_j}+\sum_i e^{-i(\mathbf{q}_1+\mathbf{q}_2)\cdot\mathbf{X}_i}\\ &= N^2\delta_{\mathbf{q}_1,0}\,\delta_{\mathbf{q}_2,0}+N\delta_{\mathbf{q}_1+\mathbf{q}_2,0}\end{aligned} \tag{5.2.9}$$

[we have approximated $N(N-1)$ by N^2 as N is very large].

For the calculation of $G(\mathbf{p})$ we need

$$\overline{\rho_{\mathbf{q}}\rho_{-\mathbf{q}}}=N^2\delta_{\mathbf{q},0}+N. \tag{5.2.10}$$

It is of crucial importance that this average includes a term, corresponding to a second-order Born scattering from a single atom (see below), which is directly proportional to N and which occurs without restriction on the momentum change $\mathbf{q}$ in the intermediate state.

Let us apply these results to the series (5.2.5). The first-order term may be represented by a diagram (Fig. 5.2) giving a contribution

$$G^0(\mathbf{p})NU(\mathbf{q}=0)G^0(\mathbf{p}) \tag{5.2.11}$$

to the expression for $G(\mathbf{p})$.

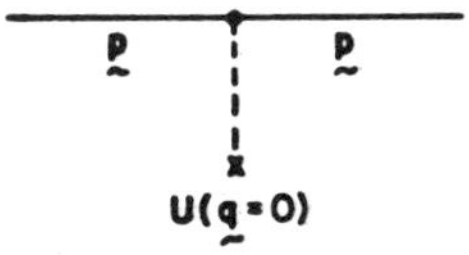

Fig. 5.2. First-order contribution to $G(\mathbf{p})$.

The second-order term may be represented by a pair of diagrams (Fig. 5.3) leading to the terms

$$G^0(\mathbf{p})NU(\mathbf{q}=0)G^0(\mathbf{p})NU(\mathbf{q}=0)G^0(\mathbf{p})$$
$$+\,G^0(\mathbf{p})\sum_{\mathbf{q}} NU(\mathbf{q})G^0(\mathbf{p}+\mathbf{q})U(-\mathbf{q})G^0(\mathbf{p}). \qquad (5.2.12)$$

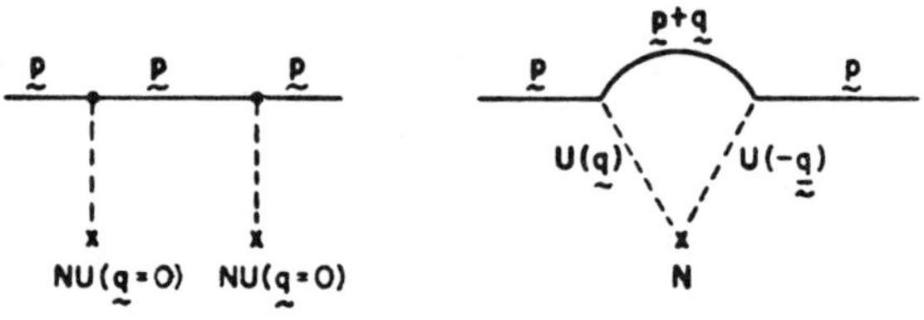

Fig. 5.3. Second-order contribution to $G(\mathbf{p})$.

In these diagrams the single dashed lines each ending in a cross may be thought of as independent scattering (in lowest Born approximation) from two different impurity atoms [i.e., the $i \neq j$ term in Eq. (5.2.9)], while the pair of dashed lines ending in a single cross corresponds to a second Born approximation scattering from the *same* atom [the $i = j$ term in (5.2.9)]. Continuing the same argument, the third-order diagrams are as in Fig. 5.4. These correspond, respectively, to (a) independent first-order Born scatterings from three distinct atoms; (b) a first-order Born scattering from $\mathbf{X}_i$ occurring together with a second-order Born

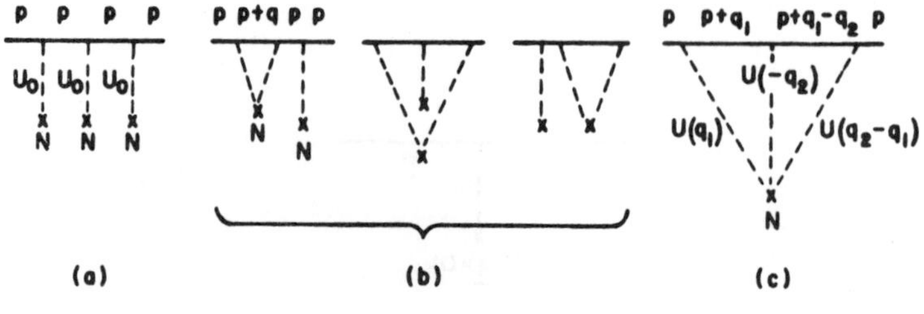

Fig. 5.4. Third-order contribution to $G(\mathbf{p})$.

scattering from a separate atom $\mathbf{X}_j$; (c) a third-order Born scattering from a single atom.

It is thus seen that the general term in the series is obtained by including all possible repeated scattering events at a single atom. Clearly the general term will involve a very large number of possible diagrams. It is not possible to sum these diagrams in closed form in the general case. However, by introducing the concept of the *irreducible diagram* (Dyson 1949), we can carry out a partial summation, in such a way that there remains only a sum over such irreducible diagrams.

5.3. DYSON'S EQUATION

An irreducible diagram is defined as a diagram which cannot be divided into two sub-diagrams joined only by a single $G^0(\mathbf{p})$ line; all other diagrams are called reducible. For example, in Fig. 5.4 the third and fifth diagrams are irreducible and the others are reducible diagrams. An *irreducible self-energy diagram* is an irreducible diagram which has the $G^0(\mathbf{p})$ lines removed from the two ends. There is a denumerable infinity $\Sigma^{(i)}(\mathbf{p}, \epsilon)$ $(i = 0, 1, 2, \ldots)$ of such diagrams, and the (total) *irreducible self-energy* is defined to be the sum

$$\Sigma(\mathbf{p}, \epsilon) = \sum_{i=0}^{\infty} \Sigma^{(i)}(\mathbf{p}, \epsilon). \tag{5.3.1}$$

We now rearrange the terms of the perturbation series by collecting together all terms from all orders of the form $G^0(\mathbf{p})\Sigma^{(i)}G^0(\mathbf{p})$, then all terms of the form $G^0(\mathbf{p})\Sigma^{(i)}G^0(\mathbf{p})\Sigma^{(j)}G^0(\mathbf{p})$, and so on. All combinations for all $i, j, \ldots$, occur; hence the sum of all contributions in each group is obtained by replacing each $\Sigma^{(i)}$ by the total self-energy Σ. The complete series may thus be written

$$\begin{aligned} G(\mathbf{p}) &= G^0(\mathbf{p}) + G^0(\mathbf{p})\,\Sigma\, G^0(\mathbf{p}) + G^0(\mathbf{p})\,\Sigma\, G^0(\mathbf{p})\,\Sigma\, G^0(\mathbf{p}) + \cdots \\ &= G^0(\mathbf{p}) + G^0(\mathbf{p})\,\Sigma \left\{G^0(\mathbf{p}) + G^0(\mathbf{p})\,\Sigma\, G^0(\mathbf{p}) + \cdots\right\} \\ &= G^0(\mathbf{p}) + G^0(\mathbf{p})\,\Sigma\, G(\mathbf{p}). \end{aligned} \tag{5.3.2}$$

This is *Dyson's equation*. Solving for $G(\mathbf{p})$ and remembering the form

(4.3.15) of $G^0(\mathbf{p})$, we have

$$G(\mathbf{p}, \epsilon) = \frac{1}{\{G^0(\mathbf{p})\}^{-1} - \sum(\mathbf{p}, \epsilon)} = \frac{1}{\epsilon - \epsilon_{\mathbf{p}} - \sum(\mathbf{p}, \epsilon)}. \tag{5.3.3}$$

We thus see that the exact one-electron Green's function is obtained from the unperturbed Green's function by adding the self-energy to the unperturbed single-particle energy.

5.4. ONE-ELECTRON GREEN'S FUNCTION IN LOW-DENSITY WEAK SCATTERING APPROXIMATION

To obtain an explicit expression for $\Sigma(\mathbf{p}, \epsilon)$ we have to approximate by selecting a sub-class of irreducible diagrams to be summed. In fact it is easy to classify the diagrams according to their dependence on the concentration $n_{\text{imp.}} = N/V$ of impurities. We note that every scattering cross in a diagram, representing an impurity atom, carries a factor N/V. We therefore obtain a low-density approximation to $\Sigma(\mathbf{p})$ by keeping only the self-energy diagrams with a single cross:

$\Sigma(\underset{\sim}{p},\epsilon) =$  $+ \dots$ (5.4.1)

These diagrams are just the same as those representing the T matrix for scattering from a single impurity [see Fig. 4.1 and Eq. (4.3.19)]; thus, except for a factor N, the sum of the self-energy diagrams (5.4.1) is given by the T matrix for single-center scattering. However, as is indicated by (5.3.3) and as will be discussed further below, in the many-center problem the singularities of $G(\mathbf{p}, \epsilon)$ lie at values of ϵ, in general complex, which differ from the unperturbed energy $\epsilon_{\mathbf{p}}$. Thus we are now really dealing with an analytic continuation in the complex ϵ-plane of the T matrix which was originally defined in Chap. 4 in terms of matrix elements between plane waves corresponding to the unperturbed energy.

We make an additional approximation which is convenient but not necessary. This is to allow the scattering potential from a given atom to be weak, so that only first and second Born scatterings from a given atom [i.e., the first two diagrams of (5.4.1)] need to be taken into account.

The first-order diagram contributes to $\Sigma(\mathbf{p}, \epsilon)$ an amount $NU(\mathbf{q} = 0)$. This simply leads to a shift in the energy of the electron states by the spatial average of the one-center potential:

$$\epsilon_{\mathbf{p}} \to \epsilon_{\mathbf{p}} + NU(\mathbf{q} = 0) = \epsilon_{\mathbf{p}} + n_{\text{imp.}} \int U(\mathbf{x})\, d^3x. \tag{5.4.2}$$

We can take care of this by a suitable redefinition of the energies $\epsilon_{\mathbf{p}}$. This leaves as our final expression for $\Sigma(\mathbf{p}, \epsilon)$, in the low-density weak-potential limit, the contribution of the second-order diagram of (5.4.1):

$$\Sigma(\mathbf{p}, \epsilon) = N \sum_{\mathbf{p}'} U^2(\mathbf{p}' - \mathbf{p}) G^0(\mathbf{p}', \epsilon). \tag{5.4.3}$$

Notice that, in adopting (5.3.3) with (5.4.3) as our approximation to $G(\mathbf{p}, \epsilon)$, we have included diagrams in the series for $G(\mathbf{p})$ of the type of diagram (a) of Fig. 5.5, but have excluded diagrams such as diagram (b).

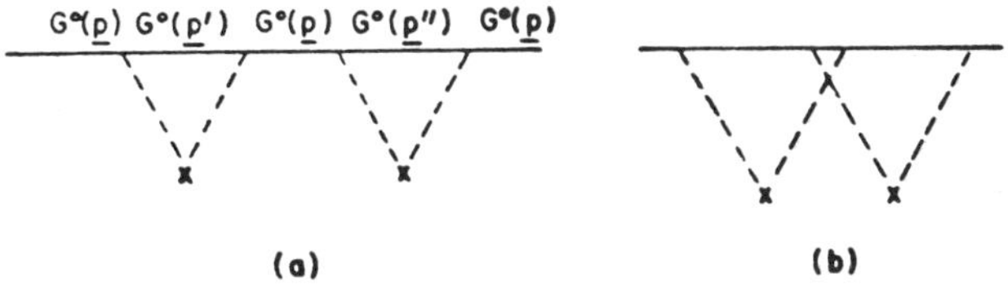

Fig. 5.5. Diagrams (a) contributing, (b) not contributing, to $G(\mathbf{p}, \epsilon)$.

Our low-density approximation thus amounts to neglecting diagrams in which the successive Born scatterings on different atoms "overlap" or interfere with each other. However, by simply shifting the energy states of the unperturbed electron through Eq. (5.4.2), we have automatically included all first-order Born independent scatterings off other atoms (Fig. 5.6). Thus we do not need to worry about these contributions as they are taken care of automatically.

We emphasize that, although $\Sigma(\mathbf{p}, \epsilon)$ is given by a single term in this approximation, we have nevertheless had to consider successive scatterings from many different atoms, i.e., the perturbation series for G had to be summed to *infinite* order to produce (5.3.3). The reason why this

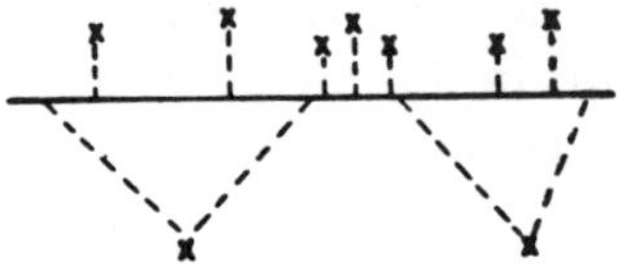

Fig. 5.6. Diagram included in $G(\mathbf{p}, \epsilon)$.

is essential may be seen from a simple argument to calculate the propagation of an electron wave through the random medium (Fig. 5.7). In any given slab of the material the wave loses some amplitude by the scattering, provided the centers are random and therefore interfere on average

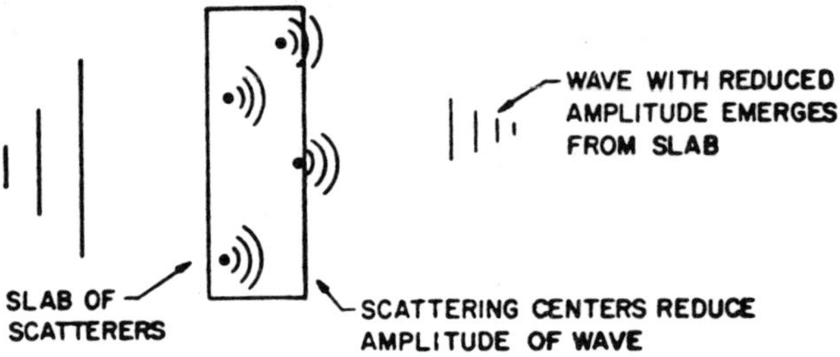

Fig. 5.7. Propagation of electron wave through the random medium.

destructively. Hence the wave will decay exponentially as $e^{-\gamma x}$ as it propagates through a succession of slabs. In quantum-mechanical terms this exponential decay can only be taken into account by summing over an infinite series of scattering events (corresponding to an infinite set of slabs).

How does irreversibility creep into the expression (5.3.3)? Physically what happens can be understood in terms of Fig. 5.7. The electron wave will only lose amplitude in an irreversible manner provided we neglect reflection from the walls of the box of the scattered wavelets in the slab being considered. Mathematically this is done by means of a limiting procedure. If we examine (5.4.3)—neglecting the q-dependence of $U(\mathbf{q})$ for convenience (this effectively implies the use of a short-range potential)-

then as a function of the complex energy variable ϵ, $\Sigma(\epsilon)$ contains a series of poles at the energy values $\epsilon_{\mathbf{p}}$ corresponding to the poles of the one-electron Green's function (4.3.15) at the electron energy levels in the large box enclosing the system. We have

$$\Sigma(\mathbf{p}, \epsilon) = N \sum_{\mathbf{p}'} \left(\frac{U}{V}\right)^2 \frac{1}{\epsilon - \epsilon_{\mathbf{p}'} + i\eta}, \tag{5.4.4}$$

where we have replaced $U(\mathbf{q})$ by U/V, U being an energy parameter.

Substituting back in (5.3.3) it may be seen that as a function of ϵ $G(\mathbf{p}, \epsilon)$ now acquires a corresponding spectrum of poles lying near the poles of (5.4.4). The poles of $G(\mathbf{p}, \epsilon)$ occur at the values of ϵ satisfying

$$\epsilon - \epsilon_{\mathbf{p}} - \Sigma(\mathbf{p}, \epsilon) = 0 \tag{5.4.5}$$

and may be determined graphically as indicated in Fig. 5.8. The spacing between the poles is of order (1/volume) and corresponds to the Poincaré

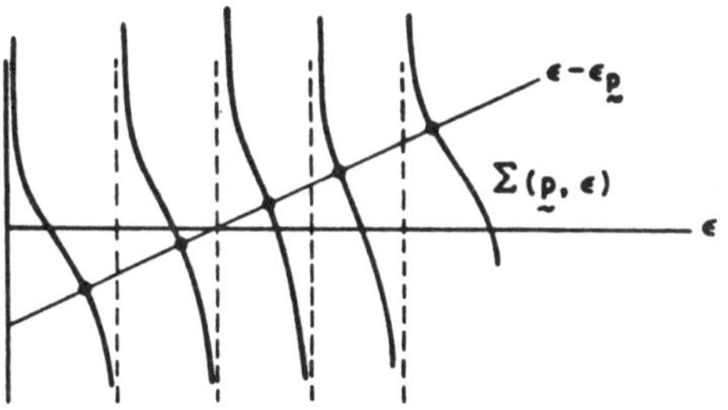

Fig. 5.8. Poles of $G(\mathbf{p}, \epsilon)$ for a finite system.

spacing discussed in Sec. 5.1. Irreversibility is now brought in by letting the volume tend to infinity *before* letting the time of observation (corresponding to the precision of taking the Fourier transform in ϵ) become very long. We are in this way ensuring that, if $\Delta\epsilon$ is the spacing between adjacent energy levels and τ is the time of observation, the inequality $\Delta\epsilon \ll \hbar/\tau$ holds. Our limiting procedure thus implies that in a macroscopic system the discrete level structure cannot be resolved and we detect only the *level density*, averaged over an energy interval

$\hbar/\tau$. The sum in (5.4.4) becomes an integral and, since the integrand depends only on $\epsilon_{\mathbf{p}}$, the integral can be written in terms of the density of unperturbed electron states per unit volume

$$\rho(\epsilon) = \frac{1}{V} \sum_{\mathbf{p}} \delta(\epsilon - \epsilon_{\mathbf{p}}), \tag{5.4.6}$$

giving

$$\Sigma(\epsilon) = n_{\text{imp.}} U^2 \int_0^\infty \frac{\rho(\epsilon')\, d\epsilon'}{\epsilon - \epsilon' + i\eta}. \tag{5.4.7}$$

Thus the infinite-volume limit alters the analytic form of the function $\Sigma(\epsilon)$ of the complex variable ϵ: the discrete poles of $\Sigma(\mathbf{p}, \epsilon)$ merge to form a *branch cut* along the real axis in the complex ϵ-plane. This results in a finite imaginary contribution to the self-energy for real ϵ: we have

$$\Sigma(\epsilon) = n_{\text{imp}} U^2 [F(\epsilon) - i\pi\rho(\epsilon)], \tag{5.4.8}$$

where $F(\epsilon)$ (as in Sec. 4.4) is the Hilbert transform of $\rho(\epsilon)$. Further, the imaginary part of $\Sigma(\epsilon)$ is negative, and this leads to an *irreversible damping* in the time-dependent Green's function $G(\mathbf{p}, t)$. The negative sign comes from the fact that, as discussed in Sec. 4.3, only the retarded $(t > 0)$ part of $G(\mathbf{p}, t)$ is non-zero for a one-electron system, leading to the prescription used in (5.4.4) and (5.4.7) for going round the poles of $G^0(\mathbf{p}', \epsilon)$. To exhibit the damping explicitly we ignore for simplicity the energy dependence of $\Sigma(\epsilon)$ and treat it as a complex constant. Then, writing $\Delta = n_{\text{imp}} U^2 F$, $\Gamma = n_{\text{imp}} U^2 \pi \rho > 0$, we have for $t > 0$

$$\begin{aligned} G(\mathbf{p}, t) &= \int_{-\infty}^{\infty} d\epsilon\, e^{-i\epsilon t} G(\mathbf{p}, \epsilon) = \int_{-\infty}^{\infty} d\epsilon \frac{e^{-i\epsilon t}}{\epsilon - \epsilon_{\mathbf{p}} - \Delta + i\Gamma} \\ &= -i\, e^{-i(\epsilon_{\mathbf{p}} + \Delta)t}\, e^{-\Gamma t}, \end{aligned} \tag{5.4.9}$$

on closing the contour of integration in the lower half-plane.

To obtain (5.4.9) we needed both the infinite-volume limiting procedure and the random-impurity averaging procedure. For a finite box, the imaginary part of the self-energy (5.4.4) consists of a discrete series of δ-functions which cannot give finite damping of $G(\mathbf{p}, t)$. If the scattering is coherent there is also no damping: in particular, if the scattering centers occupy the lattice points in a periodic lattice, the density function $\rho_{\mathbf{q}}$ is given by

$$\rho_{\mathbf{q}} = N \sum_{\mathbf{K}} \delta_{\mathbf{q},\mathbf{K}}, \tag{5.4.10}$$

where the sum goes over reciprocal lattice vectors **K**. The series for the one-electron Green's function can then be summed to give

$$G(\mathbf{p}, \epsilon) = \frac{1}{\epsilon - \epsilon_{\mathbf{p}}^{(l)} + i\eta}, \tag{5.4.11}$$

where

$$\epsilon_{\mathbf{p}}^{(l)} = \epsilon_{\mathbf{p}} + NU(\mathbf{q} = 0) + N^2 \sum_{\mathbf{K} \neq 0} \frac{|U(\mathbf{K})|^2}{\epsilon_{\mathbf{p}} - \epsilon_{\mathbf{p}+\mathbf{K}}} + \cdots \tag{5.4.12}$$

is the one-particle Bloch band energy. The poles of this Green's function lie on the real axis.

Thus in the random impurity case it is seen that the damping of the electron wave as it propagates through the medium results from an infinite sequence of *incoherent* second Born scatterings from successive impurities.

5.5. THEORY OF ELECTRICAL CONDUCTIVITY—LINEAR RESPONSE FORMULATION

Before the development of many-body theory techniques the theory of electrical conduction was based on a Boltzmann equation approach to the time dependence of the semiclassical one-particle distribution function. The essential physical assumption in this theory is the "Stosszahlansatz," or assumption of randomness after successive collision events. The understanding of the mathematical formulation of irreversibility discussed in the last section, however, allows us to dispense with this assumption which had to be imposed arbitrarily. We can now develop the theory starting from the reversible equations of motion and introduce irreversibility essentially as in the one-electron Green's function case. One can in fact proceed in two possible ways. The first is to give a more rigorous derivation, free from the Stosszahlansatz, of kinetic equations such as the Boltzmann equation, which can then be solved to give the transport coefficient of interest. This approach was initiated by van Hove (1955) and has been extensively developed by

Prigogine and his school [see the review by Chester (1963)]. A more direct approach, which we shall adopt here, starts from a formulation of the process of measurement of transport properties in general mathematical terms. This is the *linear response* or *Kubo formula* approach [Kubo (1957, 1958)], where the only assumption made is that the current is linear in the applied voltage, i.e., that Ohm's law applies. We will work within the context of the many-impurity problem, although the general approach will work for any many-body or extended hamiltonian (e.g., describing electron–phonon interactions).

To derive the Kubo formula we apply a weak time-dependent external force (e.g., electric scalar or vector potential) to the system and measure the change in some observable quantity, e.g., charge density, or electrical current, to *linear order* in the applied force. The possibility of this linear expansion implies, of course, an inherent stability in the system being tested. (One would have to be careful, for example, in the case of a superconductor.) So we write the total hamiltonian as

$$H_{\text{tot}} = H + H_{\text{ext}}, \tag{5.5.1}$$

where, if the perturbing field is described by an applied vector potential $\mathbf{A}(\mathbf{x}, t)$, we have to first order in $\mathbf{A}$

$$H_{\text{ext}} = \int d^3x\, \mathbf{A}(\mathbf{x}, t) \cdot \mathbf{j}(\mathbf{x}). \tag{5.5.2}$$

$\mathbf{j}(\mathbf{x})$ is the "paramagnetic" part of the total current operator

$$\begin{aligned} \mathbf{J}(\mathbf{x}, t) &= \tfrac{1}{2} \sum_i [\{\mathbf{p}_i - e\mathbf{A}(\mathbf{x}, t)\}\, \delta(\mathbf{x} - \mathbf{x}_i) \\ &\qquad + \delta(\mathbf{x} - \mathbf{x}_i)\{\mathbf{p}_i - e\mathbf{A}(\mathbf{x}, t)\}] \\ &= \mathbf{j}(\mathbf{x}) - ne\mathbf{A}(\mathbf{x}, t). \end{aligned} \tag{5.5.3}$$

Here n is the density of electrons and units have been chosen such that $e/m = 1$. The second-quantized form (see Appendix 1) of $\mathbf{j}(\mathbf{x})$ is

$$\mathbf{j}(\mathbf{x}) = \sum_{\mathbf{p},\mathbf{q}} e^{i\mathbf{q}\cdot\mathbf{x}} a^\dagger_{\mathbf{p}+\mathbf{q}} a_{\mathbf{p}} (\mathbf{p} + \tfrac{1}{2}\mathbf{q}). \tag{5.5.4}$$

H is the hamiltonian being "tested" by the external force.

Let us now measure the response to $\mathbf{A}$ of the electric current in the system described by H. We choose the current response function rather than the charge density or dielectric response function as it leads more directly to the electrical conductivity. (The current and charge densities are of course simply related via the continuity equation.) Further, we assume **A** to represent a *transverse* field [$\mathbf{q} \cdot \mathbf{A}(\mathbf{q}) = 0$], so that we do not have to worry about internal fields arising from induced charge densities, and **J** is the direct current response to the applied field (see Sec. 6.4).

Suppose the system is initially in some eigenstate $|E_N\rangle$ of H. Eventually we will let this be a member of a statistical mechanical (Gibbs) ensemble. Then as time evolves the state $|E_N\rangle$ will change according to both H and H_{ext}, in accordance with the Schrödinger equation

$$i\frac{\partial|E_N(t)\rangle}{\partial t} = H_{\text{tot}}|E_N(t)\rangle. \tag{5.5.5}$$

We write

$$|E_N(t)\rangle = e^{-iHt}U_{\text{ext}}(t)|E_N\rangle; \tag{5.5.6}$$

then $U_{\text{ext}}(t)$ satisfies the integral equation

$$U_{\text{ext}}(t) = 1 - i\int_0^t H_{\text{ext}}(t')U_{\text{ext}}(t')\,dt'. \tag{5.5.7}$$

The expectation value of the current density at time t is

$$\langle\mathbf{J}(\mathbf{x}, t)\rangle_N = \langle E_N(t)|\mathbf{j}(\mathbf{x})|E_N(t)\rangle - ne\,\mathbf{A}(\mathbf{x}, t). \tag{5.5.8}$$

We solve (5.5.7) to the first order in H_{ext}, obtaining

$$U_{\text{ext}}(t) = 1 - i\int_0^t H_{\text{ext}}(t')\,dt', \tag{5.5.9}$$

and, when we substitute (5.5.6) and (5.5.9) into (5.5.8) and work to the first order in H_{ext}, we obtain

$$\langle\mathbf{J}(\mathbf{x}, t)\rangle_N = \langle\mathbf{j}(\mathbf{x})\rangle_N + i\int_0^t dt'\langle E_N|[H_{\text{ext}}(t'), \mathbf{j}(\mathbf{x}, t)]|E_N\rangle - ne\,\mathbf{A}(\mathbf{x}, t), \tag{5.5.10}$$

where the time dependence of $H_{\text{ext}}(t)$ and $\mathbf{j}(\mathbf{x}, t)$ refers only to the development according to H:

$$H_{\text{ext}}(t) = e^{iHt}H_{\text{ext}}\,e^{-iHt}, \qquad \mathbf{j}(\mathbf{x}, t) = e^{iHt}\mathbf{j}(\mathbf{x})\,e^{-iHt}, \tag{5.5.11}$$

and the bracket [. . .] denotes a *commutator*. Thus, using the form

(5.5.2) of H_{ext}, we have finally

$$\langle J_\alpha(\mathbf{x}, t)\rangle = \langle j_\alpha(\mathbf{x})\rangle + \int_{-\infty}^{\infty} dt' \int d^3x' \sum_\beta R_{\alpha\beta}(\mathbf{x} - \mathbf{x}', t - t')A_\beta(\mathbf{x}', t'), \tag{5.5.12}$$

where α, β are cartesian suffixes,

$$R_{\alpha\beta}(\mathbf{x} - \mathbf{x}', t - t') = \mathscr{R}_{\alpha\beta}(\mathbf{x} - \mathbf{x}', t - t') - ne\delta_{\alpha\beta}\delta(\mathbf{x} - \mathbf{x}')\delta(t - t'), \tag{5.5.13}$$

$$\mathscr{R}_{\alpha\beta}(\mathbf{x} - \mathbf{x}', t - t') = -i\theta(t - t')\langle[j_\alpha(\mathbf{x}, t), j_\beta(\mathbf{x}', t')]\rangle, \tag{5.5.14}$$

and we assume that $\mathbf{A}(\mathbf{x}, t)$ contains the switching-on process in its time dependence. For a metal at zero temperature, the expectation value refers to the N-particle fermion ground state.

As in Chap. 2 (Sec. 2.3), this result can be extended to non-zero temperatures by considering a canonical ensemble of initial states occurring with probability $p_I = e^{-\beta E_I}/Z$. We now have

$$\begin{aligned}\langle \mathbf{J}(\mathbf{x}, t)\rangle &= \sum_I p_I\langle E_N(t)|\mathbf{j}(\mathbf{x})|E_N(t)\rangle - ne\mathbf{A}(\mathbf{x}, t)\\ &= \mathrm{Tr}\,\{e^{-\beta H}U^\dagger_{\text{ext}}(t)\mathbf{j}(\mathbf{x}, t)U_{\text{ext}}(t)\}/Z - ne\mathbf{A}(\mathbf{x}, t),\end{aligned} \tag{5.5.15}$$

($Z = \mathrm{Tr}\, e^{-\beta H}$), and the only difference is that the angular bracket in (5.5.14) now denotes a thermal average:

$$\langle[j_\alpha(\mathbf{x}, t), j_\beta(\mathbf{x}', t')]\rangle = \mathrm{Tr}\,\{e^{-\beta H}[j_\alpha(\mathbf{x}, t), j_\beta(\mathbf{x}', t')]\}/Z. \tag{5.5.16}$$

Notice that the time dependence of the operators in (5.5.14) is the general time dependence governed by the full hamiltonian H of the system being tested. This has not been approximated or restricted in any way [other than by the implied restriction that the expansion (5.5.10) is stable]. So Eqs. (5.5.12) to (5.5.14) constitute a completely general formulation of the physics of measurement of a transport coefficient. This is often referred to as the *Kubo formula* and is the analog in non-equilibrium statistical mechanics of the Gibbs formula for equilibrium statistical mechanics.

In this way the theory of electrical conduction (and other transport processes) has been reduced essentially to the mathematical problem of evaluating the response function R. (Physical reasoning is still needed, of course, as a guide in making the correct approximations during this evaluation.)

5.6. EVALUATION OF THE KUBO FORMULA FOR THE MANY-IMPURITY PROBLEM

Just as in the evaluation of the one-particle damping via a partial summation of the perturbation series for the one-particle Green's function in Sec. 5.4, we proceed by a perturbation expansion of the response function (5.5.14) to evaluate the damping effect of the impurities on the electrical response—and hence to calculate the electrical conductivity.

The response function is a retarded, real-time Green's function of the type encountered in Chap. 2 [Eq. (2.3.6)] and discussed in Appendix 2. However, on substituting for the current density operators in terms of the a and $a^\dagger$ [Eq. (5.5.4)] it may be seen that we are now dealing with a generalization of the one-particle Green's function in which both a particle *and* a hole are injected into the system at time t' and taken out again after they have propagated until time t. Thus we are dealing with a *two-particle* (i.e., one particle and one hole) *Green's function*. The evaluation of the terms in a perturbation series (in powers of an interaction hamiltonian) for a two-particle Green's function in the case of an interacting electron gas is an important problem which will be discussed in subsequent chapters. However, in the case of an electron gas in which the electrons *only* interact with the external impurity potential and not with each other, the problem simplifies considerably. This is because the interaction hamiltonian is now only quadratic in the a-operators. For a fixed array of impurities the hamiltonian can therefore in principle be diagonalized by a linear transformation of the a-operators (corresponding to the formation of scattering states out of plane waves). Let this transformation be written

$$A_m{}^\dagger = \sum_{\mathbf{p}} U_{m\mathbf{p}} a_{\mathbf{p}}{}^\dagger, \tag{5.6.1}$$

where m denotes a new set of quantum numbers. Then

$$[H, A_m{}^\dagger] = \epsilon_m A_m{}^\dagger, \tag{5.6.2}$$

where H is the impurity-scattering hamiltonian (5.2.2). Hence the operator part of the response function can be written

$$\langle a^\dagger_{\mathbf{p}_1}(t)a_{\mathbf{p}_2}(t)a^\dagger_{\mathbf{p}_3}(t')a_{\mathbf{p}_4}(t')\rangle$$
$$= \sum_{\substack{m_1m_2\\m_3m_4}} U^\dagger_{\mathbf{p}_1m_1}U_{\mathbf{p}_2m_2}U^\dagger_{\mathbf{p}_3m_3}U_{\mathbf{p}_4m_4}\langle A^\dagger_{m_1}(t)A_{m_2}(t)A^\dagger_{m_3}(t')A_{m_4}(t')\rangle. \tag{5.6.3}$$

But $A^\dagger_{m_1}(t)$ is now diagonal in time by (5.6.2), i.e., it satisfies

$$A^\dagger_{m_1}(t) = e^{i\epsilon_{m_1}t}A^\dagger_{m_1}, \tag{5.6.4}$$

so that (5.6.3) may be expressed in terms of a diagonal matrix element (with respect to the fermi ground state in the presence of the impurities) of a product of creation and annihilation operators. Hence by Wick's theorem it may be written as a sum of contractions. The equal-time contractions will eventually contribute nothing [on account of the commutator in (5.5.14)], so the only term which survives is

$$\langle A^\dagger_{m_1}(t)A_{m_2}(t)A^\dagger_{m_3}(t')A_{m_4}(t')\rangle = \langle A^\dagger_{m_1}(t)A_{m_4}(t')\rangle\langle A_{m_2}(t)A^\dagger_{m_3}(t')\rangle. \tag{5.6.5}$$

Now transform each A or $A^\dagger$ back to the a-representation using the U's in (5.6.3), and we have the result

$$\langle a^\dagger_{\mathbf{p}_1}(t)a_{\mathbf{p}_2}(t)a^\dagger_{\mathbf{p}_3}(t')a_{\mathbf{p}_4}(t')\rangle = \langle a^\dagger_{\mathbf{p}_1}(t)a_{\mathbf{p}_4}(t')\rangle\langle a_{\mathbf{p}_2}(t)a^\dagger_{\mathbf{p}_3}(t')\rangle. \tag{5.6.6}$$

Thus in the case of *a non-interacting electron gas* the two-particle Green's function for a given external potential—for a particular member of the random potential ensemble—can simply be written as a product of one-particle Green's functions. This reduction is only possible for a quadratic hamiltonian, and for general many-body problems there is no simple relation between one- and two-particle Green's functions.

For the purpose of the present calculation we can work either with the retarded or the time-ordered form of the response function[1]. We choose the time-ordered form, since this is simply related to the time-ordered one-particle functions studied earlier, and we can then obtain the retarded form by using the fluctuation-dissipation theorem [Appendix 2, Eqs. (A.2.14) to (A.2.16)]. At $T = 0$ we have

$$\left.\begin{aligned} \operatorname{Re} \mathscr{R}_{\alpha\beta}(\mathbf{x}-\mathbf{x}',\omega) &= \operatorname{Re} \mathscr{R}^T_{\alpha\beta}(\mathbf{x}-\mathbf{x}',\omega),\\ \operatorname{Im} \mathscr{R}_{\alpha\beta}(\mathbf{x}-\mathbf{x}',\omega) &= (\omega/|\omega|)\operatorname{Im} \mathscr{R}^T_{\alpha\beta}(\mathbf{x}-\mathbf{x}',\omega), \end{aligned}\right\} \tag{5.6.7}$$

[1] Actually in deriving Eq. (5.6.6) above we assumed that we were dealing with the time-ordered function.

where

$$\mathscr{R}^{T}_{\alpha\beta}(\mathbf{x}-\mathbf{x}', t-t') = -i\langle T[j_\alpha(\mathbf{x}, t)j_\beta(\mathbf{x}', t')]\rangle, \tag{5.6.8}$$

and Eq. (5.6.7) refers to the Fourier transforms (in time) of the time-dependent Green's functions. (As we are dealing with two-particle Green's functions, the operator T does not involve a change of sign, i.e., T is a boson T product.) Introducing next the Fourier transform in $\mathbf{x}$-space via the definition (5.5.4), we then have for the time-ordered function on using the factorization theorem (5.6.6) (valid only in this case)

$$\begin{aligned}\mathscr{R}^{T}_{\alpha\beta}(\mathbf{q}, t-t') &= -i\langle T[j_\alpha(\mathbf{q}, t)j_\beta(-\mathbf{q}, t')]\rangle \\ &= -i\sum_{\mathbf{p},\mathbf{p}'}(\mathbf{p}+\tfrac{1}{2}\mathbf{q})_\alpha F(\mathbf{p}, \mathbf{p}'-\mathbf{q}; t-t') \\ &\quad \times F(\mathbf{p}', \mathbf{p}+\mathbf{q}; t'-t)(\mathbf{p}'-\tfrac{1}{2}\mathbf{q})_\beta,\end{aligned} \tag{5.6.9}$$

where $F(\mathbf{p}, \mathbf{p}'; t)$ is the off-diagonal one-particle Green's function (4.3.9). Taking the Fourier transform in time Eq. (5.6.9) becomes a convolution

$$\begin{aligned}\mathscr{R}^{T}_{\alpha\beta}(\mathbf{q}, \omega) = -\frac{i}{2\pi}\int_{-\infty}^{\infty} d\epsilon \sum_{\mathbf{p},\mathbf{p}'}(\mathbf{p}+\tfrac{1}{2}\mathbf{q})_\alpha F(\mathbf{p}, \mathbf{p}'-\mathbf{q}; \epsilon+\omega) \\ \times F(\mathbf{p}', \mathbf{p}+\mathbf{q}; \epsilon)(\mathbf{p}'-\tfrac{1}{2}\mathbf{q})_\beta.\end{aligned} \tag{5.6.10}$$

Finally the ensemble average must be taken to give

$$\begin{aligned}\overline{\mathscr{R}^{T}_{\alpha\beta}}(\mathbf{q}, \omega) = -\frac{i}{2\pi}\int_{-\infty}^{\infty} d\epsilon \sum_{\mathbf{p},\mathbf{p}'}(\mathbf{p}+\tfrac{1}{2}\mathbf{q})_\alpha \\ \times \overline{F(\mathbf{p}, \mathbf{p}'-\mathbf{q}; \epsilon+\omega)F(\mathbf{p}', \mathbf{p}+\mathbf{q}; \epsilon)}(\mathbf{p}'-\tfrac{1}{2}\mathbf{q})_\beta.\end{aligned} \tag{5.6.11}$$

Now let us have a look at the perturbation expansion of this expression in powers of the potential. As in Sec. 5.2 each F in the product in Eq. (5.6.11) may be expanded in a power series. So we now get the *product* of two series. This is indicated by the diagram of Fig. 5.9, which may be

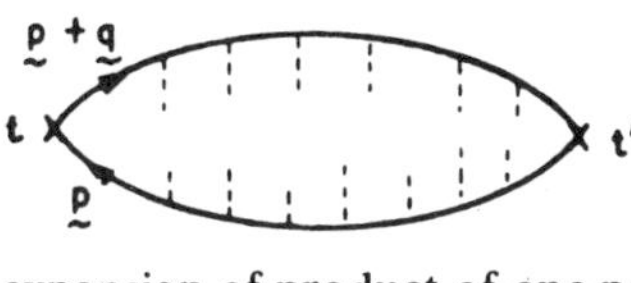

Fig. 5.9. Perturbation expansion of product of one-particle Green's functions.

interpreted in terms of an electron-hole pair injected at time t' and removed at t. The dangling dashed lines each represent a factor $U(\mathbf{q})\rho_{\mathbf{q}}$ for various values of $\mathbf{q}$. What happens on performing the ensemble average? As in (5.2.9) this results in tying together pairs (or higher powers) of dashed lines corresponding to scattering off particular atoms. However, there will now be a new class of diagrams in which simultaneous scattering from a single atom occurs both by the electron and by the hole [Fig. 5.10(b)]. In Fig. 5.10(a) the one-electron scattering diagrams are indicated.

(a) (b)

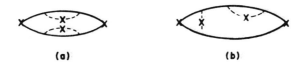

Fig. 5.10. One-electron and electron-hole scattering diagrams.

As in the one-electron case we will again make the low-density hypothesis that only successive scatterings from different atoms need be considered (together with the weak-coupling hypothesis that only pairs of dashed lines will be included). In this way we are led to sums of "ladders" of the form shown in Fig. 5.11. Now note that momentum

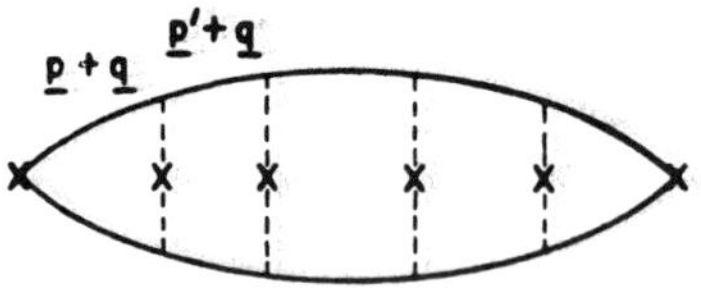

Fig. 5.11. Ladder approximation to the two-particle Green's function.

conservation along a dashed line means that the particle and hole will receive the same change of momentum at each scattering.

Let us consider an extended two-particle Green's function $\mathbf{\Pi}$, defined before summing over internal momentum $\mathbf{p}$ and integrating over internal energy ϵ in (5.6.11):

$$\mathbf{\Pi}(\mathbf{p}, \mathbf{q}; \epsilon, \omega) = \overline{\sum_{\mathbf{p}'} F(\mathbf{p}, \mathbf{p}' - \mathbf{q}; \epsilon + \omega) F(\mathbf{p}', \mathbf{p} + \mathbf{q}; \epsilon)}(\mathbf{p}' - \tfrac{1}{2}\mathbf{q}). \tag{5.6.12}$$

Then by examination of Fig. 5.11 it may be seen that the ladder sum is obtained by iteration of the following integral equation for $\mathbf{\Pi}$:

$$\mathbf{\Pi}(\mathbf{p}, \mathbf{q}; \epsilon, \omega) = G^0(\mathbf{p}, \epsilon + \omega)G^0(\mathbf{p} + \mathbf{q}, \epsilon) \times \{\mathbf{p} + \tfrac{1}{2}\mathbf{q} + \sum_{\mathbf{p}'} NU^2(\mathbf{p} - \mathbf{p}')\mathbf{\Pi}(\mathbf{p}', \mathbf{q}; \epsilon, \omega)\}. \quad (5.6.13)$$

Finally, all the one-particle diagrams previously summed in calculating the one-particle lifetime in Sec. 5.3 may be included by replacing $G^0(\mathbf{p})$ in (5.6.13) by the damped G of Eqs. (5.3.3) and (5.4.3). In this way all diagrams of type Fig. 5.10(a) are automatically inserted in Fig. 5.11. We therefore have as our final equation

$$\mathbf{\Pi}(\mathbf{p}, \mathbf{q}; \epsilon, \omega) = G(\mathbf{p}, \epsilon + \omega)G(\mathbf{p} + \mathbf{q}, \epsilon) \times \{\mathbf{p} + \tfrac{1}{2}\mathbf{q} + \sum_{\mathbf{p}'} NU^2(\mathbf{p} - \mathbf{p}')\mathbf{\Pi}(\mathbf{p}', \mathbf{q}; \epsilon, \omega)\}. \quad (5.6.14)$$

In terms of the solution of this equation we have

$$\overline{\mathscr{R}^T_{\alpha\beta}}(\mathbf{q}, \omega) = -\frac{i}{2\pi} \int_{-\infty}^{\infty} d\epsilon \sum_{\mathbf{p}} (\mathbf{p} + \tfrac{1}{2}\mathbf{q})_\alpha \Pi_\beta(\mathbf{p}, \mathbf{q}; \epsilon, \omega). \quad (5.6.15)$$

From this we can obtain the retarded function $\mathscr{R}_{\alpha\beta}(\mathbf{q}, \omega)$ via (5.6.7), and the complete current response to an applied vector potential is then [see (5.5.13)] given by

$$R_{\alpha\beta}(\mathbf{q}, \omega) = \mathscr{R}_{\alpha\beta}(\mathbf{q}, \omega) - ne\delta_{\alpha\beta}. \quad (5.6.16)$$

(To convert to c.g.s. units, the first term on the right-hand side must be multiplied by e^2/m^2 and the second term by e/m.)

5.7. SOLUTION OF THE INTEGRAL EQUATION FOR THE RESPONSE FUNCTION

In order to calculate the d.c. conductivity we require the response function $R_{\alpha\beta}(\mathbf{q}, \omega)$ in the limit $(\mathbf{q} \to 0, \omega \to 0)$ of a uniform static field. Care is needed in the approach to the $\omega = 0$ limit. At frequency ω the conductivity $\sigma(\omega)$ is defined by Ohm's law

$$J_\alpha(\omega) = \sigma_{\alpha\beta}(\omega)E_\beta(\omega) = i\omega\sigma_{\alpha\beta}(\omega)A_\beta(\omega). \quad (5.7.1)$$

A *constant* electric field $\mathbf{E}(0)$ corresponds to a *divergent* vector potential

$$\mathbf{A}(\omega) = \frac{\mathbf{E}(\omega)}{i\omega}, \qquad \omega \to 0,$$

and the d.c. conductivity is obtained from the response function as the limit

$$\sigma_{\alpha\beta}(0) = \lim_{\omega \to 0} \frac{R_{\alpha\beta}(\omega)}{i\omega}. \tag{5.7.2}$$

How can the expression (5.6.16) for $R_{\alpha\beta}$ lead to a finite result? Clearly, for $\omega = 0$, $\mathscr{R}_{\alpha\beta}(0)$ must precisely *cancel* the "diamagnetic" term $-ne\delta_{\alpha\beta}$, and the d.c. conductivity will come from the term of order ω in the expansion of $\mathscr{R}_{\alpha\beta}(\omega)$. Our calculation will verify this behavior for the random impurity case.

It is just the absence of this cancellation of the $1/\omega$ terms in the conductivity which distinguishes the electromagnetic response of superconductors from that of normal metals. It was first pointed out by F. London in 1935 that the interactions in superconductors produce a "rigidity" in the electronic wave functions which prevents the "paramagnetic" current from adjusting itself to an applied field. Thus, to a first approximation, $\mathscr{R}_{\alpha\beta}$ remains zero in a superconductor when a vector potential is applied, and (5.6.16) then leads to a finite diamagnetic current (London's equation)

$$\mathbf{J} = -\frac{ne^2}{m}\mathbf{A}. \tag{5.7.3}$$

We now consider the solution of the integral equation (5.6.14). The mathematical structure of the equation depends strongly on the fact that it is expressed in terms of the T-product one-particle Green's functions G (as opposed to a retarded form) which are rather singular for the N-particle fermion system. What is their analytic form? To see this we go back to Eq. (5.4.3) expressing the one-particle self-energy $\Sigma(\mathbf{p}, \epsilon)$ in terms of the unperturbed Green's function $G^0(\mathbf{p}, \epsilon)$. As shown in Chap. 4, [Eq. (4.6.6)], the electron-hole function $G^0(\mathbf{p}, \epsilon)$ has a discontinuous imaginary part at the fermi level ($\epsilon = 0$ in our energy scale)

$$G^0(\mathbf{p}, \epsilon) = \frac{f_{\mathbf{p}}^+}{\epsilon - \epsilon_{\mathbf{p}} + i\eta} + \frac{f_{\mathbf{p}}^-}{\epsilon - \epsilon_{\mathbf{p}} - i\eta}, \tag{5.7.4}$$

from which at zero temperature

$$\operatorname{Im} G^0(\mathbf{p}, \epsilon) = -\pi \operatorname{sgn}(\epsilon)\delta(\epsilon - \epsilon_{\mathbf{p}}) \tag{5.7.5}$$

[sgn (x) = sign of $x = x/|x|$]. Note that in Eq. (5.4.7) we were interested in the retarded as opposed to the T-product Green's function (see footnote on p. 86). Inserting (5.7.5) into the self-energy (5.4.3) and neglecting the energy dependence of $F(\epsilon)$ and $\rho(\epsilon)$ [Eq. (5.4.8)], we have

$$\begin{aligned}\Sigma(\mathbf{p}, \epsilon) &= \Delta - i\pi \operatorname{sgn}(\epsilon) N \sum_{\mathbf{p}'} U^2(\mathbf{p}-\mathbf{p}')\delta(\epsilon - \epsilon_{\mathbf{p}'}) \\ &= \Delta - i\pi\rho n_{\text{imp}} \operatorname{sgn}(\epsilon) \frac{1}{4\pi} \int d\Omega_{\mathbf{p}'} U^2(\mathbf{p}-\mathbf{p}') \\ &= \Delta - \frac{i \operatorname{sgn}(\epsilon)}{2\tau}, \qquad \text{say,} \end{aligned} \tag{5.7.6}$$

where $\mathbf{p}'$ is considered to have magnitude of order p_f, $d\Omega_{\mathbf{p}'}$ is the solid angle integral over the vector direction of $\mathbf{p}'$, and Δ is the energy shift. τ is the relaxation time which describes the damping of the one-particle Green's function. We now see that the integral equation for $\mathbf{\Pi}(\mathbf{p}, \mathbf{q}; \epsilon, \omega)$ contains a discontinuous function of both ϵ and ω. [Our treatment follows the one given by Abrikosov, Gor'kov, and Dzyaloshinskii (1963).]

The other variable we have to worry about in solving the integral equation (5.6.14) is that associated with the directionality of the electron-hole momenta $\mathbf{p}$ and $\mathbf{p} + \mathbf{q}$. As we are only interested in the limit of a uniform applied field, we can now set $\mathbf{q} = 0$. [For general $\mathbf{q}$ there are two regions to distinguish, that of the "normal" skin effect where $|\mathbf{q}|l \ll 1$, and that of the "anomalous" skin effect where $|\mathbf{q}|l \gg 1$. Here $l = v_f\tau$ is a mean free path. See Pippard (1965) for a review.] However, the evaluation of $\mathbf{\Pi}$ will still depend on the directions of $\mathbf{p}$ and $\mathbf{p}'$. This is because electric current is a directional quantity (as represented here through the tensor character of $R_{\alpha\beta}$), and the relative angle of scattering during a collision turns out to weight the corresponding response time. This may be seen physically by examining the form of Eq. (5.6.14). The first term on the right-hand side just represents the decay of the response-current, via the damping in $G(\mathbf{p}, \epsilon)$ [see Eq. (5.7.6)], due to electrons scattering out of their excited states $\mathbf{p}$ and $\mathbf{p} + \mathbf{q}$ as initially excited by the applied potential $\mathbf{A}$. This corresponds to the "scattering out" term in the Boltzmann equation formalism. The second, integral equation, term represents a scattering *back* into the excited states from electron and hole states $\mathbf{p}'$ and $\mathbf{p}' + \mathbf{q}$, hence depends on how many electrons were scattered out into these states [i.e., depends on $\mathbf{\Pi}(\mathbf{p}', \mathbf{q}; \epsilon, \omega)$] previously. This corresponds to the "scattering in" term in the usual Boltzmann equation.

However, the nice thing about our Green's function approach is that we have not had to assume anything about the randomizing of the distribution after each time interval, but get the same results *automatically* by the irreversible limiting procedures discussed above.

What the tensor character of Eq. (5.6.14) (letting $\mathbf{q} \to 0$) does is to project out the first spherical harmonic (in the angle between the vectors $\mathbf{p}$ and $\mathbf{p}'$) of the function $\mathbf{\Pi}(\mathbf{p}', 0; \epsilon, \omega)$. So we must solve the integral equation for $\mathbf{\Pi}$ in this spherical harmonic component. We assume that the rate of variation of $\mathbf{\Pi}$ with the magnitude $|\mathbf{p}|$ is slow [as in (5.7.6)] and set $|\mathbf{p}| = |\mathbf{p}'| = p_f$. We therefore insert in (5.6.14) the $l = 1$ term

$$\Pi_\beta(\mathbf{p}, 0; \epsilon, \omega) = p_f P_1(\cos\theta_\beta)\Lambda_1(\epsilon, \omega) = p_\beta \Lambda_1(\epsilon, \omega) \tag{5.7.7}$$

of the expansion of Π_β in spherical harmonics of $\cos\theta_\beta = p_\beta/p_f$. We now assume that the potential $U(\mathbf{p} - \mathbf{p}')$ is spherically symmetric, and thus depends only on the angle between $\mathbf{p}$ and $\mathbf{p}'$. We can then write

$$\sum_{\mathbf{p}'} NU^2(\mathbf{p} - \mathbf{p}')\Pi_\beta(\mathbf{p}', 0; \epsilon, \omega) = p_\beta \overline{\Lambda}_1(\epsilon, \omega), \tag{5.7.8}$$

and (5.6.14) gives

$$\Lambda_1(\epsilon, \omega) = G(\mathbf{p}, \epsilon + \omega)G(\mathbf{p}, \epsilon)\{1 + \overline{\Lambda}_1(\epsilon, \omega)\}. \tag{5.7.9}$$

We can obtain an equation for $\overline{\Lambda}_1$ from this by multiplying by $p_\beta NU^2(\mathbf{k} - \mathbf{p})$ and summing over $\mathbf{p}$. This gives

$$k_\beta \overline{\Lambda}_1(\epsilon, \omega) = \sum_{\mathbf{p}} NU^2(\mathbf{k} - \mathbf{p}) p_\beta G(\mathbf{p}, \epsilon + \omega)G(\mathbf{p}, \epsilon)\{1 + \overline{\Lambda}_1(\epsilon, \omega)\}.$$

We again, as in (5.7.6), separate the angular integration in the sum over $\mathbf{p}$ from the integral over energy. The angular integration projects out the component of $\mathbf{p}$ parallel to $\mathbf{k}$, and we obtain

$$\overline{\Lambda}_1(\epsilon, \omega) = \frac{1}{2\pi\tau_1} \int d\epsilon_{\mathbf{p}} G(\mathbf{p}, \epsilon + \omega)G(\mathbf{p}, \epsilon)\{1 + \overline{\Lambda}_1(\epsilon, \omega)\}, \tag{5.7.10}$$

where

$$\frac{1}{\tau_1} = 2\pi\rho n_{\text{imp}} \frac{1}{4\pi} \int U^2(\theta) \cos\theta \, d\Omega. \tag{5.7.11}$$

[Note the difference from $1/\tau$ as defined via Eq. (5.7.6) $-1/\tau_1$ is an $l = 1$ spherical harmonic projection of the transition probability.]

Now let us examine (5.7.10) in ϵ and ω space. From Eq. (5.7.6) we see that

$$G(\mathbf{p}, \epsilon) = \frac{1}{\epsilon - \bar{\epsilon}_{\mathbf{p}} + (i/2\tau)\,\mathrm{sgn}\,(\epsilon)} \tag{5.7.12}$$

(where $\bar{\epsilon}_{\mathbf{p}} = \epsilon_{\mathbf{p}} + \Delta$) has a pole which switches from positive to negative imaginary as ϵ changes sign. Therefore the integral over $\epsilon_{\mathbf{p}}$ in (5.7.10) is non-zero only if $\epsilon + \omega$ is >0 while ϵ is simultaneously <0 (or vice versa), otherwise both poles are on one side of the real $\epsilon_{\mathbf{p}}$ axis and the integral vanishes. Hence $\bar{\Lambda}_1$ is proportional to

$$\mathscr{D}(\epsilon, \omega) = \theta(\epsilon + \omega)\theta(-\epsilon) + \theta(\epsilon)\theta(-\epsilon - \omega) \tag{5.7.13}$$

(θ is the unit step function). On performing the contour integral we find

$$\frac{\bar{\Lambda}_1}{1 + \bar{\Lambda}_1} = \frac{i}{\tau_1} \frac{\mathscr{D}(\epsilon, \omega)}{|\omega| + i/\tau}, \tag{5.7.14}$$

from which

$$\bar{\Lambda}_1(\epsilon, \omega) = \frac{i}{\tau_1} \frac{1}{|\omega| + i/\tau_{\mathrm{tr}}} \mathscr{D}(\epsilon, \omega), \tag{5.7.15}$$

where

$$\frac{1}{\tau_{\mathrm{tr}}} = \frac{1}{\tau} - \frac{1}{\tau_1} \tag{5.7.16}$$

is a "transport relaxation rate" and is an integral of the form (5.7.11) with factor $(1 - \cos\theta)$ in place of $\cos\theta$.

We can now use this solution to evaluate the response function. The function Π_β is given by (5.7.7) with (5.7.9), and substituting this into (5.6.15) we have, in the limit $\mathbf{q} = 0$,

$$\begin{aligned}\overline{\mathscr{R}^{T}_{\alpha\beta}(\omega)} &= -\frac{i}{2\pi} \int_{-\infty}^{\infty} d\epsilon \sum_{\mathbf{p}} p_\alpha p_\beta G(\mathbf{p}, \epsilon + \omega) G(\mathbf{p}, \epsilon) \\ &\quad \times \left\{1 + \frac{i}{\tau_1} \frac{\mathscr{D}(\epsilon, \omega)}{|\omega| + i/\tau_{\mathrm{tr}}}\right\} \\ &= -\delta_{\alpha\beta} \frac{i}{2\pi} \frac{\rho p_f^2}{3} \int d\epsilon \int d\epsilon_{\mathbf{p}} G(\mathbf{p}, \epsilon + \omega) G(\mathbf{p}, \epsilon) \\ &\quad \times \left\{1 + \frac{i}{\tau_1} \frac{\mathscr{D}(\epsilon, \omega)}{|\omega| + i/\tau_{\mathrm{tr}}}\right\}.\end{aligned} \tag{5.7.17}$$

This double integral is formally divergent; to evaluate it consistently we must be careful to perform the integration over the Fourier frequency ϵ

first. (This is because of our earlier restriction to energies very close to the fermi energy; integration over $\epsilon_{\mathbf{p}}$ first includes contributions from energy regions where our approximations do not hold and hence leads to a spurious divergence.)

Using the explicit form (5.7.12) of $G(\mathbf{p}, \epsilon)$ the integrations are elementary, and we find after some calculation that

$$\int_{-\infty}^{\infty} d\epsilon G(\mathbf{p}, \epsilon + \omega) G(\mathbf{p}, \epsilon) \left\{ 1 + \frac{i}{\tau_1} \frac{\mathscr{D}(\epsilon, \omega)}{|\omega| + i/\tau_{tr}} \right\}$$

$$= \frac{i/\tau_{tr}}{|\omega|(|\omega| + i/\tau_{tr})} \log \frac{\bar{\epsilon}_{\mathbf{p}}^2 + (1/2\tau - i|\omega|)^2}{\bar{\epsilon}_{\mathbf{p}}^2 + 1/4\tau^2} . \tag{5.7.18}$$

The integral over $\epsilon_{\mathbf{p}}$ now converges, and

$$\int_{-\infty}^{\infty} d\epsilon_{\mathbf{p}} \log \frac{\bar{\epsilon}_{\mathbf{p}}^2 + (1/2\tau - i|\omega|)^2}{\bar{\epsilon}_{\mathbf{p}}^2 + 1/4\tau^2} = 2\pi i |\omega|. \tag{5.7.19}$$

Thus

$$\overline{\mathscr{R}^T_{\alpha\beta}}(\omega) = \delta_{\alpha\beta} \frac{\rho p_f^2}{3} \frac{1}{1 - i|\omega|\tau_{tr}} . \tag{5.7.20}$$

Using $\rho = mp_f/2\pi^2$, $n = p_f^3/6\pi^2$ (for electrons of one spin direction), and converting to c.g.s. units of current, this becomes

$$\overline{\mathscr{R}^T_{\alpha\beta}}(\omega) = \delta_{\alpha\beta} \frac{ne^2}{m} \frac{1}{1 - i|\omega|\tau_{tr}}$$

$$= \delta_{\alpha\beta} \frac{ne^2}{m} (1 + i|\omega|\tau_{tr} + \cdots). \tag{5.7.21}$$

We finally go over to the retarded function $\mathscr{R}_{\alpha\beta}$ by means of Eq. (5.6.7), and we see that, in the limit $\omega \to 0$, the leading term in the complete response function (5.6.16) has the expected form [compare (5.7.2)]

$$R_{\alpha\beta}(\omega) = \delta_{\alpha\beta} i\omega\sigma, \tag{5.7.22}$$

where

$$\sigma = \frac{ne^2\tau_{tr}}{m} . \tag{5.7.23}$$

This is just Ohm's law, with the expression for the d.c. conductivity obtained from the elementary Boltzmann equation approach to the problem.

Chapter 6

The Interacting Electron Gas

We now come to consider systems of particles which interact with each other through some sort of potential. We will be interested in systems where quantum effects dominate, i.e. (in principle) electrons, nucleons and He atoms below their degeneracy temperature ($\sim 2^\circ$K), as it is in such systems that the Green's function formalism which is geared to wave-like quantum phenomena proves vitally important.

The most basic physical effect of interactions between particles is of course the phenomenon of phase transitions to condensed states of matter. A classical treatment will account for condensation and the critical phenomena near the phase transition, but it predicts that all systems solidify if made cold enough. It is a very remarkable property of quantum systems that fluid-like states can persist even at the absolute zero of temperature.

The understanding of the properties of such "quantum fluids" relies on two important concepts: (a) the idea of an elementary excitation, or quasiparticle state; (b) the possibility of instabilities of the quantum fluid which occur as either the temperature is lowered or the interaction strength is increased, in which the ground state of the fluid makes a transition to a new state which is still a fluid but has lower symmetry than the original state of the system. Examples of such instabilities are superconductivity and ferromagnetism in metals. These will be discussed in later chapters.

The mathematical formulation of these concepts in terms of the original microscopic description of the system is best carried out using Green's function techniques. However, in the case of many-body problems of the above type we cannot hope to obtain exact solutions of the dynamical equations, and it is necessary to develop suitable methods of approximation. These can be formulated in terms of the equation of motion approach such as was used in Chap. 1, in which one devises approximations for breaking off the hierarchy of equations, or one can use the perturbation theory for the Green's functions in the many-body system and obtain approximate solutions by summing appropriate subsets of diagrams.

In two special limiting cases it turns out that one can obtain asymptotically exact solutions of many-body problems by the use of approximation methods of the above type. These are: the problem of a system of fermions interacting through long-range Coulomb forces in the limit of high density of the particles, and the solution for the superconducting state of a system of fermions interacting through weakly attractive forces in the so-called "pairing hamiltonian" model. In the Coulomb case a particular type of approximation, the "random phase approximation" (or r.p.a.) does the trick, while in the superconducting case a particular version of the Hartree–Fock approximation, the so-called B.C.S. approximation (named after Bardeen, Cooper, and Schrieffer) provides a solution. In other cases such as the medium- or low-density electron gas and its attendant instabilities (magnetic or insulating) and in liquid helium there are no known methods of getting a mathematically good (or even reasonable) solution. However, apart from the instability problem the concept of a quasiparticle excitation still appears to work and was formulated as a semimicroscopic phenomenology by Landau in his theory of fermi liquids [Landau (1956–8)]. Phenomenology of this type can be understood in microscopic terms using the Green's function approach, and some of Landau's assumptions can be proved in this way by making other assumptions about the convergence of the perturbation series [Luttinger (1960)]. In the region of instabilities, however, such concepts no longer apply; the phenomenology becomes entangled with the classical appearance of critical phenomena, and the subject is still wide open theoretically.

In the present chapter we use the Green's function formalism to study the effect of Coulomb interactions on the properties of an electron gas. As indicated above this is a case where, in the high density limit, r.p.a. works increasingly well. So it is useful to develop r.p.a. in this context as a prelude to the case (Chap. 7) of ferromagnetism in metals, where it can only be applied heuristically.

6.1. THE HARTREE–FOCK APPROXIMATION

The central physical phenomenon in the Coulomb gas is the effect of screening. The Hartree–Fock approximation allows calculation of this screening effect in the presence of an external non-uniform potential (such as that due to an ion) and is still the basic approximation used in the band theory of metals. However, when it comes to the discussion of the screening of the potential due to one of the electrons in the gas by the other electrons, the approximation is inadequate as it does not take sufficient account of the resulting correlations of the electronic motions. In this section we develop the Hartree–Fock approximation in the latter case, i.e., neglecting the ionic potentials, and show that it leads to unphysical results.

We consider a system of N electrons in a large volume V and write the hamiltonian in the form $H = H_0 + H_1$, where

$$H_0 = \sum_i \frac{\mathbf{p}_i^{\,2}}{2m}, \qquad H_1 = \tfrac{1}{2} \sum_{i \neq j} V(\mathbf{x}_i - \mathbf{x}_j). \tag{6.1.1}$$

The interaction potential $V(\mathbf{x}_i - \mathbf{x}_j)$ is in general taken to be an instantaneous two-body potential. We again use the plane waves $u_{\mathbf{p}}(\mathbf{x})$ of Eq. (4.2.2) to provide a complete orthonormal set of one-particle functions. In second quantization the form of the interaction term H_1 is then

$$\tfrac{1}{2} \sum_{\mathbf{pqst}} \langle \mathbf{pq}|V|\mathbf{st}\rangle a_{\mathbf{p}}{}^{\dagger} a_{\mathbf{q}}{}^{\dagger} a_{\mathbf{s}} a_{\mathbf{t}},$$

where

$$\langle \mathbf{pq}|V|\mathbf{st}\rangle = \int\int_V u_{\mathbf{p}}{}^*(\mathbf{x}) u_{\mathbf{q}}{}^*(\mathbf{x}') V(\mathbf{x} - \mathbf{x}') u_{\mathbf{s}}(\mathbf{x}') u_{\mathbf{t}}(\mathbf{x})\, d^3x\, d^3x'. \tag{6.1.2}$$

Evaluating this with (4.2.2), we obtain the second-quantized form of $H_0 + H_1$

$$H_0 = \sum_{\mathbf{p}\sigma} \epsilon_{\mathbf{p}} a^{\dagger}_{\mathbf{p}\sigma} a_{\mathbf{p}\sigma}, \qquad H_1 = \tfrac{1}{2} \sum_{\mathbf{q}} V(\mathbf{q}) \sum_{\substack{\mathbf{pp}' \\ \sigma\sigma'}} a^{\dagger}_{\mathbf{p+q},\sigma} a^{\dagger}_{\mathbf{p}'-\mathbf{q},\sigma'} a_{\mathbf{p}'\sigma'} a_{\mathbf{p}\sigma}. \tag{6.1.3}$$

Here $\epsilon_{\mathbf{p}} = \mathbf{p}^2/2m$, $V(\mathbf{q})$ is the Fourier transform of the potential $V(\mathbf{x})$, and the electron spin has been included explicitly through the subscripts σ, σ' which take the values $\pm\frac{1}{2}$. As in Chap. 4 we proceed to study the one-electron Green's function $G_\sigma(\mathbf{p}, t)$ for the present problem, defined as in (4.2.15) by

$$G_\sigma(\mathbf{p}, t) = -i\langle T[\tilde{a}_{\mathbf{p}\sigma}(t)\tilde{a}^{\dagger}_{\mathbf{p}\sigma}(0)]\rangle, \tag{6.1.4}$$

where

$$\tilde{a}_{\mathbf{p}\sigma}(t) = e^{iHt} a_{\mathbf{p}\sigma}\, e^{-iHt},$$

and the expectation value $\langle \ldots \rangle$ refers to the N-electron ground state $|\Psi_G^{(N)}\rangle$. The idea of this study is that $G_\sigma(\mathbf{p}, \epsilon)$ (Fourier transform) will turn out to have a pole with a reasonably small imaginary part. This means that the low-lying excitations in the gas will behave rather like free electrons—apart from the finite lifetime measured by this imaginary part. This is the essence of Landau's approach to the phenomenology of fermi liquids. In fact it may be shown that, in contrast to the impurity case where the width is not strongly energy dependent, the quasiparticle scattering actually vanishes (at $T = 0$) for particles at the fermi surface, even in an interacting system. The qualitative reason for this lies in the operation of the Pauli exclusion principle: lifetime effects result from electron–electron scattering (as opposed to electron-impurity scattering in the impurity problem). Such scattering can only take place to unoccupied states. It turns out that for an electron right at the fermi surface the phase space region (number of available scattering states) for purely elastic scattering events goes to zero, so that such electrons have truly infinite lifetime. Nearby electrons also have long lifetimes. Thus the fact that most properties of metals can be understood from a single-electron picture is explained. For the scattering lengths involved to be finite, the effective electron–electron scattering potential must be short-ranged. This results from screening, which is the second vital component in the justification of the one-electron theory of metals.

The approach of the present section to the calculation of $G_\sigma(\mathbf{p}, t)$ will not yet include these scattering processes. We will work with the equation of motion of G and treat the potentials due to the other electrons in the system in an average way only.

The equation of motion reads

$$i\,\frac{\partial G_\sigma(\mathbf{p})}{\partial t} = \delta(t)\langle\{a_{\mathbf{p}\sigma}(0), a_{\mathbf{p}\sigma}^\dagger(0)\}\rangle - i\langle T[\,[\tilde{a}_{\mathbf{p}\sigma}(t), H]a_{\mathbf{p}\sigma}^\dagger(0)]\rangle \qquad (6.1.5)$$

Now

$$[\tilde{a}_{\mathbf{p}\sigma}(t), H] = e^{iHt}[a_{\mathbf{p}\sigma}, H]\, e^{-iHt} = e^{iHt}[a_{\mathbf{p}\sigma}, H_0 + H_1]\, e^{-iHt}. \qquad (6.1.6)$$

The anticommutation rules (4.2.3), generalized to take the spin suffixes into account, lead to the general rule

$$[a_\mu{}^\dagger a_\nu, a_\rho{}^\dagger a_\sigma] = \delta_{\nu\rho} a_\mu{}^\dagger a_\sigma - \delta_{\mu\sigma} a_\rho{}^\dagger a_\nu,$$

from which we easily obtain

$$[a_{\mathbf{p}\sigma}, H_0] = \epsilon_{\mathbf{p}} a_{\mathbf{p}\sigma}, \tag{6.1.7}$$

and

$$[a_{\mathbf{p}\sigma}, H_1] = \tfrac{1}{2} \sum_{\mathbf{q}} V(\mathbf{q}) \sum_{\substack{\mathbf{p}'\mathbf{p}'' \\ \sigma'\sigma''}} a^{\dagger}_{\mathbf{p}'+\mathbf{q},\sigma'} a_{\mathbf{p}'\sigma'} a_{\mathbf{p}''\sigma''} \delta_{\sigma\sigma''} \delta_{\mathbf{p}''-\mathbf{q},\mathbf{p}}$$

$$+ \text{ an identical term}$$

$$= \sum_{\mathbf{q}} V(\mathbf{q}) \sum_{\mathbf{p}'\sigma'} a^{\dagger}_{\mathbf{p}'+\mathbf{q},\sigma'} a_{\mathbf{p}'\sigma'} a_{\mathbf{p}+\mathbf{q},\sigma}. \tag{6.1.8}$$

The commutator (6.1.6) thus leads to a product of three operators at time t, and, when we substitute this into (6.1.5), we obtain on the right-hand side of the equation of motion a higher-order Green's function of the form

$$\langle T[\tilde{a}^{\dagger}(t)\tilde{a}(t)\tilde{a}(t)\tilde{a}^{\dagger}(0)]\rangle.$$

If we work out the equation of motion for this, we obtain a Green's function of still higher order, and so on; thus we are led to a hierarchy of equations which continues indefinitely.

The Hartree–Fock approximation consists in linearizing the equation of motion for G by replacing products $a^{\dagger}a$ of pairs of operators in (6.1.8) by their expectation value in the fermi sea. This will be a good approximation if the *fluctuations* $\langle a^{\dagger}a\rangle - a^{\dagger}a$ of the system are sufficiently small to be negligible. Only averages of the type $\langle a^{\dagger}a\rangle$ occur; terms like $\langle a\rangle$, $\langle aa\rangle$, etc., which do not conserve particle numbers, are all zero for normal systems.

Since two combinations of a and $a^{\dagger}$ operators are possible in (6.1.8), we replace

$$a^{\dagger}_{\mathbf{p}'+\mathbf{q},\sigma'} a_{\mathbf{p}'\sigma'} a_{\mathbf{p}+\mathbf{q},\sigma}$$

by

$$\langle a^{\dagger}_{\mathbf{p}'+\mathbf{q},\sigma'} a_{\mathbf{p}'\sigma'}\rangle a_{\mathbf{p}+\mathbf{q},\sigma} - \langle a^{\dagger}_{\mathbf{p}'+\mathbf{q},\sigma'} a_{\mathbf{p}+\mathbf{q},\sigma}\rangle a_{\mathbf{p}'\sigma'}; \tag{6.1.9}$$

the minus sign in the second term comes from the change in the order of two anticommuting factors. We further assume that we are dealing with a homogeneous system whose size is so large that boundary effects are negligible. Such a system is translationally invariant, so that the single-particle eigenfunctions must be plane waves (as assumed) and the ground state is the fermi sea $|\Psi_G{}^{(N)}\rangle$. It may be shown that the above approximation is equivalent to a variational derivation in which the ground state is assumed to be a Slater determinant of single-electron wave functions. (The present case assumes these wave functions are plane waves.) In this way more general one-electron equations are obtained which, in the case where the ion potentials are included, lead to energy bands for the electron states. The effect of the bands can be simulated by replacing $\mathbf{p}^2/2m$ in the hamiltonian (6.1.3) by a more general dependence of $\epsilon_\mathbf{p}$ on wave-vector.

For complex band structures the effect of interactions, treated through the Hartree–Fock approximation, can lead to instability of the fermi sea ground state and the appearance of states of lower symmetry (ferromagnetic states or spin density wave states) which would not satisfy spin-rotation invariance or translation invariance. In this case the symmetry of expectation values $\langle a^\dagger a\rangle$ would be altered. An example of such an instability, leading to an antiferromagnetic transition, is discussed in Sec. 7.4. We stay here with the simple paramagnetic solutions, for which there is momentum conservation in both orbital and spin variables. We thus put

$$\langle\Psi_G{}^{(N)}|a^\dagger_{\mathbf{p}\sigma}a_{\mathbf{p}'\sigma'}|\Psi_G{}^{(N)}\rangle = f_{\mathbf{p}\sigma}\delta_{\mathbf{p}\mathbf{p}'}\delta_{\sigma\sigma'}, \tag{6.1.10}$$

where $f_{\mathbf{p}\sigma}$ is the distribution function (there denoted by $f_\mathbf{p}{}^-$) of Eq. (4.6.3) for electrons of energy $\epsilon_\mathbf{p}$ and spin σ. In the paramagnetic state in the absence of an external field, $f_{\mathbf{p}\sigma}$ is of course independent of σ. For ground states of lower symmetry one may find that the number $f_{\mathbf{p}\sigma}$ depends on σ (ferromagnetism), that it contains a wave-vector $\mathbf{Q}$: $\langle a^\dagger_{\mathbf{p}+\mathbf{Q}\uparrow}a_{\mathbf{p}\downarrow}\rangle \neq 0$ (spin density waves), or that it violates particle conservation: $\langle a_\mathbf{p}{}^\dagger a^\dagger_{-\mathbf{p}}\rangle \neq 0$ (superconducting case).

Putting (6.1.7–10) into (6.1.5), we now obtain the equation of motion of G in the diagonal form

$$i\,\frac{\partial G_\sigma(\mathbf{p}, t)}{\partial t} = \delta(t) + \epsilon^{\mathrm{HF}}_{\mathbf{p}\sigma}G_\sigma(\mathbf{p}, t), \tag{6.1.11}$$

where

$$\epsilon^{\mathrm{HF}}_{\mathbf{p}\sigma} = \epsilon_\mathbf{p} + V(\mathbf{q}=0)\sum_{\mathbf{p}'\sigma'} f_{\mathbf{p}'\sigma'} - \sum_{\mathbf{p}'} V(\mathbf{p}'-\mathbf{p})f_{\mathbf{p}'\sigma}. \tag{6.1.12}$$

It is seen that, in this approximation, the interaction between the particles simply produces a shift in the unperturbed energy $\epsilon_{\mathbf{p}}$.

The first term in the energy shift (6.1.12) may be written

$$\frac{N}{V}\int V(\mathbf{x})\, d^3x$$

and represents the (infinite) self-energy of an electron charge distribution of uniform density N/V. This term may be cancelled out by introducing a uniform positive background charge, of the same density, which reduces the net charge of the system to zero; this may be regarded as simulating the static ionic lattice in real metals.

The last term in (6.1.12) is the *exchange term* $\epsilon_{\mathbf{p}}^{\mathrm{ex}}$. It is easily evaluated for the case of an unscreened Coulomb potential, $V(\mathbf{x}) = e^2/|\mathbf{x}|$. Here we have

$$V(\mathbf{q}) = \frac{1}{V}\frac{4\pi e^2}{|\mathbf{q}|^2}, \tag{6.1.13}$$

and, replacing the sum over $\mathbf{p}'$ by an integral over the fermi sphere, we obtain the well-known result (independent of spin)

$$\begin{aligned}\epsilon_{\mathbf{p}}^{\mathrm{ex}} &= -\frac{4\pi e^2}{(2\pi)^3}\int_{|\mathbf{p}'|<p_f}\frac{d^3p'}{|\mathbf{p}-\mathbf{p}'|^2}\\ &= -\frac{e^2 p_f}{\pi}\left\{1+\frac{p_f^2-p^2}{2pp_f}\log\left|\frac{p_f+p}{p_f-p}\right|\right\},\end{aligned} \tag{6.1.14}$$

where p_f is the fermi momentum. This formula contains an important deficiency which is shown through the fact that the derivative $d\epsilon_{\mathbf{p}}^{\mathrm{ex}}/dp$ of the function (6.1.14) has a logarithmic singularity at $p = p_f$; hence the density of states, which is proportional to $dp/d\epsilon_{\mathbf{p}}$, vanishes at the fermi surface. This would lead to anomalous effects in properties which depend on the one-particle density of states. For example, the electronic specific heat would vary as $T/\log T$ instead of being linear in T. Such effects have not been observed (they would in fact be quite difficult to detect), but, more important, they neglect screening which both on heuristic and on mathematical grounds provides an essential contribution we will discuss below.

6.2. CALCULATION OF THE EXCHANGE CONTRIBUTION TO THE GROUND STATE ENERGY OF THE ELECTRON GAS

The results of the Hartree–Fock calculation of Sec. 6.1 now allow us to make a calculation of the effects of the Coulomb interaction on the ground state energy of the electron gas. The shift of one-electron energies calculated in Sec. 6.1 leads to a lowering of the positive kinetic energy, $\frac{3}{5}\epsilon_f$ per particle, of the non-interacting gas. It turns out that for an electron gas of high density (measured in units of the ratio of the inter-particle spacing to the Bohr radius), the effects of the potential energy become relatively weak, compared to the kinetic energy, as the density is increased.

To see this we introduce dimensionless variables. Suppose the gas contains N particles in a volume V; let r_s be the radius of a sphere of volume V/N, measured in units of the Bohr radius $a_0 = 1/(me^2)$ $(\hbar = 1)$. Thus

$$\tfrac{4}{3}\pi(r_s a_0)^3 = \frac{V}{N}, \tag{6.2.1}$$

and the fermi momentum is

$$p_f = \left(\frac{3\pi^2 N}{V}\right)^{1/3} = \left(\frac{9\pi}{4}\right)^{1/3} \frac{1}{r_s a_0}. \tag{6.2.2}$$

The unit of energy is the Rydberg $E_0 = e^2/2a_0$. In terms of reduced position and momentum variables

$$\mathbf{x}' = \mathbf{x}/(a_0 r_s), \qquad \mathbf{p}' = (a_0 r_s)\mathbf{p}, \tag{6.2.3}$$

the hamiltonian in configuration space is

$$H = \sum_i \frac{\mathbf{p}_i^2}{2m} + \tfrac{1}{2}\sum_{i\neq j}\frac{e^2}{x_{ij}} = \frac{E_0}{r_s^2}\left\{\sum_i \mathbf{p}_i'^2 + r_s \sum_{i\neq j}\frac{1}{x_{ij}'}\right\}, \tag{6.2.4}$$

and it is seen that the second term is of higher order in r_s than the first. Thus as the density is increased (corresponding to $r_s \to 0$) the effects of the potential energy become relatively weak. In order to calculate this effect within the Hartree–Fock approximation we need an expression relating the total energy shift

$$\Delta E_{\text{tot}} = \langle H_1 \rangle, \tag{6.2.5}$$

where the brackets denote the ground state expectation value and H_1 is given in (6.1.3), to the one-electron Green's function calculated in Sec. 6.1. This may be done by means of the following identity: using (6.1.8) it may be seen that

$$\lim_{t\to 0-} \sum_{\mathbf{p}\sigma} \left\{ \frac{\partial G_\sigma(\mathbf{p}, t)}{\partial t} + i\delta(t) \right\} = \langle H_0 \rangle + 2\langle H_1 \rangle, \tag{6.2.6}$$

where the time limit has been chosen so that the $a_\mathbf{p}^\dagger$ stands to the left of the $a_\mathbf{p}$ in the formula for G.

We thus have

$$\Delta E_{\text{tot}} = \tfrac{1}{2} \lim_{t\to 0-} \sum_{\mathbf{p}\sigma} \left\{ \frac{\partial G_\sigma(\mathbf{p}, t)}{\partial t} - \frac{\partial G_\sigma{}^0(\mathbf{p}, t)}{\partial t} \right\}, \tag{6.2.7}$$

where G^0 is the Green's function of the non-interacting system. Introducing the Fourier transform of $G_\sigma(\mathbf{p}, t)$,

$$G_\sigma(\mathbf{p}, t) = \frac{1}{2\pi} \int_{-\infty}^{\infty} d\epsilon\, G_\sigma(\mathbf{p}, \epsilon)\, e^{-i\epsilon t}, \tag{6.2.8}$$

ΔE_{tot} can then be evaluated in terms of $G_\sigma(\mathbf{p}, \epsilon)$ as

$$\Delta E_{\text{tot}} \doteq -\frac{i}{4\pi} \sum_{\mathbf{p}\sigma} \int_{-\infty}^{\infty} d\epsilon\, e^{-i\epsilon 0^-} \epsilon \{ G_\sigma(\mathbf{p}, \epsilon) - G_\sigma{}^0(\mathbf{p}, \epsilon) \}. \tag{6.2.9}$$

Using (6.1.11), $G_\sigma(\mathbf{p}, \epsilon)$ in the Hartree–Fock approximation is easily seen to have the form [analogous to Eq. (4.6.6) for the non-interacting Green's function]

$$G_\sigma^{\text{HF}}(\mathbf{p}, \epsilon) = \frac{f_{\mathbf{p}\sigma}^+}{\epsilon - \epsilon_{\mathbf{p}\sigma}^{\text{HF}} + i\eta} + \frac{f_{\mathbf{p}\sigma}^-}{\epsilon - \epsilon_{\mathbf{p}\sigma}^{\text{HF}} - i\eta}. \tag{6.2.10}$$

Inserting this in (6.2.9), the integral over ϵ may be evaluated by closing the contour in the upper half-plane. This picks out the $f_{\mathbf{p}\sigma}^-$ contribution from $G_\sigma(\mathbf{p}, \epsilon)$ and leads to

$$\Delta E_{\text{tot}}^{\text{HF}} = \tfrac{1}{2} \sum_{\mathbf{p}\sigma} f_{\mathbf{p}\sigma}^- (\epsilon_{\mathbf{p}\sigma}^{\text{HF}} - \epsilon_\mathbf{p}), \tag{6.2.11}$$

from which we get an exchange energy contribution of

$$E^{\text{ex}} = \tfrac{1}{2} \sum_{\mathbf{p}\sigma} f_{\mathbf{p}\sigma}^- \epsilon_\mathbf{p}^{\text{ex}}, \tag{6.2.12}$$

where $\epsilon_{\mathbf{p}}^{\mathrm{ex}}$ is given in (6.1.14). On performing the integral over **p** this reduces to an energy per particle of

$$\frac{E^{\mathrm{ex}}}{N} = -\frac{3\,e^2 p_f}{4\pi} = -\frac{3}{2}\left(\frac{3}{2\pi}\right)^{2/3}\frac{1}{r_s}\,E_0. \tag{6.2.13}$$

This is negative because the exclusion principle tends to keep apart particles of parallel spin and thus reduces the effect of the Coulomb repulsion.

We went through the above somewhat lengthy derivation in order to obtain a result consistent with the Hartree–Fock approximation for the single-particle excitation energies. However, having done this, one may also see that the result is equivalent to a simple calculation of the expectation value of the Coulomb term in the hamiltonian with respect to the unperturbed Slater determinant ground state $|\Phi_0\rangle$,

$$E^{\mathrm{ex}} = \langle \Phi_0 | H_1 | \Phi_0 \rangle. \tag{6.2.14}$$

This also follows from the fact that, for an electron gas in a uniform positive background, the variational Slater determinant wave function is simply a determinant of plane waves. Thus what the above result demonstrates is that the Slater determinant wave function has built into it a minimal amount of particle–particle correlation such that around any particle there is a reduction of density of particles of the same spin (the so-called fermi hole). However, (6.2.13) still overestimates the repulsive energy between particles of opposite spin. This is because screening has not been included, so that the effect of the bare Coulomb potential has too long a range (is too highly singular for small **q**). In order to take these screening effects into account we need to go beyond the Hartree–Fock approximation. Before we go on to this, however, we return to the calculation of Sec. 6.1 in terms of perturbation theory.

6.3. DERIVATION OF THE HARTREE–FOCK APPROXIMATION FROM A PERTURBATION EXPANSION OF *G*

As an alternative way of deriving the result (6.1.12) we apply the Feynman–Dyson expansion technique of Chap. 3 to the expansion of G. Then it will appear that the Hartree–Fock approximation corresponds to the sum of a particular set (the simplest) of the diagrams.

We start from the expression for G which corresponds to (3.1.16),

$$G(\mathbf{p}, t) = -i \frac{\langle \Phi_0 | T[a_\mathbf{p}(t) a_\mathbf{p}^\dagger(0) S] | \Phi_0 \rangle}{\langle \Phi_0 | S | \Phi_0 \rangle}, \tag{6.3.1}$$

where $|\Phi_0\rangle$ is the unperturbed N-particle ground state and the time dependence of the operators corresponds to the interaction picture, so that

$$a_\mathbf{p}(t) = e^{iH_0 t} a_\mathbf{p}\, e^{-iH_0 t}. \tag{6.3.2}$$

Also

$$S = U(\infty, -\infty) = T \exp\left\{ -i \int_{-\infty}^{\infty} dt_1 H_1(t_1) \right\}. \tag{6.3.3}$$

The spin suffixes have been omitted for simplicity.

The T operator, when applied to an arbitrary number of fermion operators, orders the operators according to the order of their time arguments (with the largest t on the left), with a *minus sign* if the new ordering is obtained from the old by an *odd* number of permutations.

To evaluate the terms in the perturbation series we must generalize Wick's theorem so that it applies to the ground state average of a time-ordered product of fermion operators. We note that, with the definition (6.3.2), the time dependence of the operators is again simply

$$a_\mathbf{p}(t) = e^{-i\epsilon_\mathbf{p} t} a_\mathbf{p}, \qquad a_\mathbf{p}^\dagger(t) = e^{i\epsilon_\mathbf{p} t} a_\mathbf{p}^\dagger, \tag{6.3.4}$$

so that the time factors can again be factored out. Wick's theorem is then of the same form as before (Sec. 3.1), except that, because of the anticommutation of the fermion operators, each fully paired product in the sum over all contractions appears with a factor $(-1)^p$, where p is the number of interchanges of fermion operators required to obtain the fully paired product from the original product.

The contractions

$$\overbrace{a(t_1) a(t_2)}, \qquad \overbrace{a^\dagger(t_1) a^\dagger(t_2)}$$

which do not conserve particle numbers vanish, and the contraction

$$\overbrace{a_\mathbf{p}(t_1) a_{\mathbf{p}'}^\dagger(t_2)} = \langle \Phi_0 | T[a_\mathbf{p}(t_1) a_{\mathbf{p}'}^\dagger(t_2)] | \Phi_0 \rangle$$

has been evaluated already in Sec. 4.6. It is

$$i\delta_{\mathbf{pp}'}G^0(\mathbf{p}, t_1 - t_2), \tag{6.3.5}$$

where $G^0(\mathbf{p}, t)$ is the free *electron-hole propagator* or Green's function, evaluated for the unperturbed fermi ground state and given by Eq. (4.6.5). Thus, by Wick's theorem, the expansion of $G(\mathbf{p})$ is in terms of the function $G^0(\mathbf{p})$.

The expanded form of (6.3.3) is

$$S = \sum_{n=0}^{\infty} \frac{(-i)^n}{n!} \int_{-\infty}^{\infty} dt_1 \ldots \int_{-\infty}^{\infty} dt_n T[H_1(t_1) \ldots H_1(t_n)]. \tag{6.3.6}$$

We now evaluate the first-order term $G^{(1)}(\mathbf{p})$ in $G(\mathbf{p})$, contributed by the term $n = 1$ in the sum (6.3.6). We have

$$G^{(1)}(\mathbf{p}, t) = (-i)(-i)\langle\Phi_0|T[a_{\mathbf{p}}(t)a_{\mathbf{p}}^{\dagger}(0)\tfrac{1}{2}\sum_{\mathbf{qkk}'} V(\mathbf{q}) \int_{-\infty}^{\infty} dt_1$$
$$\times\, a_{\mathbf{k}+\mathbf{q}}^{\dagger}(t_1)a_{\mathbf{k}'-\mathbf{q}}^{\dagger}(t_1)a_{\mathbf{k}'}(t_1)a_{\mathbf{k}}(t_1)]|\Phi_0\rangle. \tag{6.3.7}$$

Here we have ignored for the moment the denominator $\langle\Phi_0|S|\Phi_0\rangle$ of (6.3.1). Application of Wick's theorem gives two types of paired terms. First there are *unlinked* diagrams in which $a_{\mathbf{p}}(t)$, $a_{\mathbf{p}}^{\dagger}(0)$ are contracted with each other but not with the operators in H_1. These terms are cancelled by identical contributions from $\langle\Phi_0|S|\Phi_0\rangle$ (see below). Secondly we have the *linked* terms in which $a_{\mathbf{p}}(t)$ and $a_{\mathbf{p}}^{\dagger}(0)$ are each contracted with an operator in H_1; these correspond to physical effects. One such contraction of the operators in (6.3.7) is

$$a_{\mathbf{p}}(t)a_{\mathbf{p}}^{\dagger}(0)a_{\mathbf{k}+\mathbf{q}}^{\dagger}(t_1)a_{\mathbf{k}'-\mathbf{q}}^{\dagger}(t_1)a_{\mathbf{k}'}(t_1)a_{\mathbf{k}}(t_1).$$

This contributes

$$-(-1)\tfrac{1}{2}\sum_{\mathbf{k}'} \int_{-\infty}^{\infty} dt_1 V(\mathbf{q}=0)\langle T[a_{\mathbf{p}}(t)a_{\mathbf{p}}^{\dagger}(t_1)]\rangle$$
$$\times\,\langle T[a_{\mathbf{k}'}(t_1)a_{\mathbf{k}'}^{\dagger}(t_1+0)]\rangle\langle T[a_{\mathbf{p}}(t_1)a_{\mathbf{p}}^{\dagger}(0)]\rangle. \tag{6.3.8}$$

The extra factor -1 arises from an odd permutation of the factors in (6.3.7). The notation used for the time arguments shows that $a_{\mathbf{k}'}^{\dagger}$ stands to the left of $a_{\mathbf{k}'}$ in H_1. Adding to this an identical term in which **k** and k′ are interchanged, we obtain

$$i^3 \sum_{\mathbf{k}'} \int_{-\infty}^{\infty} dt_1 V(\mathbf{q} = 0) G^0(\mathbf{p}, t - t_1) G^0(\mathbf{k}', 0^-) G^0(\mathbf{p}, t_1). \tag{6.3.9}$$

This expression is represented by the diagram shown in Fig. 6.1. We say that the propagator $G^0(\mathbf{p}, t_1)$ describes a particle created in state **p** at time $t = 0$ which propagates freely to time t_1. Here it interacts with a

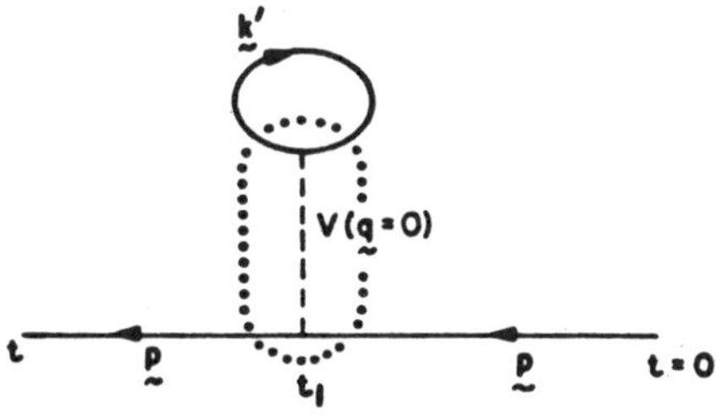

Fig. 6.1. First-order direct contribution to $G(\mathbf{p})$.

particle in state **k**′, undergoes forward scattering (**q** = 0), propagates on in state **p** and is annihilated at time t. The dotted line in Fig. 6.1 encloses a so-called *vertex*, which corresponds to a product of four a, $a^\dagger$ operators at time t_1; a creation operator $a^\dagger$ is represented by a line leaving the vertex and an annihilation operator a by a line entering the vertex. Fig. 6.2 shows a more general vertex in which there is a non-zero

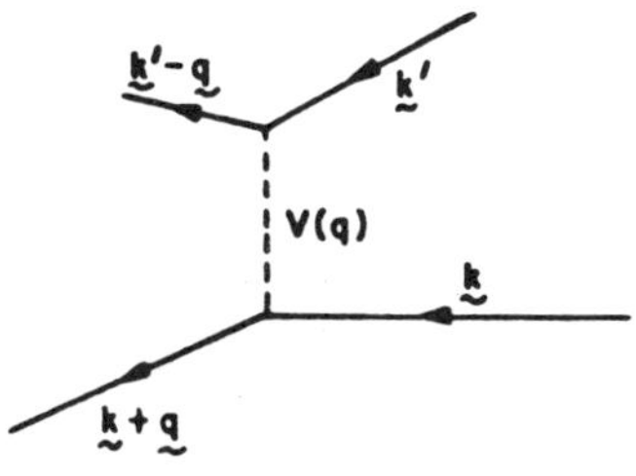

Fig. 6.2. A vertex with non-zero momentum transfer.

momentum transfer; this represents the expression

$$\sum_{\mathbf{q}} V(\mathbf{q}) a^{\dagger}_{\mathbf{k}+\mathbf{q}}(t_1) a^{\dagger}_{\mathbf{k}'-\mathbf{q}}(t_1) a_{\mathbf{k}'}(t_1) a_{\mathbf{k}}(t_1).$$

There are two further contributions to (6.3.7), one of which arises from the contractions

$$a_{\mathbf{p}}(t) a_{\mathbf{p}}^{\dagger}(0) a^{\dagger}_{\mathbf{k}+\mathbf{q}}(t_1) a^{\dagger}_{\mathbf{k}'-\mathbf{q}}(t_1) a_{\mathbf{k}'}(t_1) a_{\mathbf{k}}(t_1),$$

the other has k and $\mathbf{k}'$ interchanged. These give the *exchange term*

$$-i^3 \sum_{\mathbf{k}'} V(\mathbf{k}' - \mathbf{p}) \int_{-\infty}^{\infty} dt_1 G^0(\mathbf{p}, t - t_1) G^0(\mathbf{k}', 0^-) G^0(\mathbf{p}, t_1), \qquad (6.3.10)$$

represented by the diagram of Fig. 6.3. In this case there is a momentum transfer $\mathbf{q} = \mathbf{k}' - \mathbf{p}$ at the bottom and top of the vertex.

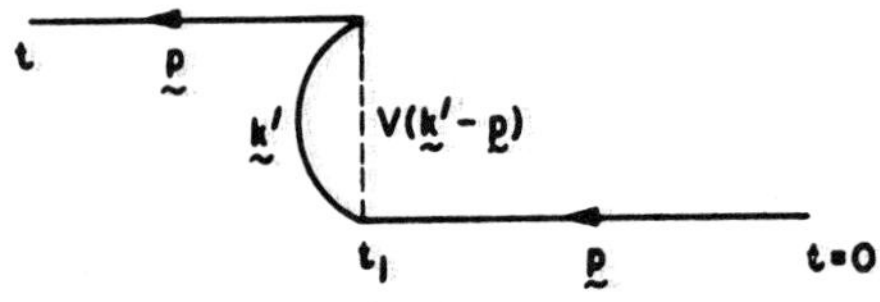

Fig. 6.3. First-order exchange contribution to $G(\mathbf{p})$.

Because of the convolution form of (6.3.9) and (6.3.10) we can again simplify the expressions by going over to Fourier transforms $G(\mathbf{p}, \epsilon)$. The total first-order contribution to the Green's function is thus obtained as

$$G^{(1)}(\mathbf{p}, \epsilon) = -iG^0(\mathbf{p}, \epsilon)\{V(\mathbf{q} = 0) \sum_{\mathbf{k}'} G^0(\mathbf{k}', t = 0^-)\} G^0(\mathbf{p}, \epsilon)$$

$$+ iG^0(\mathbf{p}, \epsilon)\{\sum_{\mathbf{k}'} V(\mathbf{k}' - \mathbf{p}) G^0(\mathbf{k}', t = 0^-)\} G^0(\mathbf{p}, \epsilon).$$

(6.3.11)

The higher-order terms in the perturbation series for G can be discussed and evaluated similarly. To formulate general rules for writing down any term it is convenient to replace the momentum sums by integrals (assuming normalization in a box of unit volume) and to introduce the four-vector momentum p associated with 3-momentum $\mathbf{p}$ and energy ϵ. Then, for example, the expression

$$\sum_{\mathbf{k}'} G^0(\mathbf{k}', t = 0^-) = \int \frac{d^3k'}{(2\pi)^3} \int \frac{d\epsilon'}{2\pi} G^0(\mathbf{k}', \epsilon')\, e^{-i\epsilon' 0^-}$$

$$= \int \frac{d^4k'}{(2\pi)^4} G^0(k')\, e^{-i\epsilon' 0^-} \tag{6.3.12}$$

becomes an integral over the four components of k'. The diagrams expressed in terms of 4-momentum variables have the same form as the diagrams in terms of $\mathbf{p}$ and t but have no time labels, and a 4-momentum p is to be associated with each line.

The general rules which result from the application of Wick's theorem for writing down the term in the perturbation series corresponding to any diagram are as follows. [An nth-order diagram contains n vertices and $(2n + 1)$ fermion lines.]

(i) The 4-momentum is conserved at each vertex. A factor $V(q)$ is associated with each vertex, where q is the 4-momentum transfer. [In the case of an instantaneous two-body interaction $V(q)$ is a function of $|\mathbf{q}|$ only.]

(ii) A factor $G^0(\mathbf{p}, \epsilon) = G^0(p)$ is associated with each fermion line of 4-momentum p.

(iii) There are integrations over the 4-momenta of all internal lines.

(iv) There is a factor (-1) for each *closed loop* in the diagram, and, in nth order, an overall factor i^n. [The factor $1/n!$ in (6.3.6) cancels out if we allow only *topologically distinct* diagrams, i.e., treat all diagrams as equivalent which are obtained from each other by relabeling the variables.]

To these rules must be added a prescription, of the type used in (6.3.12), for dealing with the limit $t' \to t$ in diagrams such as Figs. 6.1 and 6.3, in which (in the time-dependent representation) a fermion line begins and ends at the same point in time. Such diagrams occur when the interaction is instantaneous in time.

Two second-order diagrams are shown in Fig. 6.4. The first contributes to the perturbation series the term

$$(-1)i^2G^0(p)\int\frac{d^4q}{(2\pi)^4}\,V^2(q)G^0(p-q)\int\frac{d^4p'}{(2\pi)^4}\,G^0(p'+q)G^0(p')G^0(p).$$

The second is an example of an *unlinked* (or disconnected) diagram. As in the lattice dynamics example (Sec. 3.2) it turns out that the sum of the contributions of all unlinked diagrams separates out from the

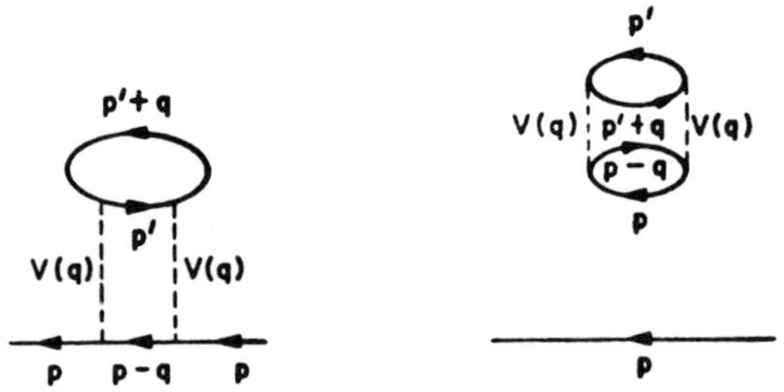

Fig. 6.4. Second-order contributions to $G(\mathbf{p})$.

numerator of (6.3.1) as a factor $\langle\Phi_0|S|\Phi_0\rangle$ and cancels with the denominator. We therefore obtain the complete perturbation series by including only the linked diagrams and ignoring the denominator of (6.3.1).

Eq. (6.3.13) shows explicitly all connected and topologically distinct contributions to $G(p)$ up to the second order.

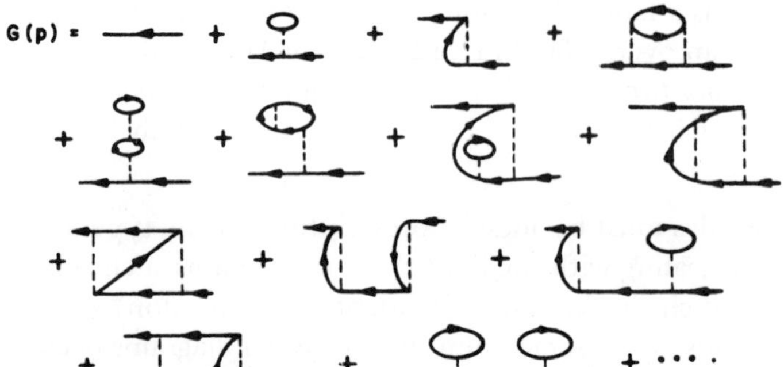

(6.3.13)

We can again sum the perturbation series formally by introducing the irreducible self-energy $\Sigma(p)$. This is defined, as in Sec. 5.3, as the sum of all distinct self-energy diagrams which cannot be divided into two sub-diagrams connected only by a single solid line carrying the momentum p of the incoming line. [It is seen that the first six of the ten second-order diagrams in (6.3.13) are irreducible.] We are thus led to Dyson's equation for the present problem,

$$G(p) = G^0(p) + G^0(p)\Sigma(p)G(p), \tag{6.3.14}$$

where

$\Sigma(p) =$ 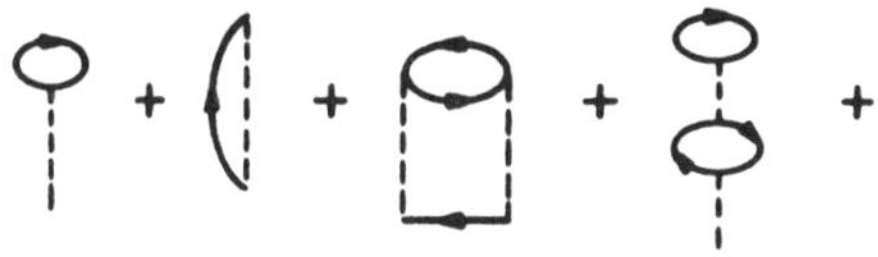 $+ \cdots\cdots$ (6.3.15)

Solving for $G(p)$ we have

$$G(p) = \frac{G^0(p)}{1 - \Sigma(p)G^0(p)}, \quad \text{where, for } |\mathbf{p}| > p_f,$$

$$G^0(p) \equiv G^0(\mathbf{p}, \epsilon) = \frac{1}{\epsilon - \epsilon_{\mathbf{p}} + i\eta}, \tag{6.3.16}$$

[see Eq. (4.6.6)]. Hence, for $|\mathbf{p}| > p_f$,

$$G(\mathbf{p}, \epsilon) = \frac{1}{\epsilon - \epsilon_{\mathbf{p}} - \Sigma(\mathbf{p}, \epsilon) + i\eta} \qquad (\eta > 0), \tag{6.3.17}$$

with a corresponding expression valid for $|\mathbf{p}| < p_f$.

The singularities of $G(\mathbf{p}, \epsilon)$ in the complex ϵ-plane, and hence the form of $G(\mathbf{p}, t)$, are thus determined by the properties of the self-energy function $\Sigma(\mathbf{p}, \epsilon)$. In particular, the asymptotic form of $G(\mathbf{p}, t)$ for large t is determined by the pole of $G(\mathbf{p}, \epsilon)$ in the lower half-plane which lies nearest to the real axis. If this pole occurs at $\epsilon = \bar{\epsilon}_{\mathbf{p}} - i\Gamma_{\mathbf{p}}$, then, for sufficiently large t,

$$G(\mathbf{p}, t) \sim -i\, e^{-i\bar{\epsilon}_{\mathbf{p}} t}\, e^{-\Gamma_{\mathbf{p}} t} \qquad (|\mathbf{p}| > p_f). \tag{6.3.18}$$

Thus $G(\mathbf{p}, t)$ corresponds to the propagation of a quasiparticle with energy $\bar{\epsilon}_{\mathbf{p}}$, damped in time with damping constant $\Gamma_{\mathbf{p}}$.

We now evaluate $\Sigma(\mathbf{p}, \epsilon)$ in the lowest non-vanishing order and show that this leads to the Hartree–Fock results of Sec. 6.1. We require the contributions of the two first-order diagrams in the series (6.3.15) for Σ; these are given by Eq. (6.3.11) if we remove the factors $G^0(\mathbf{p}, \epsilon)$ from the beginning and end of each of the two terms. Thus, remembering the definition (6.1.4) of $G(\mathbf{p}, t)$, we obtain from the first term of (6.3.11)

$$-iV(\mathbf{q}=0) \sum_{\mathbf{k}} G^0(\mathbf{k}, t=0^-)$$

$$= -V(\mathbf{q}=0) \sum_{\mathbf{k}} \lim_{\epsilon \to 0+} \langle T[a_{\mathbf{k}}(-\epsilon) a_{\mathbf{k}}^{\dagger}(0)] \rangle$$

$$= V(\mathbf{q}=0) \sum_{\mathbf{k}} \langle a_{\mathbf{k}}^{\dagger} a_{\mathbf{k}} \rangle = V(\mathbf{q}=0) \sum_{\mathbf{k}} f_{\mathbf{k}},$$

and the second term gives similarly

$$i \sum_{\mathbf{k}} V(\mathbf{k}-\mathbf{p}) G^0(\mathbf{k}, t=0^-) = -\sum_{\mathbf{k}} V(\mathbf{k}-\mathbf{p}) f_{\mathbf{k}}.$$

Hence we have, in this approximation,

$$G(\mathbf{p}, \epsilon) = \frac{1}{\epsilon - \bar{\epsilon}_{\mathbf{p}} + i\eta}, \qquad \text{where } \bar{\epsilon}_{\mathbf{p}} = \epsilon_{\mathbf{p}} + V(\mathbf{q}=0) \sum_{\mathbf{k}} f_{\mathbf{k}} - \sum_{\mathbf{k}} V(\mathbf{k}-\mathbf{p}) f_{\mathbf{k}}. \tag{6.3.19}$$

This is in exact agreement with the energy shift (6.1.12) obtained in the Hartree–Fock approximation. Note that $\Sigma(p)$ is real in this approximation, so that the damping is zero. Note also that the approximation (6.3.19) involves the summation of a particular infinite subset of diagrams (corresponding to repeated direct and exchange interactions) in the perturbation series for $G(\mathbf{p}, \epsilon)$:

$G(\mathbf{p}, \epsilon) =$

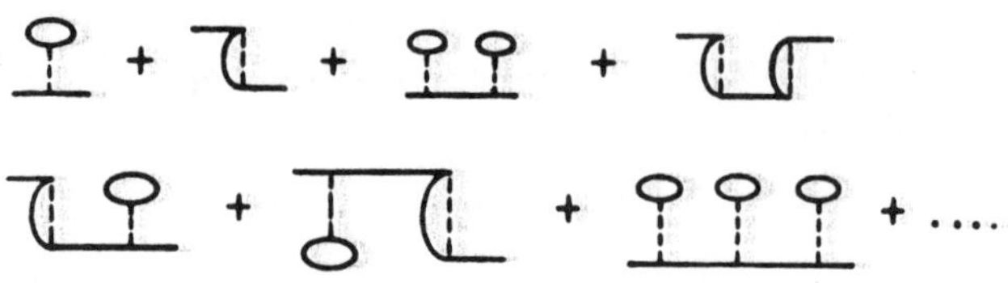

(6.3.20)

6.4. THE DIELECTRIC RESPONSE FUNCTION OF A DENSE ELECTRON GAS

We have already noted that the Hartree–Fock approximation, because of its neglect of screening, does not give an adequate account of the properties of the electron gas. The next level of approximation is the random phase approximation which provides a more satisfactory description. In the high-density limit this may be shown to lead to the next significant correction to the ground state energy after the exchange term [Gell-Mann and Brueckner (1957)].

Historically, the random phase approximation was first introduced in a heuristic fashion by Bohm and Pines (1953). They analyzed the possible dynamical degrees of freedom of an interacting electron gas and argued that most of the Coulomb correlation will be absorbed in a *plasma mode* of collective oscillation which, because of its high zero-point energy, will not be excited at low temperatures. The remaining modes can be regarded as electrons, moving in a weak screened potential, which may be described to a good approximation by an independent-particle model. This approach can be reformulated in a systematic way, following the work of Nozières and Pines (1958), by which the properties of the electron gas are described by a complex, wave-number and frequency dependent, dielectric response function. In this section we obtain the general Kubo formula expression for the dielectric response which is used later to discuss the possible modes of excitation of the system.

We can obtain the dielectric function by studying the response of the electron gas to an applied external charge density $\rho_{ext}(\mathbf{x}, t)$ [we assume that $\rho_{ext}(\mathbf{x}, t)$ contains an adiabatic switching-on process in its time dependence]. The interaction between the external charge and the electron gas is given by the hamiltonian

$$H_{ext} = \int d^3x \int d^3x' \rho(\mathbf{x}) V(\mathbf{x} - \mathbf{x}') \rho_{ext}(\mathbf{x}', t), \qquad (6.4.1)$$

where $\rho(\mathbf{x})$ is the charge density operator for the electrons. In the absence of the external charge the expectation value of ρ is zero because of translational symmetry, but as a result of the perturbation we get a non-zero induced charge density $\rho_{ind}(\mathbf{x}, t)$. We calculate $\langle \rho_{ind} \rangle$ by linear response theory, proceeding exactly as in the calculation of the electrical conductivity in Sec. 5.5. We then introduce the *dielectric response function* $\epsilon(\mathbf{q}, \omega)$ by the definition

$$\langle \rho_{tot}(\mathbf{q}, \omega) \rangle = \rho_{ext}(\mathbf{q}, \omega) + \langle \rho_{ind}(\mathbf{q}, \omega) \rangle = \frac{\rho_{ext}(\mathbf{q}, \omega)}{\epsilon(\mathbf{q}, \omega)}. \qquad (6.4.2)$$

[Here $\rho(\mathbf{q}, \omega)$ is the Fourier transform with respect to space and time variables of $\rho(\mathbf{x}, t)$.] No approximations are made in this calculation except that ρ_{ext} is regarded as a weak perturbation, so that only linear terms in ρ_{ext} need to be retained. We thus obtain an expression for $\epsilon(\mathbf{q}, \omega)$ in which the Coulomb interactions are still formally included to all orders. Once again, however, the assumption that the linear expansion is possible implies a stability in the system, so that the true ground state of the interacting system evolves continuously from the ground state of the non-interacting gas as the Coulomb interaction increases from zero to its full value (this would not be true for a superconductor). Although the complete evaluation of the expression for $\epsilon(\mathbf{q}, \omega)$ of course requires the exact solution of the many-body problem, the formulation in terms of the dielectric response is useful in that it leads in a natural way to approximations suggested by physical considerations. This formulation also focuses attention on the roots of the equation

$$\epsilon(\mathbf{q}, \omega) = 0. \tag{6.4.3}$$

When this condition is satisfied, we see from (6.4.2) that $\langle\rho_{\text{tot}}\rangle$ can be non-zero when $\rho_{\text{ext}} = 0$. The free modes of oscillation of the electron gas thus correspond to the frequencies and wave-numbers satisfying (6.4.3).

The definition (6.4.2) corresponds to replacing $\mathbf{E}$ by $\mathbf{D} = \epsilon\mathbf{E}$ in Maxwell's equations for the electric field produced by ρ_{ext}. In general, Maxwell's equations relating the total electric and magnetic fields in the medium to source densities ρ_{tot}, $\mathbf{J}_{\text{tot}}$ are

$$i\mathbf{q} \times \mathbf{B} = \frac{4\pi}{c}\mathbf{J}_{\text{tot}} - \frac{i\omega}{c}\mathbf{E}, \tag{6.4.4}$$

$$i\mathbf{q} \cdot \mathbf{E} = 4\pi\rho_{\text{tot}}. \tag{6.4.5}$$

With $\rho_{\text{tot}} = \rho_{\text{ext}}/\epsilon$, (6.4.5) becomes

$$i\mathbf{q} \cdot \epsilon\mathbf{E} = 4\pi\rho_{\text{ext}}. \tag{6.4.6}$$

Note that (6.4.6) contains only the longitudinal part (parallel to $\mathbf{q}$) of the electric field, so that ϵ is a *longitudinal* dielectric constant. One can formally define a longitudinal conductivity σ_l associated with ϵ: to do this put $\mathbf{J}_{\text{tot}} = \mathbf{J}_{\text{ext}} + \mathbf{J}_{\text{ind}}$, where

$$i\omega\rho_{\text{ind}} + \mathbf{q} \cdot \mathbf{J}_{\text{ind}} = 0 \tag{6.4.7}$$

(the continuity equation), and $\mathbf{J}_{\text{ind}} = \sigma_l \mathbf{E}$. Then (6.4.4) can be written

$$i\mathbf{q} \times \mathbf{B} = \frac{4\pi}{c} \mathbf{J}_{\text{ext}} - \frac{i\omega}{c} \epsilon \mathbf{E}, \qquad \text{with } \epsilon = 1 + \frac{4\pi i \sigma_l}{\omega}. \tag{6.4.8}$$

But σ_l is *not* in general the conductivity calculated in Chap. 5. There we were concerned with the direct current response to an applied field, and we therefore took the fields and currents to be *transverse* to $\mathbf{q}$. There is then no induced charge density modifying the electric field in the medium.

We now proceed with the linear response calculation of the dielectric function. The total hamiltonian is

$$H_{\text{tot}} = H + H_{\text{ext}}, \tag{6.4.9}$$

where H is the exact hamiltonian (6.1.3) of the interacting gas. We have to evaluate the expectation value

$$\langle \rho_{\text{ind}}(\mathbf{x}, t) \rangle = \langle E(t) | \rho(\mathbf{x}) | E(t) \rangle, \tag{6.4.10}$$

where $|E(t)\rangle$ is the state of the system at time t. $|E(t)\rangle$ has evolved in time according to both H and H_{ext} from the initial state $|E\rangle$ which is taken to be the ground state of H. For simplicity we consider the zero-temperature theory only, but there is no difficulty in generalizing the formalism to the finite T case in order to study thermodynamic properties. The time development of $|E(t)\rangle$ is given by Eqs. (5.5.6) and (5.5.9) and we obtain, working to the first order in H_{ext},

$$\langle \rho_{\text{ind}}(\mathbf{x}, t) \rangle = i \int_0^t dt' \langle [H_{\text{ext}}(t'), \rho(\mathbf{x}, t)] \rangle, \tag{6.4.11}$$

where the time dependence of $H_{\text{ext}}(t)$, $\rho(\mathbf{x}, t)$ refers only to the development according to H. H_{ext} is given by (6.4.1), and thus

$$\langle \rho_{\text{ind}}(\mathbf{x}, t) \rangle = \int_{-\infty}^{\infty} dt' \int d^3x' \int d^3x'' \mathscr{K}(\mathbf{x} - \mathbf{x}', t - t') V(\mathbf{x}' - \mathbf{x}'') \times \rho_{\text{ext}}(\mathbf{x}'', t'), \tag{6.4.12}$$

where

$$\mathscr{K}(\mathbf{x} - \mathbf{x}', t - t') = -i\theta(t - t') \langle [\rho(\mathbf{x}, t), \rho(\mathbf{x}', t')] \rangle. \tag{6.4.13}$$

This result is a Kubo formula entirely analogous to the expressions (5.5.12) and (5.5.14) for the electrical conductivity response.

Introducing Fourier transforms in x-space, (6.4.12) becomes

$$\langle \rho_{\text{ind}}(\mathbf{q}, t) \rangle = \int_{-\infty}^{\infty} dt' \mathscr{K}(\mathbf{q}, t - t') V(\mathbf{q}) \rho_{\text{ext}}(\mathbf{q}, t'), \tag{6.4.14}$$

where

$$\mathscr{K}(\mathbf{q}, t - t') = -i\theta(t - t') \langle [\rho(\mathbf{q}, t), \rho(-\mathbf{q}, t')] \rangle \tag{6.4.15}$$

and [see Eq. (4.2.14)]

$$\rho(\mathbf{q}) = \sum_{\mathbf{p}} a^{\dagger}_{\mathbf{p}+\mathbf{q}} a_{\mathbf{p}} \tag{6.4.16}$$

is the Fourier transform of the charge density operator. Next, taking Fourier transforms in time, (6.4.14) becomes a product

$$\langle \rho_{\text{ind}}(\mathbf{q}, \omega) \rangle = V(\mathbf{q}) \mathscr{K}(\mathbf{q}, \omega) \rho_{\text{ext}}(\mathbf{q}, \omega), \tag{6.4.17}$$

in which $\mathscr{K}(\mathbf{q}, \omega)$ is the time Fourier transform of (6.4.15). Finally from (6.4.2) and (6.4.17) we now obtain the dielectric response function in the form

$$\epsilon^{-1}(\mathbf{q}, \omega) = 1 + V(\mathbf{q}) \mathscr{K}(\mathbf{q}, \omega). \tag{6.4.18}$$

6.5. THE PARTICLE-HOLE GREEN'S FUNCTION AND THE RANDOM PHASE APPROXIMATION FOR THE DIELECTRIC RESPONSE

According to (6.4.15) and (6.4.18) the dielectric response function is determined by the time Fourier transform of

$$\mathscr{K}(\mathbf{q}, t - t') = -i\theta(t - t') \langle [\rho(\mathbf{q}, t), \rho(-\mathbf{q}, t')] \rangle. \tag{6.5.1}$$

This is a two-particle Green's function of the same type as those studied in Chap. 5 [see Eq. (5.5.14)] in the case of the electrical conductivity. One way to proceed with the calculation of this function for the interacting electron gas is to use the equation-of-motion method (as in Sec. 6.1) to generate an approximation scheme by means of truncation of the hierarchy of coupled equations of motion. This is a natural way to extend the Hartree–Fock approximation to a time-dependent version which in fact leads directly to the r.p.a.

An alternative approach is to use perturbation theory to pick out classes of diagrams which are of leading order for small r_s (high density). We follow this approach in this section. For this purpose we again consider, instead of the retarded function (6.5.1), the time-ordered form

$$\mathscr{K}^T(\mathbf{q}, t - t') = -i\langle T[\rho(\mathbf{q}, t)\rho(-\mathbf{q}, t')]\rangle, \tag{6.5.2}$$

the two being related by formulas given in Appendix 2.

The diagrammatic perturbation expansion of $\mathscr{K}^T(\mathbf{q}, \omega)$ can be obtained by applying the rules given in Sec. 6.3. In the interaction picture we have

$$\mathscr{K}^T(\mathbf{q}, t - t') = -i\frac{\langle\Phi_0|T[\rho(\mathbf{q}, t)\rho(-\mathbf{q}, t')S]|\Phi_0\rangle}{\langle\Phi_0|S|\Phi_0\rangle}, \tag{6.5.3}$$

where S is given by (6.3.3) and is to be expanded in powers of $V(\mathbf{q})$. $\rho(\mathbf{q})$ is the operator (6.4.16). It follows from the rules for calculating diagrams in momentum space that the zero-order term in the expansion of $\mathscr{K}^T(\mathbf{q}, \omega)$ is

$\mathscr{P}_0(\mathbf{q}, \omega) =$ 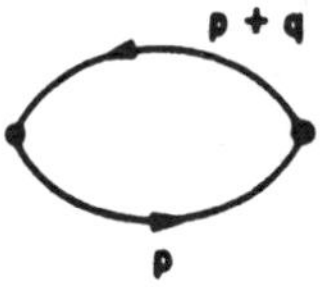 , (6.5.4)

evaluated for fixed $q = (\mathbf{q}, \omega)$. The complete perturbation series includes all the higher-order diagrams associated with (6.5.4), and may be represented as

$\mathscr{K}^T(\mathbf{q}, \omega) =$ 

 (6.5.5)

where each bubble carries a net 4-momentum q.

The series (6.5.5) can be summed formally by introducing the *irreducible polarization propagator* $\mathscr{P}(\mathbf{q}, \omega)$, defined as the sum of all those diagrams in (6.5.5) which cannot be divided into two diagrams connected only by a single interaction line carrying momentum $\mathbf{q}$. In terms of $\mathscr{P}$ the series (6.5.5) can be rearranged as

$$\begin{aligned}\mathscr{K}^T(\mathbf{q}, \omega) &= \mathscr{P} + \mathscr{P}V\mathscr{P} + \mathscr{P}V\mathscr{P}V\mathscr{P} + \cdots \\ &= \mathscr{P} + \mathscr{P}V(\mathscr{P} + \mathscr{P}V\mathscr{P} + \cdots) \\ &= \mathscr{P} + \mathscr{P}V\mathscr{K}^T,\end{aligned}$$

so that

$$\mathcal{K}^T(\mathbf{q}, \omega) = \frac{\mathcal{P}(\mathbf{q}, \omega)}{1 - \mathcal{P}(\mathbf{q}, \omega)V(\mathbf{q})}. \tag{6.5.6}$$

If we consider (6.5.6) as a power series in e^2, it may be seen that for small $\mathbf{q}$ the leading terms in each order will be those in which $V(\mathbf{q})$ occurs to maximum power. This is because the singularity $1/\mathbf{q}^2$ will be strongest in such terms. Diagrams in which the Coulomb interaction occurs with a momentum different from $\mathbf{q}$, the input momentum for the response, will be less singular as $\mathbf{q} \to 0$. This suggests that the leading corrections in the $\mathbf{q} \to 0$ limit will be obtained by including only the ring diagrams of the form of Fig. 6.5. These are summed exactly by (6.5.6)

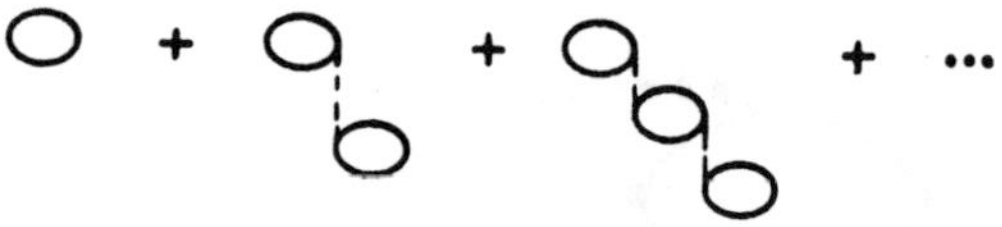

Fig. 6.5. Diagrams giving leading contribution in $\mathbf{q} \to 0$ limit.

if $\mathcal{P}(\mathbf{q}, \omega)$ is replaced by the unperturbed electron-hole propagator $\mathcal{P}_0(\mathbf{q}, \omega)$, defined by Eq. (6.5.4). This sum is what is usually referred to as the r.p.a.

We now evaluate $\mathcal{P}_0(\mathbf{q}, \omega)$. We have, using the expression (4.6.6) for the free electron-hole propagator G^0,

$$\begin{aligned}\mathcal{P}_0(\mathbf{q}, \omega) &= 2(-1)i \int \frac{d^4p}{(2\pi)^4} G^0(p+q)G^0(p) \\ &= -2i \int \frac{d^4p}{(2\pi)^4} \frac{1}{(p_0 + \omega - \epsilon_{\mathbf{p}+\mathbf{q}} + i\eta_{\mathbf{p}+\mathbf{q}})} \\ &\quad \times \frac{1}{(p_0 - \epsilon_{\mathbf{p}} + i\eta_{\mathbf{p}})}.\end{aligned} \tag{6.5.7}$$

Here p_0, ω are the 4-components of the 4-vector momenta p, q; $\eta_{\mathbf{p}}$ is an infinitesimal which is positive for $|\mathbf{p}| > p_f$ and negative for $|\mathbf{p}| < p_f$; and the factor 2 comes from the summation over two spin directions in the closed loop.

We now integrate over p_0 in (6.5.7). There are several cases to distinguish. If $|\mathbf{p}| < p_f$, $|\mathbf{p} + \mathbf{q}| > p_f$, the integrand has poles at $\epsilon_{\mathbf{p}}$ above the real axis and at $\epsilon_{\mathbf{p}+\mathbf{q}} - \omega$ below the real axis. Closing the contour in the upper half-plane we obtain, in the notation of Sec. 4.6, a contribution

$$-2i\,\frac{2\pi i}{2\pi}\int\frac{d^3p}{(2\pi)^3}\left[\frac{f_{\mathbf{p}}^{-}f_{\mathbf{p}+\mathbf{q}}^{+}}{p_0+\omega-\epsilon_{\mathbf{p}+\mathbf{q}}+i\eta}\right]_{p_0=\epsilon_{\mathbf{p}}}$$

$$= 2\int\frac{d^3p}{(2\pi)^3}\,\frac{f_{\mathbf{p}}^{-}f_{\mathbf{p}+\mathbf{q}}^{+}}{\omega-\omega_{\mathbf{q}}(\mathbf{p})+i\eta}\quad(\eta>0), \tag{6.5.8}$$

where

$$\omega_{\mathbf{q}}(\mathbf{p}) = \epsilon_{\mathbf{p}+\mathbf{q}} - \epsilon_{\mathbf{p}}. \tag{6.5.9}$$

If $|\mathbf{p}| > p_f$, $|\mathbf{p} + \mathbf{q}| < p_f$, we close the contour in the lower half-plane and obtain similarly

$$-2\int\frac{d^3p}{(2\pi)^3}\,\frac{f_{\mathbf{p}+\mathbf{q}}^{-}f_{\mathbf{p}}^{+}}{\omega-\omega_{\mathbf{q}}(\mathbf{p})-i\eta}\quad(\eta>0). \tag{6.5.10}$$

Finally, if both $\mathbf{p}$ and $\mathbf{p} + \mathbf{q}$ are on the same side of the fermi surface, both poles are on the same side of the real axis; we may then close the contour on the opposite side and obtain zero. Thus

$$\mathscr{P}_0(\mathbf{q},\omega) = 2\int\frac{d^3p}{(2\pi)^3}\left\{\frac{f_{\mathbf{p}}^{-}f_{\mathbf{p}+\mathbf{q}}^{+}}{\omega-\omega_{\mathbf{q}}(\mathbf{p})+i\eta}-\frac{f_{\mathbf{p}+\mathbf{q}}^{-}f_{\mathbf{p}}^{+}}{\omega-\omega_{\mathbf{q}}(\mathbf{p})-i\eta}\right\},$$

and this can, with a change of integration variable in the second term, be written as

$$\mathscr{P}_0(\mathbf{q},\omega) = 2\int\frac{d^3p}{(2\pi)^3}\,f_{\mathbf{p}}^{-}f_{\mathbf{p}+\mathbf{q}}^{+}\left\{\frac{1}{\omega-\omega_{\mathbf{q}}(\mathbf{p})+i\eta}-\frac{1}{\omega+\omega_{\mathbf{q}}(\mathbf{p})-i\eta}\right\}. \tag{6.5.11}$$

In order to obtain the retarded form (6.4.18) needed in the response function formulation, we again have to make use of the fluctuation-dissipation theorem. As shown in Appendix 2, for $T = 0$, this gives

$$\begin{aligned}\operatorname{Re}\mathscr{K}(\mathbf{q},\omega) &= \operatorname{Re}\mathscr{K}^T(\mathbf{q},\omega),\\ \operatorname{Im}\mathscr{K}(\mathbf{q},\omega) &= \operatorname{sgn}(\omega)\operatorname{Im}\mathscr{K}^T(\mathbf{q},\omega).\end{aligned} \tag{6.5.12}$$

From (6.5.6) we have in the r.p.a.

$$\operatorname{Re}\mathscr{K}^T(\mathbf{q},\omega)=\frac{\operatorname{Re}\mathscr{P}_0(\mathbf{q},\omega)-|\mathscr{P}_0(\mathbf{q},\omega)|^2V(\mathbf{q})}{\{1-\operatorname{Re}\mathscr{P}_0(\mathbf{q},\omega)V(\mathbf{q})\}^2+\{\operatorname{Im}\mathscr{P}_0(\mathbf{q},\omega)V(\mathbf{q})\}^2},$$

$$\operatorname{Im}\mathscr{K}^T(\mathbf{q},\omega)=\frac{\operatorname{Im}\mathscr{P}_0(\mathbf{q},\omega)}{\{1-\operatorname{Re}\mathscr{P}_0(\mathbf{q},\omega)V(\mathbf{q})\}^2+\{\operatorname{Im}\mathscr{P}_0(\mathbf{q},\omega)V(\mathbf{q})\}^2}. \tag{6.5.13}$$

Hence it may be seen that, if we define a retarded electron-hole propagator by

$$\begin{aligned}\operatorname{Re}\mathscr{P}_0{}^R(\mathbf{q},\omega)&=\operatorname{Re}\mathscr{P}_0(\mathbf{q},\omega),\\ \operatorname{Im}\mathscr{P}_0{}^R(\mathbf{q},\omega)&=\operatorname{sgn}(\omega)\operatorname{Im}\mathscr{P}_0(\mathbf{q},\omega),\end{aligned} \tag{6.5.14}$$

we have

$$\mathscr{K}(\mathbf{q},\omega)=\frac{\mathscr{P}_0{}^R(\mathbf{q},\omega)}{1-\mathscr{P}_0{}^R(\mathbf{q},\omega)V(\mathbf{q})}. \tag{6.5.15}$$

Finally, $\mathscr{P}_0{}^R(\mathbf{q},\omega)$ may be expressed, via (6.5.11) and (6.5.14), as

$$\mathscr{P}_0{}^R(\mathbf{q},\omega)=\sum_{\mathbf{p}}\frac{(f_{\mathbf{p}}^- - f_{\mathbf{p}+\mathbf{q}}^-)}{\omega-\omega_{\mathbf{q}}(\mathbf{p})+i\eta}. \tag{6.5.16}$$

From (6.4.18), (6.5.15) and (6.5.16) we can now discuss the propertie of

$$\epsilon^{-1}(\mathbf{q},\omega)=\{1-\mathscr{P}_0{}^RV(\mathbf{q})\}^{-1} \tag{6.5.17}$$

as a function of $\mathbf{q}$, ω.

We consider first the results obtained in the *static limit* $\omega = 0$. Here $\epsilon^{-1}(\mathbf{q},0)$ is purely real and, with $V(\mathbf{q}) = 4\pi e^2/\mathbf{q}^2$ and $\epsilon_{\mathbf{p}} = \mathbf{p}^2/2m$ we have, on again replacing the sum over p by an integration,

$$\epsilon^{-1}(\mathbf{q},0)=\left\{1+\frac{4me^2}{\pi^2q^2}\int_R\frac{d^3p}{2\mathbf{p}\cdot\mathbf{q}+q^2}\right\}^{-1}. \tag{6.5.18}$$

The region of integration R is the volume inside the sphere $|\mathbf{p}| = p_f$ which is exterior to the sphere $|\mathbf{p}+\mathbf{q}| = p_f$. A straightforward calculation gives

$$\begin{aligned}\epsilon^{-1}(\mathbf{q},0)&=\left\{1+\frac{4me^2p_f}{\pi q^2}\,u\left(\frac{q}{2p_f}\right)\right\}^{-1}\\&=\left\{1+\left(\frac{4}{9\pi^4}\right)^{1/3}r_s\,\frac{u(x)}{x^2}\right\}^{-1},\end{aligned} \tag{6.5.19}$$

where $q = |\mathbf{q}|$, $x = q/2p_f$, r_s is defined by (6.2.2), and

$$u(x) = \frac{1}{2}\left\{1 + \frac{1}{2x}(1 - x^2)\log\left|\frac{1+x}{1-x}\right|\right\}. \tag{6.5.20}$$

The function $u(x)$ decreases steadily from 1 to 0 as x increases from 0 to infinity. Thus, in the long-wavelength limit $q \to 0$, we have $u = 1$ and

$$\epsilon(\mathbf{q}, 0) = 1 + \frac{\lambda^2}{q^2}, \qquad \text{with } \lambda = \left(\frac{4}{\pi} me^2 p_f\right)^{1/2} = \left(\frac{16}{3\pi^2}\right)^{1/3} r_s^{1/2} p_f. \tag{6.5.21}$$

This is just the result obtained in the semiclassical *Thomas–Fermi approximation* [Thomas (1927); Fermi (1928)]. The function $\epsilon^{-1}(x)$ is shown in Fig. 6.6 for $\lambda/2p_f = 1$ ($r_s \simeq 6$) and compared with the Thomas-Fermi result.

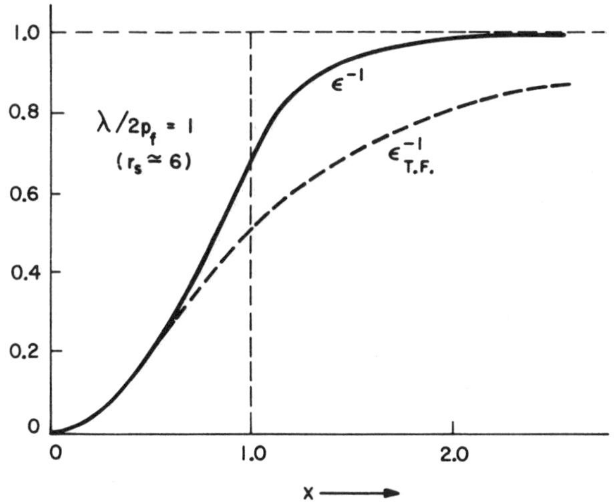

Fig. 6.6. Static dielectric constant in the r.p.a. [The broken curve is the Thomas-Fermi formula (6.5.21).]

We can use the static dielectric function $\epsilon(\mathbf{q}, 0)$ to define an *effective potential* $V(\mathbf{q})/\epsilon(\mathbf{q}, 0)$. If (6.5.21) is assumed to hold for all values of q, we obtain an effective potential $4\pi e^2/(q^2 + \lambda^2)$ which is the Fourier transform of an exponentially screened Coulomb potential $e^{-\lambda x}/x$ in real space, with a constant screening length $1/\lambda$. The more exact formula (6.5.19) shows, however, that the screening length in fact increases with

q; thus the electrons are less effective in screening the potential components of shorter wavelengths.

The function $u(x)$ not only decreases with increasing x, but it has an infinite first derivative when $x = 1$. This singularity has a marked effect on the form of the screened potential; at large distances it is in fact not damped exponentially, but exhibits long-range oscillations of the form $\cos(2p_f x)/x^3$ [Friedel (1958)]. The screening charge density is found to show a similar behavior. The singularity occurs when q equals the diameter $2p_f$ of the fermi sphere, and comes from the fact that transitions between states $\mathbf{p}$ and $\mathbf{p} + \mathbf{q}$ on the fermi sphere are possible only if $q < 2p_f$. Thus, as q increases through $2p_f$, the large contributions to $\epsilon(\mathbf{q}, 0)$ from processes in which energy is approximately conserved disappear, and there is a sharp drop in ϵ. Kohn (1959) has pointed out that, since the interaction between the ions in a metal is governed by the screened potential, the sudden change in ϵ at $q = 2p_f$ will be reflected in a corresponding sharp change at this value of q in the frequency $\omega(\mathbf{q})$ of phonons of wave-vector $\mathbf{q}$. Although the effect is small, it can be observed, and phonon spectra can in this way be used to investigate the dimensions of fermi surfaces in metals [Woll and Kohn (1962)].

For arbitrary $\mathbf{q}$ and ω explicit expressions for the real and imaginary parts of $\epsilon(\mathbf{q}, \omega)$ can easily be obtained from (6.5.16) by integration. The full results, first given by Lindhard (1954), are somewhat complicated and will not be quoted here.

We have already noted in Sec. 6.4 that the physical excitation energies of the system are determined by the condition $\epsilon(\mathbf{q}, \omega) = 0$. In the random phase approximation this gives us the equation

$$V(\mathbf{q})\mathscr{P}_0(\mathbf{q}, \omega) = 1,$$

where

$$\mathscr{P}_0(\mathbf{q}, \omega) = \sum_{\substack{p<p_f \\ |\mathbf{p}+\mathbf{q}|>p_f}} \left\{ \frac{1}{\omega - \omega_{\mathbf{q}}(\mathbf{p}) + i\eta} - \frac{1}{\omega + \omega_{\mathbf{q}}(\mathbf{p}) + i\eta} \right\}. \quad (6.5.22)$$

This is the *dispersion relation* for the eigenfrequencies $\omega(\mathbf{q})$ of the system. The function $\mathscr{P}_0(\mathbf{q}, \omega)$ has poles at the unperturbed frequencies $\pm\omega_{\mathbf{q}}(\mathbf{p})$, and, if we plot $V(\mathbf{q})\mathscr{P}_0(\mathbf{q}, \omega)$ against ω (for fixed $\mathbf{q}$), we obtain curves of the type shown in Fig. 6.7 (for $\omega > 0$). The roots of (6.5.22) are given by the intersections of these curves with unity. There is only a

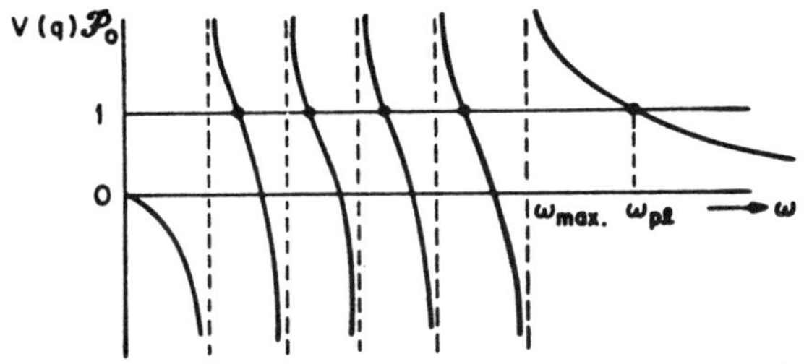

Fig. 6.7. Eigenfrequencies of electron gas in r.p.a.

slight shift in the unperturbed frequencies, by an amount, of the order of the level spacing, which goes to zero as the volume of the system tends to infinity. These roots correspond to the scattering eigenstates of a particle-hole pair. The scattering states have a maximum frequency ω_m, given by the largest value of $\omega_{\mathbf{q}}(\mathbf{p})$ such that $|\mathbf{p}| \leqslant p_f$, $|\mathbf{p}+\mathbf{q}| \geqslant p_f$. This occurs when $|\mathbf{p}| = p_f$ and $\mathbf{p}$, $\mathbf{q}$ are parallel, so that

$$\omega_m = \frac{p_f q}{m} + \frac{q^2}{2m}. \tag{6.5.23}$$

The continuum of scattering states extends over the frequency range $0 \leqslant \omega \leqslant \omega_m$ which shrinks to zero as $q \to 0$.

Fig. 6.7 shows that the dispersion relation (6.5.22) has an additional root $\omega = \omega_{pl}$, split off from the top of the continuum. This is the collective *plasmon* mode of excitation. The root $\omega = \omega_{pl}$ alone survives in the long-wavelength limit $q = 0$ when the scattering states disappear. We can obtain the approximate form of the dielectric constant in this limit by retaining only the leading terms in the expansion of $\mathscr{P}_0(\mathbf{q}, \omega)$ in powers of $\mathbf{q}$. We find that (considering only real parts)

$$\mathscr{P}_0(\mathbf{q}, \omega) = \sum_{p<p_f} \frac{q^2}{m\omega^2} + O(q^4) = \frac{q^2 N}{m\omega^2} + O(q^4),$$

and hence, with $V(\mathbf{q}) = 4\pi e^2/\mathbf{q}^2$, we have for small q

$$\epsilon^{-1}(\mathbf{q}, \omega) = 1 \bigg/ \left(1 - \frac{4\pi N e^2}{m\omega^2}\right) = 1 \bigg/ \left(1 - \frac{\omega_{pl}^2}{\omega^2}\right), \tag{6.5.24}$$

where

$$\omega_{pl} = \left(\frac{4\pi Ne^2}{m}\right)^{1/2} \tag{6.5.25}$$

is the classical *plasma frequency*. These results hold for frequencies such that $\omega \simeq \omega_{pl} \gg \omega_m$.

We can now map out the spectrum of poles of $\epsilon^{-1}(\mathbf{q}, \omega)$ in the $\omega - q$ plane. For each value of q there will be a continuum of poles from $\omega = 0$ up to qv_f, followed by a discrete pole at $\omega = \omega_{pl}$. As q increases there comes a critical value beyond which the plasma branch merges with the continuum or "independent particle-hole" branch (Fig. 6.8). Above this value of q the concept of a sharp collective mode is no longer useful.

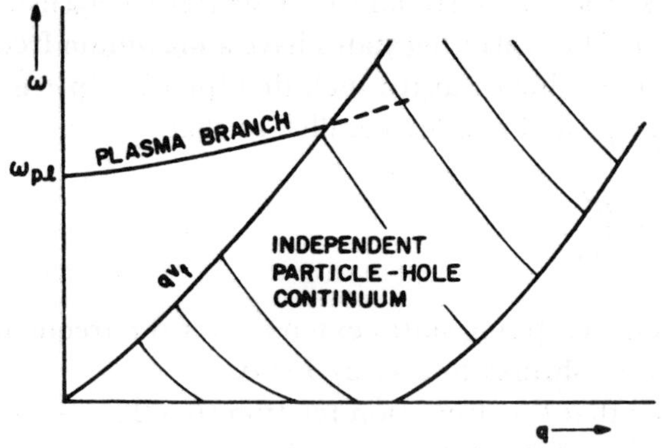

Fig. 6.8. The spectrum of poles of ϵ^{-1} (q, ω) for the Coulomb gas.

6.6. THE R.P.A. CALCULATION OF THE CORRELATION ENERGY

We now see that, for small q (long waves), the electron gas behaves in many ways like a set of oscillators, or plasmons. This analogy extends also to the calculation of the "correlation terms" in the total ground state energy of the interacting gas.

Although the correlation energy is only a small fraction of the total electronic energy, it is of the order of magnitude of the binding energy of a metal, and an accurate determination is thus important. In this section we show that in the r.p.a. the energy in fact has a contribution which is essentially the zero-point energy of the plasmon oscillators in the gas.

This result follows from a general expression for the ground state energy in terms of the dielectric response function of the system. The formulation is analogous to the corresponding treatment in our lattice dynamics example of Sec. 1.2.

We use the fact that the Coulomb interaction

$$H_1 = \tfrac{1}{2} \sum_{\mathbf{q} \neq 0} V(\mathbf{q})\{\rho^\dagger(\mathbf{q})\rho(\mathbf{q}) - N\} \tag{6.6.1}$$

may be written in terms of $\mathscr{K}$ defined in (6.5.1), using again the fluctuation-dissipation theorem. Using the T-product form (6.5.2) the expectation value of H_1 becomes explicitly

$$\langle H_1 \rangle = \tfrac{1}{2} \sum_{\mathbf{q}} V(\mathbf{q})\{ \lim_{t \to 0^-} i\mathscr{K}^T(\mathbf{q}, t) - N\}. \tag{6.6.2}$$

In terms of the Fourier transform of $\mathscr{K}^T$ this becomes

$$\langle H_1 \rangle = \tfrac{1}{2} \sum_{\mathbf{q}} V(\mathbf{q}) \int_{-\infty}^{\infty} \frac{d\omega}{2\pi} e^{-i\omega 0^-} i\mathscr{K}^T(\mathbf{q}, \omega) - \tfrac{1}{2}N \sum_{\mathbf{q}} V(\mathbf{q}). \tag{6.6.3}$$

This may be calculated in terms of the spectral density function for $\mathscr{K}$, via the definition (6.4.18) of $\epsilon^{-1}(\mathbf{q}, \omega)$, as

$$\langle H_1 \rangle = -\tfrac{1}{2} \sum_{\mathbf{q}} \int_{-\infty}^{\infty} \frac{d\omega}{2\pi} \operatorname{Im} \epsilon^{-1}(\mathbf{q}, \omega) - \tfrac{1}{2}N \sum_{\mathbf{q}} V(\mathbf{q}). \tag{6.6.4}$$

If the full hamiltonian is $H_0 + \lambda H_1$, the ground state energy is [see Eq. (1.2.3)]

$$E_G = E_0 + \int_0^1 \frac{d\lambda}{\lambda} \langle \lambda H_1 \rangle_\lambda, \tag{6.6.5}$$

where $\langle \ldots \rangle_\lambda$ refers to the ground state for coupling constant λ and E_0 is the ground state energy of H_0. Hence, if $\epsilon_\lambda^{-1}(\mathbf{q}, \omega)$ is the dielectric response for coupling constant λ,

$$E_G = E_0 - \tfrac{1}{2} \int_0^1 \frac{d\lambda}{\lambda} \sum_{\mathbf{q}} \left\{ \int_{-\infty}^{\infty} \operatorname{Im} \epsilon_\lambda^{-1}(\mathbf{q}, \omega) \frac{d\omega}{2\pi} + \lambda N V(\mathbf{q}) \right\}. \tag{6.6.6}$$

A knowledge of the dielectric response function for all values of λ is thus sufficient to determine the exact ground state energy.

Formula (6.6.6) is quite general. Within the r.p.a. for $\epsilon^{-1}(\mathbf{q}, \omega)$ we can now substitute the expression (6.5.17) and discuss the value of the integral. We have

$$E_G = E_0 - \tfrac{1}{2} \int_0^1 \frac{d\lambda}{\lambda} \sum_{\mathbf{q}} \int_{-\infty}^{\infty} \frac{d\omega}{2\pi} \operatorname{Im} \left\{ \frac{\lambda V(\mathbf{q}) \mathscr{P}_0(\mathbf{q}, \omega)}{1 - \lambda V(\mathbf{q}) \mathscr{P}_0(\mathbf{q}, \omega)} + 1 \right\}$$

$$- \tfrac{1}{2} N \sum_{\mathbf{q}} V(\mathbf{q}). \tag{6.6.7}$$

The integral over λ can be done explicitly to give

$$E_G = E_0 + \tfrac{1}{2} \sum_{\mathbf{q}} \int_{-\infty}^{\infty} \frac{d\omega}{2\pi} \operatorname{Im} \log\{1 - V(\mathbf{q}) \mathscr{P}_0(\mathbf{q}, \omega)\}$$

$$- \tfrac{1}{2} N \sum_{\mathbf{q}} V(\mathbf{q}). \tag{6.6.8}$$

This formula was first studied by Gell–Mann and Brueckner. It is of the same form as the energy formula (3.4.8) for the lattice oscillators.

To obtain the contribution to the ground state energy arising from the plasma mode we need the imaginary part of ϵ^{-1} which, near $|\omega| = \omega_{pl}$, is the sum of delta-functions at $\omega = \pm\omega_{pl}$. From the r.p.a. expression (6.5.17) we obtain, in the long-wavelength approximation corresponding to (6.5.24),

$$\operatorname{Im} \epsilon^{-1}(\mathbf{q}, \omega) = -\tfrac{1}{2}\pi\omega_{pl}\{\delta(\omega - \omega_{pl}) + \delta(\omega + \omega_{pl})\}. \tag{6.6.9}$$

The ground state energy is obtained from (6.6.6). For the λ integration it is here important to note that the interaction $V(\mathbf{q})$ (and hence the coupling constant) is proportional to e^2, whereas (6.6.9) is proportional to $|e|$. Hence, for fixed q, the contribution to E_G of the plasmon mode is

$$\tfrac{1}{2} \cdot \tfrac{1}{2}\pi\omega_{pl} \int_0^1 \frac{d\lambda}{\lambda} \lambda^{1/2} \int_{-\infty}^{\infty} \{\delta(\omega - \omega_{pl}) + \delta(\omega + \omega_{pl})\} \frac{d\omega}{2\pi}$$

$$= \tfrac{1}{2}\omega_{pl}, \tag{6.6.10}$$

which is just the zero-point energy of the plasma oscillation.

For small q it may be shown that (6.6.10) is the major contribution to the integral over ω in (6.6.7) or (6.6.8). However, as q increases the spectral weight in the continuum region (Fig. 6.8) starts to become more important, until for large q the plasma mode disappears completely. Thus, in contrast to the purely harmonic lattice dynamics case, both the collective character and the single-particle character show up in their contribution to the ground state energy. The energy per particle, expressed in terms of the dimensionless variables of Sec. 6.2, finally comes out to be

$$\frac{E}{E_0} = \frac{3}{5}\left(\frac{9\pi}{4}\right)^{2/3}\frac{1}{r_s^2} - \frac{3}{2}\left(\frac{3}{2\pi}\right)^{2/3}\frac{1}{r_s} + \frac{2}{\pi^2}(1 - \log 2)\log r_s - 0{\cdot}094 + O(r_s \log r_s)$$

$$= \frac{2{\cdot}21}{r_s^2} - \frac{0{\cdot}916}{r_s} + 0{\cdot}0622 \log r_s - 0{\cdot}094 + O(r_s \log r_s). \quad (6.6.11)$$

The first term is the kinetic energy, the second term is the exchange energy discussed in Sec. 6.2, and the remaining terms represent the correlation energy. [The logarithmic term was first obtained by Macke (1950) and the constant term by Gell–Mann and Brueckner (1957). The constant term requires numerical integration; an improved evaluation was given by Onsager *et al.* (1966). Higher-order terms have been considered by DuBois (1959) and Carr and Maradudin (1964).]

Unfortunately r_s is not small for metallic densities (in general $2 \leqslant r_s \leqslant 5$), and thus (6.6.11) cannot be expected to give quantitatively correct results for real metals. Improvements over the r.p.a. have more recently been made by Hubbard (1967) and by Singwi *et al.* (1970), in which the polarization part $\mathscr{P}(\mathbf{q}, \omega)$ of (6.5.6) is calculated taking into account short-range repulsive correlations in the electron gas. As r_s increases in real materials, the effects of the ionic potential become more and more important and the uniform-positive-background model becomes less realistic.

As an academic problem, the latter model is again soluble in the opposite limit of very *low* densities. Here the interaction energy dominates the kinetic energy, and the state of minimum energy is expected to be one in which the electrons are arranged on a regular lattice [Wigner (1938)]. This is a non-conducting state in which the electrons vibrate like ions in an insulating crystal. The ground state energy can then be obtained as an expansion in powers of $1/r_s^{1/2}$, the leading terms being the electrostatic lattice energy and the zero-point energy of the lattice vibrations. More realistically, the limit $r_s \to \infty$ is the "atomic limit" in which the ionic potentials dominate and bind the electrons to form a set of weakly coupled hydrogenic atoms.

This brings up the interesting problem emphasized by Mott of the transition from an insulating, atom-like state (with one electron per atom) to a metallic band-like state at small r_s. This so-called "insulator-metal transition" is still under active study and becomes quite dependent on the detailed chemistry of the systems involved. We return to this topic in Chap. 8.

Chapter 7

The Magnetic Instability of the Interacting Electron Gas

The properties of the high-density Coulomb gas discussed in Chap. 6 are based on one important assumption: that the basic fermi-sea picture of the ground state is a stable one and that the interactions, while introducing electron correlations, do not cause qualitative changes. However, it is well known that, in circumstances where the correlations are sufficiently strong, the "normal" gas of interacting fermions can become unstable and undergo a spontaneous transition to a "condensed" phase exhibiting long-range order. One well-established example of such an instability, induced by the attractive effective potential resulting from electron–phonon interactions, is superconductivity. This will be discussed in Chap. 10.

In this chapter we shall be concerned with instability resulting from the repulsive Coulomb interactions. This is particularly likely to occur when the ionic potentials lead to the formation of narrow energy bands, and it can take various forms which, for a rigid lattice, are all magnetic in character. [In a compressible lattice instabilities involving charge density oscillations can also occur, in particular the "excitonic insulator" state studied by Kohn (1968) and reviewed by Halperin and Rice (1968).] The effect of the instability is to produce a fundamental change in the symmetry of the ground state of the interacting system. We will concentrate here on the simplest form of the magnetic instability, namely ferromagnetism.

A more subtle form of instability occurs when the Coulomb interaction is localized at a point impurity inserted in the metal. Here the instability is not a sharp one as in the uniform ferromagnetic case, but a gradual one which leads to new types of low-lying excitations in the system associated with the formation of a local magnetic moment. In Sec. 7.3 we consider the simplest aspects of this localized moment problem as formulated by P. W. Anderson.

7.1. THE HUBBARD MODEL

As mentioned above, the band structure of a metal has an important influence on the stability of the electron gas in the metal. In order to discuss band structure effects it is not enough to consider the simple Coulomb gas in a positive background, and one requires a theory of correlations which takes into account the atomic structure of the solid. In the presence of the ions the Coulomb problem becomes much more complicated owing to the multiband nature which is an essential feature in discussing, say, the *d*-electrons of transition metals. An important physical simplification was introduced by J. Hubbard (1963) who pointed out that it is the short-ranged part of the Coulomb interaction which is dominant in leading to the instabilities. The long-range part of the Coulomb interaction can be considered to be screened out as discussed in Chap. 6.

He therefore proposed a model in which the electrons are considered to be in a narrow energy band with Bloch energies $\epsilon_{\mathbf{p}}$. For such a band one can form Wannier wave functions for an electron at site i

$$\varphi_i(\mathbf{x}) = N^{-1/2} \sum_{\mathbf{p} \in \text{zone}} e^{i\mathbf{p}\cdot\mathbf{x}_i} \psi_{\mathbf{p}}(\mathbf{x}), \tag{7.1.1}$$

where $\psi_{\mathbf{p}}(\mathbf{x})$ are the Bloch states. For electrons projected onto these band states the screened Coulomb interaction can be expressed in terms of matrix elements between Wannier states on different sites

$$U_{iji'j'} = \int d^3x_1 d^3x_2\, \varphi_i^*(\mathbf{x}_1)\varphi_j^*(\mathbf{x}_2) V_{sc}(\mathbf{x}_1 - \mathbf{x}_2)\varphi_{i'}(\mathbf{x}_1)\varphi_{j'}(\mathbf{x}_2), \tag{7.1.2}$$

where V_{sc} is some sort of screened effective Coulomb interaction. Hubbard argued that the dominant term in such an interaction would be the *intra-atomic* matrix element U_{iiii}, denoted by U. His model then consists in neglecting all matrix elements other than those for which both interacting electrons are on the same site. This leads to an interaction hamiltonian

$$H_1' = U \sum_{i\sigma\sigma'} n_{i\sigma} n_{i\sigma'}, \tag{7.1.3}$$

where $n_{i\sigma}$ is the Wannier number operator

$$n_{i\sigma} = a_{i\sigma}^{\dagger} a_{i\sigma} \tag{7.1.4}$$

and $a_{i\sigma}^\dagger$ is the Wannier creation operator

$$a_{i\sigma}^\dagger = N^{-1/2} \sum_{\mathbf{p}\in \text{zone}} a_{\mathbf{p}\sigma}^\dagger \, e^{i\mathbf{p}\cdot\mathbf{X}_i}. \tag{7.1.5}$$

We now have the important identity resulting from the anticommuting properties of the $a^\dagger$:

$$(n_{i\sigma})^2 = n_{i\sigma}, \tag{7.1.6}$$

which allows us to write

$$H_1' = UN + U\sum_i n_{i\uparrow}n_{i\downarrow}. \tag{7.1.7}$$

This automatically embodies the Pauli principle by asserting that only electrons of opposite spin will interact. It is this effect which leads to ferromagnetism: by aligning their spins the electrons can lower the strength of the repulsive potential. (They do this at the cost of kinetic energy involved in piling up electrons in the fermi sea of one sign of the spin.) Thus if U is large enough it may be energetically favorable to have an overall spin in the ground state, even though the original hamiltonian is spin independent.

To study this effect we therefore start with a model hamiltonian

$$H = H_0 + H_1,$$

$$H_0 = \sum_{\mathbf{p}\sigma} \epsilon_{\mathbf{p}} a_{\mathbf{p}\sigma}^\dagger a_{\mathbf{p}\sigma},$$

$$H_1 = U\sum_i n_{i\uparrow}n_{i\downarrow} = \frac{U}{N}\sum_{\mathbf{p}\mathbf{p}'\mathbf{q}} a_{\mathbf{p}+\mathbf{q}\uparrow}^\dagger a_{\mathbf{p}\uparrow} a_{\mathbf{p}'-\mathbf{q}\downarrow}^\dagger a_{\mathbf{p}'\downarrow}, \tag{7.1.8}$$

and investigate the response of the system to an external space- and time-dependent magnetic field. A phase transition to a magnetically ordered state will then occur if the response of the system to the applied magnetic field becomes infinitely large at some stage as the strength of the interaction increases. Such a divergence indicates the appearance of a spontaneous non-zero magnetization at the transition point. In the paramagnetic–ferromagnetic transition the magnetization, which serves as a measure of long-range order, changes continuously through the transition but has a discontinuous first derivative, so that we have a second-order phase transition [Landau and Lifshitz (1959)].

7.2. THE KUBO FORMULA FOR THE SUSCEPTIBILITY TENSOR

The derivation of a general formula for the susceptibility is again a problem in linear response theory and proceeds as in Sec. 5.5. H_{ext} now represents the interaction of the spins with a classical applied magnetic field $\mathscr{H}(\mathbf{x}, t)$. This may be written as

$$H_{\text{ext}} = -\int \mathscr{H}(\mathbf{x}, t) \cdot \sigma(\mathbf{x})\, d^3x, \tag{7.2.1}$$

where the field $\mathscr{H}$ is assumed to have an arbitrary dependence on space and time variables, and $\sigma(\mathbf{x})$ is the *spin density operator*, in first quantized form

$$\sigma(\mathbf{x}) = \sum_{\text{el}} \delta(\mathbf{x} - \mathbf{x}_{\text{el}})\sigma_{\text{el}} \tag{7.2.2}$$

for a set of electrons, spin operator σ_{el} at positions $\mathbf{x}_{\text{el}}$. [We choose units for the magnetic field such that the Bohr magneton $\mu_B = |e|\hbar/2mc = 1$; in a representation with σ_z diagonal the components of the vector σ for any electron are then the Pauli matrices

$$\sigma_x = \begin{pmatrix} 0 & 1 \\ 1 & 0 \end{pmatrix}, \quad \sigma_y = \begin{pmatrix} 0 & -i \\ i & 0 \end{pmatrix}, \quad \sigma_z = \begin{pmatrix} 1 & 0 \\ 0 & -1 \end{pmatrix}, \tag{7.2.3}$$

and $\sigma(\mathbf{x})$ is also the *magnetic moment density*. In conventional units the spin $\mathbf{s}$ and magnetic moment μ of an electron are $\mathbf{s} = \frac{1}{2}\hbar\sigma$, $\mu = (e\hbar/2mc)\sigma$, with $e < 0$.]

We wish to calculate the expectation value of the *induced moment* $\sigma(\mathbf{x}, t)$ in the true ground state at time t:

$$\langle \sigma(\mathbf{x}, t) \rangle = \langle \Psi(t) | \sigma(\mathbf{x}) | \Psi(t) \rangle. \tag{7.2.4}$$

We work to the first order in $\mathscr{H}$ and find, proceeding as in Sec. 5.5, that

$$\begin{aligned} \langle \sigma_i(\mathbf{x}, t) \rangle_{\mathscr{H}} &= \langle \sigma_i(\mathbf{x}, t) \rangle_{\mathscr{H}=0} + \sum_j \int dt' \int d^3x' \chi_{ij}(\mathbf{x} - \mathbf{x}', t - t') \\ &\quad \times \mathscr{H}_j(\mathbf{x}', t'), \end{aligned} \tag{7.2.5}$$

where the space- and time-dependent susceptibility tensor is given by

$$\chi_{ij}(\mathbf{x} - \mathbf{x}', t - t') = i\theta(t - t')\langle [\tilde{\sigma}_i(\mathbf{x}, t), \tilde{\sigma}_j(\mathbf{x}', t')] \rangle. \tag{7.2.6}$$

Here the spin density operators are Heisenberg operators with respect to the hamiltonian H of the interacting electron gas in the absence of the applied field, and the expectation value refers to the ground state of H. The suffixes i, j now denote cartesian vector and tensor components.

To obtain the second-quantized form of the spin density operators, we make a Fourier transformation and write (7.2.2) in the form

$$\sigma(\mathbf{x}) = \sum_{\mathbf{q}} e^{i\mathbf{q}\cdot\mathbf{x}} \sigma(\mathbf{q}),$$

where

$$\sigma(\mathbf{q}) = \sum_{\mathrm{el}} e^{-i\mathbf{q}\cdot\mathbf{x}_{\mathrm{el}}} \sigma_{\mathrm{el}}. \tag{7.2.7}$$

$\sigma(\mathbf{q})$ is a sum of one-electron operators which act on the momentum and spin state of an electron, and the single-particle wave functions are

$$u_{\mathbf{p}\alpha}(\mathbf{x}) = V^{-1/2}\psi_{\mathbf{p}}(\mathbf{x})\chi_\alpha, \tag{7.2.8}$$

where $\psi_{\mathbf{p}}(\mathbf{x})$ is a Bloch function and χ_α is a spin function with the two components $\chi_\uparrow$ and $\chi_\downarrow$. The second-quantized form of $\sigma(\mathbf{q})$ is thus

$$\sigma(\mathbf{q}) = \sum_{\mathbf{p}\alpha\beta} a^\dagger_{\mathbf{p}+\mathbf{q},\alpha}\sigma_{\alpha\beta}a_{\mathbf{p}\beta}, \tag{7.2.9}$$

where $\sigma_{\alpha\beta}$ denotes an element of the appropriate Pauli matrix. Hence

$$\sigma(\mathbf{x}) = \sum_{\mathbf{q}} e^{i\mathbf{q}\cdot\mathbf{x}} \sum_{\mathbf{p}\alpha\beta} a^\dagger_{\mathbf{p}+\mathbf{q},\alpha}\sigma_{\alpha\beta}a_{\mathbf{p}\beta} \tag{7.2.10}$$

and

$$\tilde{\sigma}(\mathbf{x}, t) = \sum_{\mathbf{q}} e^{i\mathbf{q}\cdot\mathbf{x}} \sum_{\mathbf{p}\alpha\beta} \tilde{a}^\dagger_{\mathbf{p}+\mathbf{q},\alpha}(t)\sigma_{\alpha\beta}\tilde{a}_{\mathbf{p}\beta}(t). \tag{7.2.11}$$

The expression (7.2.6) for χ_{ij} is thus again a retarded two-particle Green's function of the type studied in Sec. 5.5.[1] Now define the spin raising and lowering operators σ^+, σ^- by

$$\sigma^\pm = \tfrac{1}{2}(\sigma_x \pm i\sigma_y); \tag{7.2.12}$$

[1] Note the change in sign compared with our standard definition of a retarded Green's function in previous chapters and in Appendix 2. This comes from the negative sign in the interaction (7.2.1).

then

$$\sigma^+(\mathbf{x}) = \sum_{\mathbf{q}} e^{i\mathbf{q}\cdot\mathbf{x}} \sum_{\mathbf{p}} a^\dagger_{\mathbf{p}+\mathbf{q}\uparrow} a_{\mathbf{p}\downarrow}, \tag{7.2.13}$$

and

$$\sigma^-(\mathbf{x}) = \sum_{\mathbf{q}} e^{i\mathbf{q}\cdot\mathbf{x}} \sum_{\mathbf{p}} a^\dagger_{\mathbf{p}+\mathbf{q}\downarrow} a_{\mathbf{p}\uparrow}. \tag{7.2.14}$$

The susceptibility tensor χ_{ij} can be obtained from a knowledge of the *transverse* susceptibility χ^{-+}, defined by

$$\chi^{-+}(\mathbf{x}-\mathbf{x}', t-t') = i\theta(t-t')\langle[\tilde{\sigma}^-(\mathbf{x}, t), \tilde{\sigma}^+(\mathbf{x}', t')]\rangle, \tag{7.2.15}$$

and the *longitudinal* susceptibility χ_{zz}, defined similarly in terms of the commutator $[\tilde{\sigma}_z(\mathbf{x}, t), \tilde{\sigma}_z(\mathbf{x}', t')]$. For an isotropic or cubic medium in the paramagnetic state above the Curie point, it follows from symmetry considerations that χ_{ij} is diagonal and isotropic, with

$$\chi_{ij} = 2\chi^{-+}\delta_{ij}; \tag{7.2.16}$$

thus χ^{-+} determines the susceptibility completely. Below the Curie point, however, the longitudinal susceptibility differs from the transverse susceptibility and must be calculated separately. In what follows we confine ourselves to discussing χ^{-+}.

Substituting for $\tilde{\sigma}^-(\mathbf{x}, t)$, we have

$$\chi^{-+}(\mathbf{x}, t) = \sum_{\mathbf{pq}} e^{i\mathbf{q}\cdot\mathbf{x}}\, \chi^{-+}(\mathbf{p}, \mathbf{q}; t), \tag{7.2.17}$$

where

$$\chi^{-+}(\mathbf{p}, \mathbf{q}; t) = i\theta(t)\langle[\tilde{a}^\dagger_{\mathbf{p}+\mathbf{q}\downarrow}(t)\tilde{a}_{\mathbf{p}\uparrow}(t), \sigma^+(0, 0)]\rangle. \tag{7.2.18}$$

7.3. EVALUATION OF THE SUSCEPTIBILITY IN THE R.P.A.

To evaluate the Green's function (7.2.18) we set up its equation of motion and solve it in a generalized Hartree–Fock approximation [Wolff (1960), Izuyama, Kim and Kubo (1963)]. For the special case of the Hubbard model interaction this turns out to lead to a separable integral equation for $\chi(\mathbf{p}, \mathbf{q})$. As in the Coulomb case, the generalized Hartree–Fock approximation can also be derived by summing a particular set of particle-hole diagrams. These turn out to be the *ladder* diagrams containing repeated interactions of electron and hole lines (Fig. 7.1), as contrasted with the bubble diagrams which were important in the long-range Coulomb case.

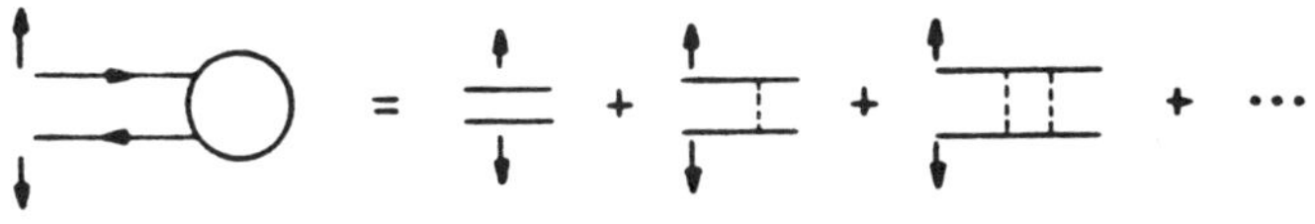

Fig. 7.1. Ladder diagrams contributing to $\chi^{-+}_{(\mathbf{p},\,\mathbf{q})}$.

The equation of motion method comes out as follows:

$$i\frac{\partial}{\partial t}\chi^{-+}(\mathbf{p},\mathbf{q};t) = -\delta(t)\langle[a^\dagger_{\mathbf{p}+\mathbf{q}\downarrow}a_{\mathbf{p}\uparrow},\sigma^+(0,0)]\rangle + i\theta(t)\langle[[\tilde{a}^\dagger_{\mathbf{p}+\mathbf{q}\downarrow}(t)\tilde{a}_{\mathbf{p}\uparrow}(t),H],\sigma^+(0,0)]\rangle. \quad (7.3.1)$$

Substituting the hamiltonian (7.1.8) we find

$$[a^\dagger_{\mathbf{p}+\mathbf{q}\downarrow}a_{\mathbf{p}\uparrow},H] = -(\epsilon_{\mathbf{p}+\mathbf{q}}-\epsilon_{\mathbf{p}})a^\dagger_{\mathbf{p}+\mathbf{q}\downarrow}a_{\mathbf{p}\uparrow} + \frac{U}{N}\sum_{\mathbf{p}'\mathbf{q}'}(a^\dagger_{\mathbf{p}+\mathbf{q}\downarrow}a_{\mathbf{p}-\mathbf{q}'\uparrow}a^\dagger_{\mathbf{p}'-\mathbf{q}'\downarrow}a_{\mathbf{p}'\downarrow} - a^\dagger_{\mathbf{p}'+\mathbf{q}'\uparrow}a_{\mathbf{p}'\uparrow} \times a^\dagger_{\mathbf{p}+\mathbf{q}-\mathbf{q}'\downarrow}a_{\mathbf{p}\uparrow}). \quad (7.3.2)$$

The generalized Hartree–Fock approximation now consists in replacing, in the products of four operators on the right-hand side of (7.3.2), all pairs of the type $a^\dagger a$ by their expectation values and taking the sum over all such averages, paying due regard to sign changes arising from changes in the order of anticommuting factors. Assuming also that there is spin and momentum uniformity in the ground state, we put

$$\langle a^\dagger_{\mathbf{p}\alpha}a_{\mathbf{p}'\beta}\rangle = \delta_{\mathbf{p}\mathbf{p}'}\delta_{\alpha\beta}f_{\mathbf{p}\alpha}. \quad (7.3.3)$$

(For the paramagnetic state $f_{\mathbf{p}\alpha}$ is also independent of α.) The second term in (7.3.2) then becomes

$$\frac{U}{N}\sum_{\mathbf{p}'}\{(f_{\mathbf{p}\uparrow}-f_{\mathbf{p}+\mathbf{q}\downarrow})a^\dagger_{\mathbf{p}+\mathbf{p}'+\mathbf{q}\downarrow}a_{\mathbf{p}+\mathbf{p}'\uparrow} + (f_{\mathbf{p}'\downarrow}-f_{\mathbf{p}'\uparrow})a^\dagger_{\mathbf{p}+\mathbf{q}\downarrow}a_{\mathbf{p}\uparrow}\}. \quad (7.3.4)$$

The first part of this corresponds to repeated exchange scattering involving a virtual electron-hole pair, represented by the ladder-type diagrams shown in Fig. 7.1. The second part is an exchange self-energy correction to each one-particle state.

Finally, noting that

$$\sigma^+(0, 0) = \sum_{\mathbf{pq}} a^\dagger_{\mathbf{p+q}\uparrow} a_{\mathbf{p}\downarrow},$$

we have

$$\langle [a^\dagger_{\mathbf{p+q}\downarrow} a_{\mathbf{p}\uparrow}, \sigma^+(0, 0)] \rangle = f_{\mathbf{p+q}\downarrow} - f_{\mathbf{p}\uparrow}. \tag{7.3.5}$$

Collecting everything together, the equation of motion (7.3.1) becomes

$$\left\{ i\frac{\partial}{\partial t} + (\bar{\epsilon}_{\mathbf{p+q}\uparrow} - \bar{\epsilon}_{\mathbf{p}\downarrow}) \right\} \chi^{-+}(\mathbf{p}, \mathbf{q}; t)$$
$$= -\delta(t)(f_{\mathbf{p+q}\downarrow} - f_{\mathbf{p}\uparrow}) - (f_{\mathbf{p+q}\downarrow} - f_{\mathbf{p}\uparrow}) \frac{U}{N} \sum_{\mathbf{p}'} \chi^{-+}(\mathbf{p}', \mathbf{q}; t), \tag{7.3.6}$$

where

$$\bar{\epsilon}_{\mathbf{p}\sigma} = \epsilon_{\mathbf{p}} - \frac{U}{N} \sum_{\mathbf{p}'} f_{\mathbf{p}',-\sigma} \qquad (\sigma = \uparrow \text{ or } \downarrow) \tag{7.3.7}$$

is the one-particle energy modified by the exchange self-energy.

Equation (7.3.6) can be solved in closed form by introducing the Fourier transform

$$\chi(\omega) = \int_{-\infty}^{\infty} \chi(t)\, e^{i\omega t}\, dt; \tag{7.3.8}$$

we find

$$\chi^{-+}(\mathbf{p}, \mathbf{q}; \omega) = \frac{(f_{\mathbf{p}\uparrow} - f_{\mathbf{p+q}\downarrow})\{1 + (U/N)\chi^{-+}(\mathbf{q}, \omega)\}}{\omega + \bar{\epsilon}_{\mathbf{p+q}\uparrow} - \bar{\epsilon}_{\mathbf{p}\downarrow}},$$

$$\text{with } \chi(\mathbf{q}) = \sum_{\mathbf{p}} \chi(\mathbf{p}, \mathbf{q}). \tag{7.3.9}$$

We can obtain $\chi^{-+}(\mathbf{q})$ from this by summing over $\mathbf{p}$. The final result (expressed as a susceptibility per atom) is

$$\chi^{-+}(\mathbf{q}, \omega) = \frac{\Gamma^{-+}(\mathbf{q}, \omega)}{1 - U\Gamma^{-+}(\mathbf{q}, \omega)}, \tag{7.3.10}$$

where

$$\Gamma^{-+}(\mathbf{q}, \omega) = \frac{1}{N} \sum_{\mathbf{p}} \frac{f_{\mathbf{p}\uparrow} - f_{\mathbf{p+q}\downarrow}}{\omega - (\tilde{\epsilon}_{\mathbf{p}\downarrow} - \tilde{\epsilon}_{\mathbf{p+q}\uparrow}) + i\eta} \tag{7.3.11}$$

is the same unperturbed particle-hole propagator which appears in the Coulomb problem [there called $\mathscr{P}_0(\mathbf{q}, \omega)$]. The infinitesimal $i\eta$, with $\eta > 0$, has been added in the denominator of (7.3.11) to characterize $\Gamma(\mathbf{q})$ as a retarded Green's function.

Equation (7.3.10), in our approximation, is the general expression for the frequency- and wave-vector-dependent susceptibility of the interacting electron gas. The general space- and time-dependent susceptibility $\chi^{-+}(\mathbf{x}, t)$ is the Fourier transform with respect to $\mathbf{q}$ and ω of the function $\chi^{-+}(\mathbf{q}, \omega)$.

The function $\Gamma(\mathbf{q}, \omega)$ represents the complex susceptibility per atom of a non-interacting electron gas, for an applied field with wave-vector $\mathbf{q}$ and frequency ω. For $\omega = \mathbf{q} = 0$, this reduces to the well-known *Pauli susceptibility* which is proportional to the one-particle density of states $\mathscr{N}(\epsilon_f)$ [see Eq. (7.4.5)].

7.4. THE INSTABILITY CRITERION

In the paramagnetic state, for vanishing external field, the populations of the "up" and "down" spin states are equal. We therefore put $f_{\mathbf{p}\uparrow} = f_{\mathbf{p}\downarrow} = f_{\mathbf{p}}$, $\tilde{\epsilon}_{\mathbf{p}\uparrow} = \tilde{\epsilon}_{\mathbf{p}\downarrow} = \epsilon_{\mathbf{p}}$ [absorbing the exchange self-energy correction in (7.3.7) into the single-particle energy], and $\Gamma^{-+}(\mathbf{q}, \omega) = \Gamma(\mathbf{q}, \omega)$. The instabilities which we are looking for now come out from a study of the response to a static spatially varying external field. This simply corresponds to the limit $\omega = 0$ in the response function $\chi(\mathbf{q}, \omega)$. Singularities in $\chi(\mathbf{q}, \omega)$ do, of course, occur for non-zero ω (in the form of a cut in the ω-plane); however, these correspond to damped propagation in the system of the type discussed in Chap. 5 in the case of the electrical conductivity. At $\omega = 0$ there can be no energy loss so that $\mathrm{Im}\, \chi(\mathbf{q}, \omega = 0) = 0$, and any instability which occurs is then a sign of an instability in the ground state of the system: the low-lying excited states (with finite magnetization) start to have the same energy as the ground state at the instability point. From (7.3.10) it may now be seen that the instability criterion is

$$U\Gamma(\mathbf{q}, \omega = 0) = 1, \tag{7.4.1}$$

where

$$\Gamma(\mathbf{q}, \omega = 0) = \frac{1}{N} \sum_{\mathbf{p}} \frac{f_{\mathbf{p}} - f_{\mathbf{p}+\mathbf{q}}}{\epsilon_{\mathbf{p}+\mathbf{q}} - \epsilon_{\mathbf{p}}}. \tag{7.4.2}$$

Thus, for arbitrary $\mathbf{q}$, the instability will occur at a critical value of the interaction strength U, corresponding to

$$U = 1/\Gamma(\mathbf{q}, \omega = 0). \tag{7.4.3}$$

For the particular case $\mathbf{q} \to 0$ we can evaluate (7.4.2) explicitly by expanding

$$f_{\mathbf{p}+\mathbf{q}} = f_{\mathbf{p}} + \mathbf{q} \cdot \frac{\partial \epsilon_{\mathbf{p}}}{\partial \mathbf{p}} \frac{\partial f_{\mathbf{p}}}{\partial \epsilon_{\mathbf{p}}} + \cdots, \quad \epsilon_{\mathbf{p}+\mathbf{q}} = \epsilon_{\mathbf{p}} + \mathbf{q} \cdot \frac{\partial \epsilon_{\mathbf{p}}}{\partial \mathbf{p}} + \cdots. \tag{7.4.4}$$

We then find

$$\lim_{\mathbf{q} \to 0} \Gamma(\mathbf{q}, \omega = 0) = \frac{1}{N} \sum_{\mathbf{p}} \left(- \frac{\partial f_{\mathbf{p}}}{\partial \epsilon_{\mathbf{p}}} \right). \tag{7.4.5}$$

At zero temperature $-\partial f/\partial \epsilon = \delta(\epsilon - \epsilon_f)$, from which

$$\Gamma(\mathbf{q} = 0, \omega = 0) = \mathcal{N}(\epsilon_f), \tag{7.4.6}$$

where $\mathcal{N}(\epsilon)$ is the single-particle density of states for the band

$$\mathcal{N}(\epsilon) = \frac{1}{N} \sum_{\mathbf{p}} \delta(\epsilon - \epsilon_{\mathbf{p}}). \tag{7.4.7}$$

The $\mathbf{q} = 0$ form of the instability criterion thus becomes

$$U \mathcal{N}(\epsilon_f) = 1, \tag{7.4.8}$$

which is known as the Stoner criterion. An instability at $\mathbf{q} = 0$ corresponds to a tendency for the system to acquire spontaneously a uniform, or ferromagnetic, spin density.

For a simple parabolic band $\Gamma(\mathbf{q})$ is a monotonically decreasing function as $|\mathbf{q}|$ is increased, so that, as U is increased, the instability will first occur at $\mathbf{q} = 0$. However, for more realistic band structures $\Gamma(\mathbf{q})$ starts to change its character as the fermi level moves up in the band, corresponding to varying the number of electrons per atom. (In a real metal there are several bands, so that the fermi level need not be symmetrically placed in a given narrow band.)

In particular, for a simple cubic band

$$\epsilon_{\mathbf{p}} = -\epsilon_0(\cos p_1 a + \cos p_2 a + \cos p_3 a), \tag{7.4.9}$$

the point at which the zone is half full leads to a strong instability at a critical value $\mathbf{Q} = \frac{1}{2}\mathbf{G}$, where G is a lattice vector $(2\pi/a)(1, 1, 1)$. This value of $\mathbf{q}$ has the special property

$$\epsilon_{\mathbf{p}+\mathbf{Q}} = -\epsilon_{\mathbf{p}} \tag{7.4.10}$$

for all $\mathbf{p}$, and hence

$$\Gamma(\mathbf{Q}, \omega = 0) = -\int_{-3\epsilon_0}^{\epsilon_f} \frac{\mathcal{N}(\epsilon)}{2\epsilon}\, d\epsilon + \int_{-\epsilon_f}^{3\epsilon_0} \frac{\mathcal{N}(-\epsilon)}{2\epsilon}\, d\epsilon$$

$$= -\int_{-3\epsilon_0}^{\epsilon_f} \frac{\mathcal{N}(\epsilon)}{\epsilon}\, d\epsilon, \tag{7.4.11}$$

where $\mathcal{N}(\epsilon)$ is the density of states in the band as a function of energy (measured from the center of the band). $\Gamma(\mathbf{Q})$ increases as the fermi level

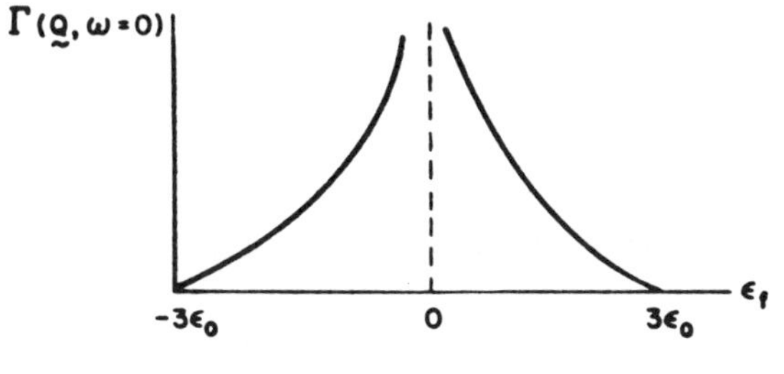

Fig. 7.2. Variation of $\Gamma(\mathbf{Q}, \omega = 0)$ with position of fermi level.

moves up in the band and has a singularity when the band is just half full ($\epsilon_f = 0$) (Fig. 7.2). Thus at this value of the electron density (one electron per atom) the instability occurs at finite $\mathbf{q}$ rather than $\mathbf{q} = 0$, and moreover for this specific case (which is somewhat unrealistic for real metals) the instability occurs for arbitrarily small values of U [Penn (1966)].

An instability with non-zero $\mathbf{q}$ corresponds to the system attempting to align itself with a spatially oscillating or antiferromagnetic spin density. This type of instability, first pointed out by Overhauser (1962), is called a spin density wave. For the example considered above, $\exp(i\mathbf{Q}\cdot\mathbf{x})$ changes sign at neighboring lattice sites, and the fermi surface for the half-filled band coincides with the boundary of the "magnetic Brillouin zone." Thus, for one conduction electron per atom, an arbitrarily small interaction causes the growth of a spin density wave of wave-number $\mathbf{Q} = (\pi/a)(1, 1, 1)$ which converts the metal into an insulating antiferromagnet.

The magnetic instabilities discussed in this section generally occur (apart from special cases of the type discussed above) only if the repulsive potential is sufficiently strong; in the approximation considered, a new bound state must be created in which electrons above the fermi surface are bound to holes (vacant states below the fermi surface) of opposite spin, and for this a minimum energy is necessary. This should be contrasted with the instability leading to the superconducting state, discussed in Chap. 10, which involves the binding of two electrons above the fermi surface and is possible for an attractive potential of any magnitude.

7.5. NEUTRON SCATTERING AND THE q,ω-DEPENDENT GENERALIZED SUSCEPTIBILITY FUNCTION

Slow neutrons scatter from solids via two forces: the nuclear force which leads to the neutron-ion scattering discussed in Secs. 1.3 and 2.3 and hence to the emission and absorption of phonons by slow neutrons traversing a solid; and the magnetic dipole interaction in which the (very weak) magnetic moment of the neutron interacts via the classical dipole-dipole interaction with the spin magnetic moment of electrons in the solid. This latter interaction leads to enhanced neutron scattering from magnetic systems.

The inelastic cross-section for magnetic neutron scattering from a magnetic system in the first Born approximation may be written in terms of spin density correlation functions [van Hove (1954)]. For a cubic paramagnet the differential cross-section $d^2\sigma/d\Omega d\omega$ is proportional to $\mathscr{S}^{-+}(\mathbf{q}, \omega)$, where

$$\mathscr{S}^{-+}(\mathbf{q}, \omega) = \int_{-\infty}^{\infty} dt\, e^{i\omega t} \langle \sigma^-(\mathbf{q}, t)\sigma^+(-\mathbf{q}, 0)\rangle$$

$$= \sum_{\mathbf{p}} \int_{-\infty}^{\infty} dt\, e^{i\omega t}\, \mathscr{S}^{-+}(\mathbf{p}, \mathbf{q}; t), \tag{7.5.1}$$

and

$$\mathscr{S}^{-+}(\mathbf{p}, \mathbf{q}; t) = \langle \tilde{a}^\dagger_{\mathbf{p}+\mathbf{q}\downarrow}(t)\tilde{a}_{\mathbf{p}\uparrow}(t)\sigma^+(0, 0)\rangle. \tag{7.5.2}$$

This formula (at finite T) is analogous to Eqs. (2.3.1) and (2.3.3) giving the cross-section for the phonon case. $\mathbf{q} = \mathbf{k}' - \mathbf{k}$ is the scattering vector ($\mathbf{k}$ being the incident and $\mathbf{k}'$ the scattered neutron wave-vector), and ω is the energy loss. As in the phonon case [Eq. (2.3.12)], the Fourier transform of the time correlation function (7.5.2), which determines the neutron cross-section, can be expressed in terms of the associated retarded Green's function which is just the susceptibility function $\chi^{-+}(\mathbf{p}, \mathbf{q}; t)$, Eq. (7.2.18). The required relation (a form of the fluctuation-dissipation theorem) is derived at finite temperature ($\beta = 1/k_B T$) in Appendix 2, and from Eq. (A.2.9) it follows that

$$\mathscr{S}^{-+}(\mathbf{q}, \omega) = \frac{2}{1-e^{-\beta\omega}} \operatorname{Im} \chi^{-+}(\mathbf{q}, \omega). \tag{7.5.3}$$

Notice that $\operatorname{Im} \chi$ is an odd function of ω so that $\mathscr{S}^{-+}(\mathbf{q}, \omega)$ is a positive function for all ω, but is asymmetric because of the Planck function factor $1/(1 - e^{-\beta\omega})$.

The general features of $\mathscr{S}(\mathbf{q}, \omega)$ are as follows: for positive ω, $1/(1 - e^{-\beta\omega})$ is approximately unity for $\omega \gg k_B T$, so that the neutron cross-section measures directly the absorptive part of the magnetic response at frequency ω for a wave-vector $\mathbf{q}$ determined by the kinematics of the neutron scattering angle. This is the "energy loss" regime in which the neutrons emit excitations into the solid. In the $\omega < 0$ regime the Planck factor is small except for $\omega \ll k_B T$. This is the "energy gain" regime in which the neutrons pick up energy from thermally excited magnetic fluctuations.

Thus neutron scattering can be used as a direct experimental probe into the spectrum of the magnetic excitations of the system. Returning to the magnetic instability problem, the approach of the instability point discussed in Sec. 7.4 also leads to a strong enhancement of the inelastic part of the scattering via Im χ^{-+}. In this regime $\Gamma(\mathbf{q}, \omega)$ may be approximated by its value for small $\mathbf{q}$ and ω. For a parabolic band ($\epsilon_{\mathbf{p}} = \mathbf{p}^2/2m$) one finds

$$\begin{aligned} \operatorname{Re} \Gamma(\mathbf{q}, \omega) &\simeq \mathcal{N}(\epsilon_f) \qquad (\mathbf{q} \to 0) \\ \operatorname{Im} \Gamma(\mathbf{q}, \omega) &\simeq C(\omega/qv_f) \end{aligned} \tag{7.5.4}$$

where C is a constant equal to $\frac{1}{4}\pi\mathcal{N}(\epsilon_f)$.

For the non-interacting case ($U = 0$) the behavior of $\mathcal{S}(\mathbf{q}, \omega)$ is thus determined at low frequencies by the slowly varying function $C\omega/qv_f$, and in this case the response extends over a frequency range on a scale measured by the fermi energy ϵ_f. For $U \neq 0$ we then have, from the expression (7.3.10) for $\chi^{-+}(\mathbf{q})$,

$$\operatorname{Im} \chi^{-+}(\mathbf{q}, \omega) = \frac{C\omega/qv_f}{\{1 - U\mathcal{N}(\epsilon_f)\}^2 + \{CU\omega/qv_f\}^2}. \tag{7.5.5}$$

Thus, as the instability point $U\mathcal{N}(\epsilon_f) = 1$ is approached, the inelastic scattering becomes strongly enhanced for small ω. Fig. 7.3 shows the

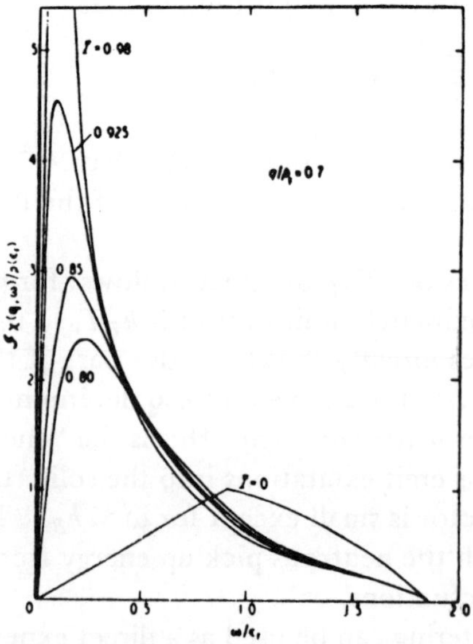

Fig. 7.3. Frequency variation of Im $\chi(\mathbf{q}, \omega)$ for fixed q. [$\bar{I} = U\mathcal{N}(\epsilon_f)$.] [After Doniach (1967).]

frequency variation in the energy loss regime at $T = 0$ for fixed q ($q/p_f = 0{\cdot}7$) and different values of $U\mathcal{N}(\epsilon_f)$ [Doniach (1967)].

The effect arises from a slowing down, because of the interaction, of the rate at which a spin turned over in a small region of the electron gas relaxes back to the equilibrium value. This enhancement of the density of low-lying excitations (so-called "paramagnons") in the system influences, apart from the neutron scattering, such physical properties as the electrical resistivity and the electronic specific heat (see below). These effects are thought to be significant in metals such as palladium and its alloys and also in the low-temperature properties of liquid He_3 (a fermi liquid with short-ranged repulsive interactions). (See the references listed at the end of the book.) An extensive investigation of the character of the magnetic excitations in metallic nickel has been conducted using slow neutron scattering by Lowde and Windsor (1970).

7.6. PARAMAGNON CONTRIBUTION TO LOW-TEMPERATURE SPECIFIC HEAT

Just as in the Coulomb gas case, the approximate form for the particle-hole response function can be used to generate an approximate formula for the free energy by integration over the coupling constant. The finite-temperature expression corresponding to the ground state energy (6.6.8) is

$$\Delta F = -\int_{-\infty}^{\infty} \frac{d\omega}{2\pi} \sum_{\mathbf{q}} n(\omega) \text{ Im } [\log \{1 - U\Gamma(\mathbf{q}, \omega)\}] \tag{7.6.1}$$

$$= \int_{-\infty}^{\infty} \frac{d\omega}{2\pi} \sum_{\mathbf{q}} n(\omega) \tan^{-1} \left[\frac{U \text{ Im } \Gamma(\mathbf{q}, \omega)}{1 - U \text{ Re } \Gamma(\mathbf{q}, \omega)} \right], \tag{7.6.2}$$

where $n(\omega) = 1/(e^{\beta\omega} - 1)$. For very small T, the integral is dominated by the $\omega \to 0$ part of the argument. In this region [where $k_B T \ll \{1 - U\mathcal{N}(\epsilon_f)\}\epsilon_f$] the inverse tangent varies, using (7.5.5), linearly with ω. Taking account only of this dependence, and differentiating once with respect to T, one has

$$\frac{\partial \Delta F}{\partial T} \simeq k_B \sum_{\mathbf{q}} \lim_{\omega \to 0} \frac{1}{\omega} \frac{U \text{ Im } \Gamma}{1 - U \text{ Re } \Gamma} \int_{-\infty}^{\infty} \frac{d\omega}{2\pi} (\beta\omega)^2 n(\omega) \{1 + n(\omega)\} \tag{7.6.3}$$

$$\propto k_B T \sum_{\mathbf{q}} \frac{(CU/qv_f)}{1 - U \operatorname{Re} \Gamma(\mathbf{q}, 0)}. \tag{7.6.4}$$

For the integral over $\mathbf{q}$ to converge one needs to take into account the q-dependence of $\operatorname{Re} \Gamma(\mathbf{q}, \omega = 0)$. This has the form

$$\operatorname{Re} \Gamma(\mathbf{q}, \omega = 0) \simeq \mathscr{N}(\epsilon_f)\{1 - \tfrac{1}{12}(q/p_f)^2\}. \tag{7.6.5}$$

Inserting this into (7.6.4) we have

$$\frac{\partial \Delta F}{\partial T} \propto k_B T \int_0^{12p_f^2} dq^2 \frac{U \mathscr{N}(\epsilon_f)}{\{1 - U \mathscr{N}(\epsilon_f)\} + \{U \mathscr{N}(\epsilon_f) q^2/12 p_f^2\}}, \tag{7.6.6}$$

where the upper limit on q is put in to take account of the fact that the approximate form for $\Gamma(\mathbf{q}, \omega)$ only holds for small q. Hence finally

$$\frac{\partial \Delta F}{\partial T} \propto k_B T \log \{1 - U \mathscr{N}(\epsilon_f)\}. \tag{7.6.7}$$

From the usual thermodynamic formula one then has for the electronic specific heat

$$C_v \propto k_B T \log \frac{1}{\{1 - U \mathscr{N}(\epsilon_f)\}}, \tag{7.6.8}$$

which is to be compared with

$$C_v \propto \gamma^\circ T \tag{7.6.9}$$

for the non-interacting electron gas. The effect of the approaching instability is thus to lead to a logarithmic singularity in the electronic effective mass.

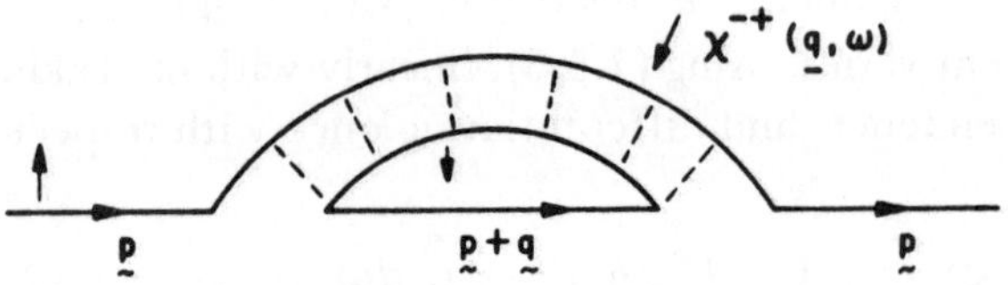

Fig. 7.4. Diagrams for electron self-energy from electron–paramagnon interaction.

An alternative way of deriving this result is to compute the electron self-energy arising from electron–paramagnon interaction (Fig. 7.4) and to use the relation

$$\frac{m^*}{m} = 1 - \left.\frac{\partial \Sigma\,(p_f, \epsilon)}{\partial \epsilon}\right|_{\epsilon=\epsilon_f}, \qquad (7.6.10)$$

which comes from examining the position of the quasiparticle pole in the general Dyson expression

$$G(\mathbf{p}, \epsilon) = \frac{1}{\epsilon - \epsilon_{\mathbf{p}} - \Sigma(\mathbf{p}, \epsilon)}. \qquad (7.6.11)$$

Near the instability point it may be shown that $\Sigma\,(\mathbf{p}, \epsilon)$ depends only weakly on $\mathbf{p}$ in the region of $p = p_f$, so that (7.6.10) gives a roughly momentum-independent value of the effective mass. This leads to the same logarithmic singularity as found in (7.6.8).

7.7. THE FERROMAGNETIC STATE

What happens to the electron gas as the interaction strength is increased beyond the instability point? It then turns out that a fundamental reorganization of the electron correlations in the ground state takes place so that the new excitation spectrum is now stable against external perturbations. For the ferromagnet this reorganization or "condensation" is a particularly simple one. More spins now occupy the fermi sea for one sign of the spin than for the other. In the place of the instability there now appears a new mode of excitation of the system which corresponds to the fact that the magnetization of the gas commutes with the total hamiltonian, hence in the $q = 0$ limit can assume arbitrary direction. For infinitesimal q it now costs very little (but finite) energy to excite a mode in which the magnetization oscillates slowly as a function of position. The existence of this mode whose frequency tends to zero as $q \to 0$ is a general result of the symmetry-breaking characteristics of the condensed ground state which was found by Goldstone (1961), so these modes are referred to as Goldstone modes. For the ferromagnet they are simply spin waves. In this section we discuss the transition to the ferromagnetic state and the existence of the spin wave modes for the metallic ferromagnet.

The ferromagnetic state is derived within the generalized Hartree-Fock scheme by simply assigning a spin-dependent expectation value to the number operators appearing in the equation of motion (7.3.6)

$$\langle n_{\mathbf{p}\sigma}\rangle = f_{\mathbf{p}\sigma}, \qquad \langle n_\sigma\rangle = \frac{1}{N}\sum_{\mathbf{p}} f_{\mathbf{p}\sigma}. \tag{7.7.1}$$

The single-particle self-energies (7.3.7) are now spin dependent:

$$\tilde{\epsilon}_{\mathbf{p}\sigma} = \epsilon_{\mathbf{p}} - U\langle n_{-\sigma}\rangle, \tag{7.7.2}$$

with a resulting *energy gap* for excitation from spin down to spin up state

$$\tilde{\epsilon}_{\mathbf{p}\uparrow} - \tilde{\epsilon}_{\mathbf{p}\downarrow} = U\langle n_\uparrow - n_\downarrow\rangle = \Delta. \tag{7.7.3}$$

The value of Δ is simply determined by the requirement that the chemical potential should be the same for both signs of the spin:

$$\langle n_\downarrow\rangle + \langle n_\uparrow\rangle = n, \tag{7.7.4}$$

where n is the number of electrons per atom, and

$$\langle n_\sigma\rangle = \frac{1}{N}\sum_{\mathbf{p}} \frac{1}{e^{\beta(\tilde{\epsilon}_{\mathbf{p}\sigma}-\mu)}+1}. \tag{7.7.5}$$

(7.7.3) and (7.7.4) together constitute a pair of coupled implicit equations to be solved self-consistently for Δ. For very small Δ they can be expanded to give

$$\left.\begin{aligned} \langle n_\uparrow\rangle &= \tfrac{1}{2}n + \frac{\Delta}{2N}\sum_{\mathbf{p}} \frac{\partial f}{\partial \epsilon_{\mathbf{p}}}, \\ \langle n_\downarrow\rangle &= \tfrac{1}{2}n - \frac{\Delta}{2N}\sum_{\mathbf{p}} \frac{\partial f}{\partial \epsilon_{\mathbf{p}}}, \end{aligned}\right\} \tag{7.7.6}$$

from which, using (7.7.3),

$$\Delta = U\{\Delta\,\mathcal{N}(\epsilon_f)\},$$

which gives the Stoner criterion

$$U\mathcal{N}(\epsilon_f) = 1 \tag{7.7.7}$$

for the point at which a ferromagnetic solution becomes possible. In the ferromagnetic regime the response function must be examined very carefully in the region $q \simeq 0$, $\omega \simeq 0$. For the *static* response at non-zero q we have

$$\Gamma^{-+}(\mathbf{q}, 0) = \frac{1}{N} \sum_{\mathbf{p}} \frac{f_{\mathbf{p}\uparrow} - f_{\mathbf{p}+\mathbf{q}\downarrow}}{\epsilon_{\mathbf{p}+\mathbf{q}} - \epsilon_{\mathbf{p}} + \Delta}. \tag{7.7.8}$$

In the limit $q \to 0$ this becomes

$$\Gamma^{-+}(0, 0) = \frac{\langle n_\uparrow \rangle - \langle n_\downarrow \rangle}{\Delta} = \frac{1}{U}. \tag{7.7.9}$$

Thus $\chi^{-+}(\mathbf{q}, \omega = 0)$ still looks singular in the limit $q = 0$. However, closer examination shows that, for $\omega \neq 0$,

$$\lim_{q \to 0} \Gamma^{-+}(\mathbf{q}, \omega) = \lim_{q \to 0} \frac{1}{N} \sum_{\mathbf{p}} \frac{f_{\mathbf{p}\uparrow} - f_{\mathbf{p}+\mathbf{q}\downarrow}}{\omega + (\epsilon_{\mathbf{p}+\mathbf{q}} - \epsilon_{\mathbf{p}} + \Delta)}$$

$$= \frac{\Delta/U}{\omega + \Delta}, \tag{7.7.10}$$

so that the full response function for $\omega \neq 0$ is given in the limit $q \to 0$ by

$$\lim_{q \to 0} \chi^{-+}(\mathbf{q}, \omega) = \frac{\Delta/U}{\omega}. \tag{7.7.11}$$

Thus it may be seen that the instability has actually been converted into a dynamic frequency-dependent response with a zero-frequency pole. Thermodynamically it may also be shown that the specific heat is no longer divergent as it is at the instability point, as discussed in Sec. 7.6, but that the spin-wave modes now give a well-defined contribution to the free energy analogous to that in the insulating ferromagnet (see Chap. 8).

At finite q the denominator of $\chi^{-+}(\mathbf{q}, \omega)$ may now be seen to have two types of singularity, somewhat like those occurring in the Coulomb problem discussed in Chap. 6. For low frequencies $\omega \ll \Delta$ the spin-wave branch acquires a q-dependence of the form (for small q)

$$\omega = Dq^2,$$

while at higher frequencies there is a spectrum of singularities along a particle-hole cut in the ω-plane, referred to as Stoner excitations. The spectrum is then as sketched in Fig. 7.5. At a critical value of q the spin-wave branch merges into the Stoner continuum, although it still retains a resonant character for some distance inside the continuum.

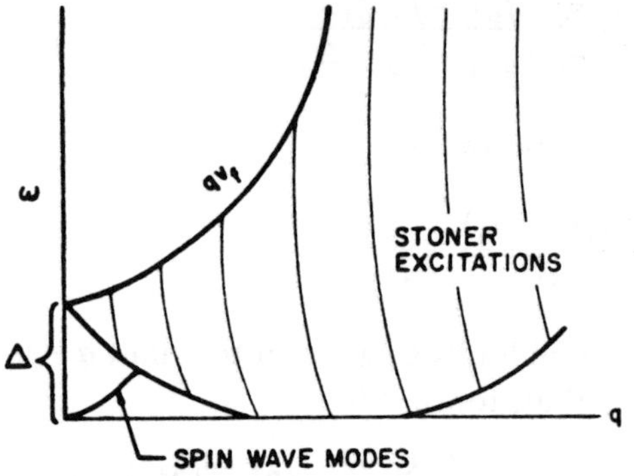

Fig. 7.5. Spectrum of excitation energies of a ferromagnet.

7.8. LOCALIZED MAGNETIC STATES IN METALS

The magnetic instabilities discussed in earlier sections occur principally in metals with narrow bands (d-bands: transition metals and compounds; f-bands: rare earth and actinide metals and compounds). When atoms of these metals are dissolved in nonmagnetic hosts to form alloys they still retain many features of magnetism—the simplest being a temperature-dependent paramagnetic susceptibility following a Curie or Curie–Weiss law. Since such a strongly temperature-dependent property could not be easily explained from a one-electron model (even with resonant scattering effects as discussed in Chap. 4), it is clear that electron–electron interactions must be playing an essential role.

In order to discuss the formation of localized magnetic states, Anderson (1961) proposed a model hamiltonian which turns out to be very closely related to the Hubbard hamiltonian of Sec. 7.1 in the case of the extended system. Two essential features of the extended system have their counterparts in the localized system: (a) the existence of a localized one-electron state (analogous to a narrow band in the extended case) which is connected with the fermi sea of conduction electrons. In Anderson's model this is described by assigning an extra d-like atomic orbital to the impurity atom with matrix elements for transfer of an electron to the conduction band of s-electrons (in states described by a wave-vector $\mathbf{k}$). Alternative descriptions [the Wolff model, Wolff (1961)] replace the d-orbital by a resonant scattering state as discussed in Chap. 4. This part of the hamiltonian becomes

$$H_0 = \sum_{\mathbf{k}} \epsilon_{\mathbf{k}} a_{\mathbf{k}}{}^{\dagger} a_{\mathbf{k}} + \epsilon_d a_d{}^{\dagger} a_d + \sum_{\mathbf{k}} V_{\mathbf{k}d}(a_{\mathbf{k}}{}^{\dagger} a_d + a_d{}^{\dagger} a_{\mathbf{k}}). \tag{7.8.1}$$

(b) The electron–electron interaction which is now considered localized to the electrons residing in the localized d-orbital. As in Sec. 7.1 the Pauli principle leads to repulsion only between electrons of opposite spin:

$$H_1 = U n_{d\uparrow} n_{d\downarrow}, \tag{7.8.2}$$

where

$$n_{d\sigma} = a_{d\sigma}^{\dagger} a_{d\sigma}. \tag{7.8.3}$$

It is assumed that all s-electron creation and annihilation operators anticommute with all d-electron operators. In the *absence* of interactions ($U = 0$), the one-electron Green's function may be obtained in closed form, as the scattering integral equation is separable just as in Sec. 4.4. The equation of motion leads to a pair of coupled equations for $G(\mathbf{k}, \mathbf{k}')$, the band Green's function, and $G(dd)$, the localized Green's function defined (as a retarded Green's function) by

$$G^{\sigma}(dd; t) = -i\theta(t)\langle[\tilde{a}_{d\sigma}(t), \tilde{a}_{d\sigma}^{\dagger}(0)]\rangle, \tag{7.8.4}$$

where $\tilde{a}(t)$ is a Heisenberg operator. The equation of motion is

$$i\frac{\partial G^{\sigma}(dd)}{\partial t} = \delta(t) - i\theta(t)\langle[[\tilde{a}_{d\sigma}(t), H], \tilde{a}_{d\sigma}^{\dagger}(0)]\rangle. \tag{7.8.5}$$

The commutator in (7.8.5) couples $G(dd)$ to a mixed propagator

$$G^{\sigma}(\mathbf{k}d; t) = -i\theta(t)\langle[\tilde{a}_{\mathbf{k}\sigma}(t), \tilde{a}_{d\sigma}^{\dagger}(0)]\rangle \tag{7.8.6}$$

via the equation

$$i\frac{\partial G^{\sigma}(dd)}{\partial t} = \delta(t) + \epsilon_d G^{\sigma}(dd) + \sum_{\mathbf{k}} V_{\mathbf{k}d} G^{\sigma}(\mathbf{k}d; t). \tag{7.8.7}$$

Fourier transforming to frequency variables we find in the same way that the equation of motion for $G^\sigma(\mathbf{k}d;\omega)$ is

$$(\omega-\epsilon_{\mathbf{k}})G^\sigma(\mathbf{k}d;\omega)-V_{d\mathbf{k}}G^\sigma(dd;\omega)=0. \tag{7.8.8}$$

Eliminating $G(\mathbf{k}d)$ we finally have

$$G(dd;\omega)=\frac{1}{\omega-\epsilon_d-\sum(d,\omega)} \tag{7.8.9}$$

(independent of spin), where

$$\sum(d,\omega)=\sum_{\mathbf{k}}\frac{|V_{\mathbf{k}d}|^2}{\omega-\epsilon_{\mathbf{k}}+i\eta}, \tag{7.8.10}$$

and the $i\eta$ has been inserted to make G a retarded function. The function $G(dd)$ behaves in many ways like the localized Green's function defined in the potential scattering problem, Eq. (4.4.3). If we write Σ in terms of its real and imaginary parts,

$$\begin{aligned}\Sigma(d,\omega)&=\sum_{\mathbf{k}}\mathscr{P}\frac{|V_{\mathbf{k}d}|^2}{\omega-\epsilon_{\mathbf{k}}}-i\pi\sum_{\mathbf{k}}|V_{\mathbf{k}d}|^2\,\delta(\omega-\epsilon_{\mathbf{k}})\\&=\Lambda(d,\omega)-i\Delta(d,\omega),\end{aligned} \tag{7.8.11}$$

we find, if we neglect the detailed $\mathbf{k}$ dependence of $V_{\mathbf{k}d}$ (equivalent to saying this is a short-ranged overlap matrix element),

$$\Delta(d,\omega)=\pi|V|^2\mathscr{N}(\epsilon), \tag{7.8.12}$$

where $\mathscr{N}(\epsilon)$ is the band density of states. Thus we can define a localized density of states as

$$\rho_d(\omega)=-\frac{1}{\pi}\operatorname{Im}G(dd;\omega)=\frac{1}{\pi}\frac{\Delta(d,\omega)}{\{\omega-\epsilon_d-\Lambda(d,\omega)\}^2+\Delta^2(d,\omega)}, \tag{7.8.13}$$

which is essentially analogous to the imaginary part of the T matrix defined in (4.4.9). The main difference is that the position of the resonance is dominated by the position of the localized orbital ϵ_d, rather than by the potential binding condition $UF(\epsilon)=1$ of Eq. (4.4.11). In this non-interacting limit ($U=0$) there is no magnetism, and the system has a temperature-independent Pauli susceptibility.

If we now include the interaction term H_1, Eq. (7.8.2), we again find enhancement of the spin fluctuation propagator as U is increased, as in the bulk ferromagnet. This may be seen by defining a localized susceptibility propagator

$$\chi^{-+}(dd;t) = i\theta(t)\langle[\sigma_d^-(t), \sigma_d^+(0)]\rangle, \qquad (7.8.14)$$

where $\sigma_d^+ = a_{d\uparrow}^\dagger a_{d\downarrow}$ as in the bulk case. To calculate $\chi^{-+}(dd;\omega)$ in the generalized Hartree–Fock approximation scheme is now a bit more complicated owing to the nature of the coupled equations. The simplest way to proceed is to sum ladder diagrams for the interaction of d-state electrons with d-state holes. To do this we notice that the solution (7.8.9) for $G(dd)$ is the result of iterating

$$G(dd;\omega) = G^0(dd;\omega) + \sum_{\mathbf{k}} G^0(dd)V_{d\mathbf{k}}G^0(\mathbf{k})V_{\mathbf{k}d}G^0(dd) + \cdots, \qquad (7.8.15)$$

which may be denoted by a composite diagram

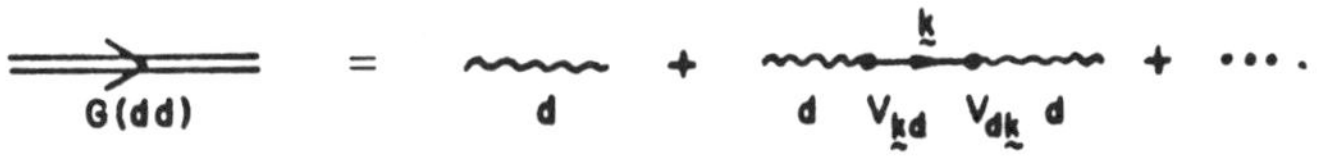

The interaction vertex now only couples a pair of d-lines so that we are finally led to

$$\chi^{-+}(dd;\omega) = \frac{\Gamma(d,\omega)}{1 - U\Gamma(d,\omega)}, \qquad (7.8.16)$$

where $\Gamma(d,\omega)$ is the unperturbed local susceptibility function

$$\Gamma(d,\omega) = \frac{1}{2\pi}\int_{-\infty}^{\infty} d\epsilon G(dd;\omega+\epsilon)G(dd;\epsilon). \qquad (7.8.17)$$

Again, within this approximation scheme, we see that there may occur an instability, for static response, when

$$U\Gamma(d,0) = 1. \qquad (7.8.18)$$

However, there is now an important distinction from the case of the extended ferromagnet—the instability of (7.8.18) in fact only concerns $1/N$ of the degrees of freedom of the system, so, it may be argued, cannot lead to a qualitative change in the ground state wave function of the metal. This suggests that the ladder diagram approximation scheme is in fact breaking down rather badly in the instability region. Despite this it appears to have some qualitative merit for describing the magnetic state of the impurity, and Anderson proceeded to use a Hartree–Fock approximation scheme to discuss the conditions for the appearance of a local moment.

Recent studies using a functional integral formalism (Schrieffer, Hamann) suggest that Anderson's scheme does incorporate some of the correct features of the model, although leaving out some of the dynamical properties of the local spin associated with the precession (spin-flip) of the local moment. The latter process itself leads to a remarkable physical phenomenon, the Kondo effect, which is most simply described in the localized moment limit (see Sec. 9.4 below).

We proceed to a description of Anderson's Hartree–Fock solution. Returning to $G(dd;t)$ the equation of motion including H_1 becomes

$$i\frac{\partial G^{\sigma}(dd)}{\partial t} = \delta(t) + \epsilon_d G^{\sigma}(dd;t) + U\Gamma^{\sigma}(dd;t) + \sum_{\mathbf{k}} V_{\mathbf{k}d} G^{\sigma}(\mathbf{k}d;t), \tag{7.8.19}$$

where

$$\Gamma^{\sigma}(dd;t) = -i\theta(t)\langle[\tilde{n}_{d,-\sigma}(t)\tilde{a}_{d\sigma}(t), \tilde{a}^{+}_{d\sigma}(0)]\rangle. \tag{7.8.20}$$

If we also write down the equation of motion for $\Gamma^{\sigma}(dd)$, we find that this involves three additional higher order Green's functions, and the system of equations of motion does not close up.

To solve the problem in the Hartree–Fock approximation, we approximate to $\Gamma^{\sigma}(dd)$ by replacing the operator $\tilde{n}_{d,-\sigma}$ by its expectation value $\langle n_{d,-\sigma}\rangle$. Then

$$\Gamma^{\sigma}(dd;\omega) = \langle n_{d,-\sigma}\rangle G^{\sigma}(dd;\omega). \tag{7.8.21}$$

With this approximation (7.8.19) becomes formally identical with the equation of motion (7.8.7) in the non-interacting case, except that ϵ_d is now modified to

$$\tilde{\epsilon}_{d\sigma} = \epsilon_d + U\langle n_{d,-\sigma}\rangle, \tag{7.8.22}$$

just as in the bulk ferromagnet problem.

The solution (7.8.9) thus becomes modified to a spin-dependent result

$$G^{\sigma}(dd;\omega) = \frac{1}{\omega - \tilde{\epsilon}_{d\sigma} - \sum(d,\omega)}. \tag{7.8.23}$$

We can now determine $\langle n_{d\sigma}\rangle$, the number of d-electrons of spin σ, by writing down the self-consistency condition from the definition of $G(dd; t)$ as

$$\langle n_{d\sigma}\rangle = \int_{-\infty}^{\epsilon_f} \rho_d{}^{\sigma}(\omega)\, d\omega = \frac{1}{\pi} \cot^{-1} \frac{\epsilon_d - \epsilon_f + U\langle n_{d,-\sigma}\rangle}{\Delta}. \tag{7.8.24}$$

This is equivalent to the two simultaneous equations

$$\langle n_{d\uparrow}\rangle = \frac{1}{\pi} \cot^{-1} \frac{\epsilon_d - \epsilon_f + U\langle n_{d\downarrow}\rangle}{\Delta}, \tag{7.8.25}$$

$$\langle n_{d\downarrow}\rangle = \frac{1}{\pi} \cot^{-1} \frac{\epsilon_d - \epsilon_f + U\langle n_{d\uparrow}\rangle}{\Delta}, \tag{7.8.26}$$

which are to be solved for the populations $\langle n_{d\uparrow}\rangle$ and $\langle n_{d\downarrow}\rangle$.

To study the solutions of these equations we introduce the parameters

$$x = \frac{\epsilon_f - \epsilon_d}{U}, \qquad y = \frac{U}{\Delta}, \tag{7.8.27}$$

which are dimensionless measures of the position of the d-level relative to the fermi level and of the strength of the d–d interaction. Writing also $\langle n_{d\uparrow}\rangle = n_1$, $\langle n_{d\downarrow}\rangle = n_2$, we thus have the equations

$$\cot \pi n_1 - y(n_2 - x) = 0, \tag{7.8.28}$$

$$\cot \pi n_2 - y(n_1 - x) = 0. \tag{7.8.29}$$

These can be solved graphically by looking for the intersections of the two curves obtained on plotting n_1 against n_2.

The *non-magnetic* solutions are such that $n_1 = n_2 = n$. These are obtained from the single equation

$$f(n) \equiv \cot \pi n - y(n - x) = 0. \tag{7.8.30}$$

This always has a unique root in the range $0 \leqslant n \leqslant 1$, corresponding to uniform partial occupation of the two d-levels. However, for suitable values of the parameters, (7.8.28) and (7.8.29) also have *magnetic* solutions for which $n_1 \neq n_2$. In particular, if y is large and $0 < x < 1$, the equations are satisfied approximately by

$$n_1 = 1 - \frac{1}{\pi x y}, \qquad n_2 = \frac{1}{\pi(1 - x)y}, \tag{7.8.31}$$

so that one d-level is nearly empty and the other is nearly full. (Since the formalism is symmetric in n_1 and n_2, each magnetic solution is accompanied by a second solution which has n_1 and n_2 interchanged.) It can be shown, from the variational formulation of the Hartree–Fock equations, that the magnetic solutions correspond to minima in the energy and are thus the stable ones energetically. It is the dynamics of spin-flip between the pair of degenerate solutions which leads to the startling Kondo effect. However, it is not possible to discuss this within the simple Hartree–Fock solution set up above.

The *transition curve* from magnetic to nonmagnetic behavior may be obtained from (7.8.28) and (7.8.29) as the relation between the parameters x and y for which the magnetic solutions coalesce into the single solution $n_1 = n_2 = n$. The conditions for this are $f(n) = 0$ and $f'(n) = 0$, where $f(n)$ is given by (7.8.30), i.e.,

$$\cot \pi n = y(n - x), \qquad \frac{\pi}{\sin^2 \pi n} = y,$$

which may be written

$$x = \frac{1}{2\pi}(\theta - \sin\theta), \qquad \frac{\pi}{y} = \tfrac{1}{2}(1 - \cos\theta), \tag{7.8.32}$$

where $\theta = 2\pi n$. If x is plotted against π/y for $0 \leqslant n \leqslant 1$, we obtain the cycloidal curve shown in Fig. 7.6. It may be seen from this that the tendency towards magnetism is strongest when the d–d interaction is large and when x is close to $\frac{1}{2}$, i.e., when the d-levels ϵ_d and $\epsilon_d + U$ are symmetrically disposed about the fermi level.

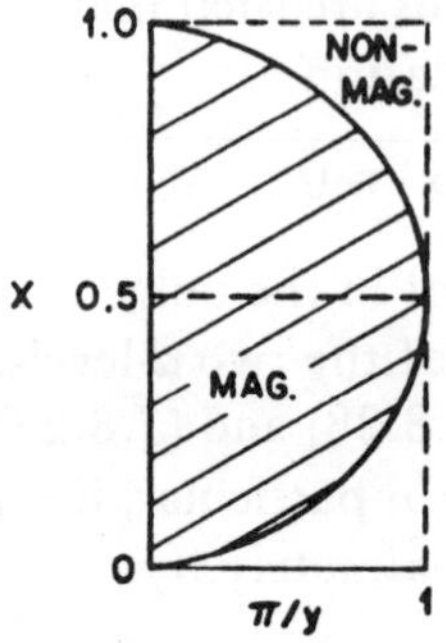

Fig. 7.6. Magnetic and nonmagnetic regions for localized states in metals. [$x = (\epsilon_f - \epsilon_d)/U$, $y = U/\Delta$.]

Figure 7.7 shows the observed magnetic moment (obtained by fitting the measured susceptibility to a Curie–Weiss law) of iron dissolved in various transition metals. The appearance and disappearance of a finite magnetic moment can be interpreted [Clogston *et al.* (1962)] in terms of changes in the position and width of the impurity level as the electron concentration is varied.

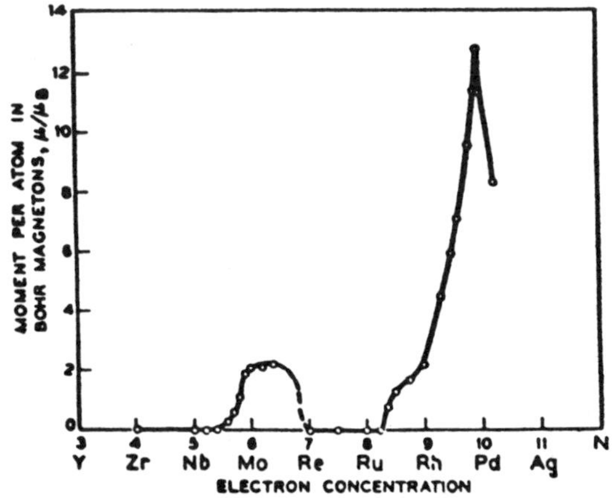

Fig. 7.7. Magnetic moment of an iron atom dissolved in second row transition metals and alloys, as a function of electron concentration. [After Clogston *et al.* (1962).]

Chapter 8

Interacting Electrons in the Atomic Limit

In Chap. 7 we noted that the existence of narrow bands or localized states for electrons in solids tends to promote instabilities in the interacting electron gas which lead to magnetic states. As the ratio of the band width to the strength of the Coulomb interaction parameter U continues to decrease we come to a regime where the generalized Hartree–Fock description we used there ceases to have much usefulness, even at a qualitative level. A good example of this is in solids containing rare earth metals. These contain very tightly bound f-shells which overlap very little with neighboring ions. If band theory is used for such materials, it gives bandwidths of the order of 10^{-2} eV. The concepts of the band model, which starts from a free-electron gas into which correlations are introduced, are then virtually meaningless; the physics is dominated by the Coulomb interactions, and the electrons are to a high degree localized on particular atoms.

In this limit it is more sensible to start from atomic wave functions into which the Coulomb interactions are introduced by the usual Hund's rules and to discuss the effects of atomic overlap as a weak perturbation. This transition from band limit to atomic limit was discussed quantum-mechanically by Hubbard in 1963. One of the characteristic phenomena which occur as U is increased is an abrupt disappearance of metallic nature at a critical value of the bandwidth—the so-called "metal-insulator transition." This transition, whose physics was first discussed by Mott (1949, 1961), occurs at the stage when the reduction of kinetic energy obtained by allowing electrons to tunnel from one atom to another is balanced by the Coulomb repulsion which occurs when two electrons occupy the same orbital of a single atom. For larger values of U the band approximation, which predicts that a substance with a partially filled band has metallic conductivity, is inapplicable, and the substance has the insulating properties predicted in the atomic limit. In this chapter we discuss this transition to the atomic limit, and we then go on to study the spin wave excitations in the resulting insulating magnetic system (the Heisenberg model).

8.1. THE ATOMIC LIMIT OF THE HUBBARD MODEL

In order to emphasize the atomic limit of the Hubbard model (Sec. 7.1) we rewrite his hamiltonian in the form

$$\left.\begin{aligned} H_{00} &= \sum_{i\sigma} t_0 n_{i\sigma}, \\ H_0 &= \sum_{\substack{i\neq j\\ \sigma}} t_{ij} a_{i\sigma}^\dagger a_{j\sigma}, \end{aligned}\right\} \tag{8.1.1}$$

$$H_U = U \sum_i n_{i\uparrow} n_{i\downarrow}, \tag{8.1.2}$$

where the a_i are the Wannier operators defined in (7.1.5). (8.1.1) is equivalent to (7.1.8) for the case of a tight-binding electron band for which one has, on diagonalizing (8.1.1),

$$\epsilon_{\mathbf{k}} = t_0 + t_{\mathbf{k}},$$

with

$$t_{\mathbf{k}} = \sum_{\alpha} t_{i,i+\alpha}\, e^{i\mathbf{k}\cdot\mathbf{X}_\alpha}, \tag{8.1.3}$$

where $\mathbf{X}_\alpha$ is a nearest-neighbor vector. The atomic limit now simply corresponds to letting $t_{ij} = 0$ in (8.1.1), so that we now have to solve

$$H_{\text{atomic}} = H_0 + H_U = \sum_i (t_0 n_i + U n_{i\uparrow} n_{i\downarrow}). \tag{8.1.4}$$

We see immediately that there are three possible configurations for each atom: $|0\rangle$ (vacuum); $a_i^\dagger|0\rangle$ and $a_{i\uparrow}^\dagger a_{i\downarrow}^\dagger|0\rangle$, with energies 0, t_0 and $t_0 + U$. Thus, if we have N_e electrons in the system of N atoms, we can form product wave functions in which N_0 atoms are unoccupied, N_1 have one electron and $\frac{1}{2}(N_e - N_1)$ have two electrons.

Suppose we now insert an extra electron into the system. The way this electron "propagates" (i.e., the time dependence of the resulting wave function) will depend on whether it lands on an empty site or on a singly occupied site (it cannot land on a doubly occupied one). This is a matter of chance so that, on the average, a test electron placed in the system will propagate in a mixture of states. If the system is in the lowest state and $n = N_e/N < 1$, there will be a fraction $\frac{1}{2}n$ of sites occupied with a given spin, and a fraction $(1 - \frac{1}{2}n)$ sites unoccupied by electrons with this spin. This mixed behavior results if we examine the electron Green's function, for which we choose the retarded form:

$$G^0(t - t') = -i\theta(t - t')\langle\{\tilde{a}_i(t), \tilde{a}_i{}^\dagger(t')\}\rangle, \tag{8.1.5}$$

where $\{\ldots\}$ denotes the anticommutator. Suppose the ground state was that of a single atom; then if it was unoccupied we would have

$$\begin{aligned} G_{\text{empty}}(t - t') &= -i\theta(t - t')\langle 0|\{\tilde{a}_i(t), \tilde{a}_i{}^\dagger(t')\}|0\rangle \\ &= -i\theta(t - t') \exp\{-it_0(t - t')\}, \end{aligned} \tag{8.1.6}$$

which has a Fourier transform

$$G_{\text{empty}}(\epsilon) = \frac{1}{\epsilon - t_0 + i\eta}. \tag{8.1.7}$$

Similarly, for an occupied ground state $a^\dagger_{i,-\sigma}|0\rangle$ we have

$$\begin{aligned} G^{\sigma\sigma'}_{\text{full}}(t - t') &= -i\theta(t - t')\langle 0|a_{i,-\sigma'}\{\tilde{a}_{i\sigma}(t), \tilde{a}^\dagger_{i\sigma}(t')\}a^\dagger_{i,-\sigma'}|0\rangle \\ &= -\delta_{\sigma\sigma'}i\theta(t - t') \exp\{-i(t_0 + U)(t - t')\}; \end{aligned}$$

therefore

$$G^{\sigma\sigma'}_{\text{full}}(\epsilon) = \frac{\delta_{\sigma\sigma'}}{\epsilon - (t_0 + U) + i\eta}. \tag{8.1.8}$$

Thus for the complete ground state of N_e electrons, we expect a test electron to propagate as a mixture of G_{empty} and G_{full} if we average over the possible configurations of occupied atomic states. (This averaging will result below from switching on t_{ij}, so allowing electrons to tunnel through and make the ground state into a mixture of different atomic configurations.)

The above averaging results automatically if we look at the equation of motion of G^0. We have

$$-i\frac{\partial G_\sigma{}^0(t)}{\partial t} = -\delta(t)\langle\{a_{i\sigma}, a^\dagger_{i\sigma}\}\rangle + i\theta(t)\langle\{[\tilde{a}_{i\sigma}(t), H], \tilde{a}^\dagger_{i\sigma}(0)\}\rangle. \tag{8.1.9}$$

Now use

$$\left.\begin{aligned} [a_{i\sigma}, H_{00}] &= t_0 a_{i\sigma}, \\ [a_{i\sigma}, H_U] &= U n_{i,-\sigma} a_{i\sigma}, \end{aligned}\right\} \tag{8.1.10}$$

so that

$$\left(-i\frac{\partial}{\partial t}+t_0\right)G_\sigma{}^0(t)=-\delta(t)-U\Gamma^0_{i\sigma}(t), \tag{8.1.11}$$

where

$$\Gamma^0_{i\sigma}(t)=-i\theta(t)\langle\{\tilde{n}_{i,-\sigma}(t)\tilde{a}_{i\sigma}(t),a^\dagger_{i\sigma}(0)\}\rangle. \tag{8.1.12}$$

From this we find a Fourier-transformed equation

$$(\epsilon-t_0)G_\sigma{}^0(\epsilon)=1+U\Gamma^0_{i\sigma}(\epsilon). \tag{8.1.13}$$

On differentiating $\Gamma(t)$ we similarly find an equation involving $n^2_{i,-\sigma}$. However, in contrast to the procedure used in the generalized Hartree-Fock approximation, we can now use the important identity

$$n^2_{i,-\sigma}=n_{i,-\sigma}$$

to break the chain of equations of motion and allow an exact solution in this limit:

$$(\epsilon-t_0)\Gamma=\langle n_{i,-\sigma}\rangle+U\Gamma, \tag{8.1.14}$$

leading to

$$\Gamma^0_{i\sigma}=\frac{\langle n_{-\sigma}\rangle}{\epsilon-(t_0+U)} \tag{8.1.15}$$

(assuming the ground state is spatially invariant), and hence from (8.1.13)

$$\begin{aligned}G_\sigma{}^0(\epsilon)&=G_{\text{empty}}+G_{\text{empty}}U\Gamma\\&=\frac{1}{\epsilon-t_0}+\frac{1}{\epsilon-t_0}U\frac{\langle n_{-\sigma}\rangle}{\epsilon-(t_0+U)}.\end{aligned} \tag{8.1.16}$$

We can rewrite $G_\sigma{}^0$, using $\langle n_{-\sigma}\rangle=\frac{1}{2}n$ for the ground state:

$$G_\sigma{}^0(\epsilon)=\frac{1-\frac{1}{2}n}{\epsilon-t_0}+\frac{\frac{1}{2}n}{\epsilon-(t_0+U)}, \tag{8.1.17}$$

so that G^0 represents a mixed amplitude for the electron to propagate either in empty or occupied states.

8.2. THE TRANSITION BETWEEN THE ATOMIC LIMIT AND THE BAND LIMIT

We now want to discuss what happens as the atoms are brought closer together and the overlap between atoms becomes significant. We will do this by expanding in a power series in t_{ij}. Let us consider first the case $U = 0$, i.e., the usual tight-binding model of band theory:

$$H_{\mathrm{TB}} = H_{00} + H_0. \tag{8.2.1}$$

The Green's function is now

$$G_{ij}^{\sigma}(t - t') = -i\theta(t - t')\langle\{\tilde{a}_{i\sigma}(t), \tilde{a}_{j\sigma}^{\dagger}(t')\}\rangle, \tag{8.2.2}$$

and satisfies the equation of motion

$$(\epsilon - t_0)G_{ij} = \delta_{ij} + \sum_{l \neq i} t_{il}G_{lj}. \tag{8.2.3}$$

For a periodic system the equation is solved by going to k-space and writing

$$G_{ij}(\epsilon) = \frac{1}{N}\sum_{\mathbf{k}} e^{i\mathbf{k}\cdot(\mathbf{x}_i - \mathbf{x}_j)}G_{\mathbf{k}}(\epsilon) \tag{8.2.4}$$

(where the sum over **k** is confined to the first Brillouin zone), to give

$$G_{\mathbf{k}}(\epsilon) = \frac{1}{\epsilon - t_0 - t_{\mathbf{k}} + i\eta} = \frac{1}{\epsilon - \epsilon_{\mathbf{k}} + i\eta}, \tag{8.2.5}$$

where $\epsilon_{\mathbf{k}}$ is given in (8.1.3). Eq. (8.2.3) generates a perturbation series on iterating in t_{ij}, given by

$$G_{ij} = G_{\mathrm{empty}} + G_{\mathrm{empty}}t_{ij}G_{\mathrm{empty}}$$
$$+ G_{\mathrm{empty}}\sum_{l} t_{il}G_{\mathrm{empty}}t_{lj}G_{\mathrm{empty}} + \cdots, \tag{8.2.6}$$

which we may represent by diagrams:

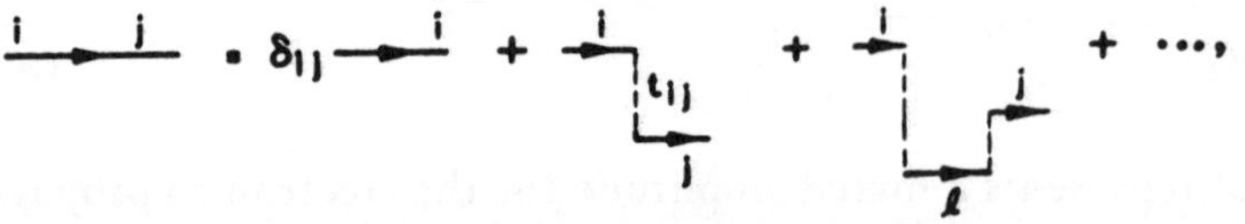

where the dotted lines represent the interatomic tunneling and $G_{\mathrm{empty}}(\epsilon)$ is the propagator for the empty atom, $1/(\epsilon - t_0 + i\eta)$. It may be seen that this is the usual one-particle Green's function for a band electron. Hubbard now combines features of both the limiting forms (8.2.5) and (8.1.17) by a decoupling procedure for the equation of motion of the full hamiltonian

$$H = H_{00} + H_0 + H_U$$

which turns out to be equivalent to a generalization of Eq. (8.2.6) in which G_{empty} is replaced by the *atomic limit* solution $G_\sigma{}^0$, i.e., we replace the equation of motion (8.2.3) by

$$G_{ij}^\sigma(\epsilon) = G_\sigma{}^0(\epsilon)\delta_{ij} + G_\sigma{}^0(\epsilon) \sum_l t_{il} G_{lj}^\sigma. \tag{8.2.7}$$

On Fourier-transforming to k space (8.2.7) becomes

$$G_{\mathbf{k}}{}^\sigma(\epsilon) = G_\sigma{}^0 + G_\sigma{}^0 t_{\mathbf{k}} G_{\mathbf{k}}{}^\sigma(\epsilon); \tag{8.2.8}$$

therefore

$$G_{\mathbf{k}}{}^\sigma(\epsilon) = \frac{1}{\{G_\sigma{}^0(\epsilon)\}^{-1} - t_{\mathbf{k}}}, \tag{8.2.9}$$

where $G_\sigma{}^0(\epsilon)$ is given in (8.1.17). Substituting this we have finally

$$G_{\mathbf{k}}{}^\sigma(\epsilon) = \frac{1}{(\epsilon - t_0)\{[\epsilon - t_0 - U]/[\epsilon - t_0 - U(1 - \frac{1}{2}n)]\} - t_{\mathbf{k}}}. \tag{8.2.10}$$

This function has two discrete poles, which in the limit $t_{\mathbf{k}} \to 0$ just reduce to the energies t_0, $t_0 + U$ of the atomic states. For finite $t_{\mathbf{k}}$ they are given as the solutions of the quadratic equation

$$(\epsilon - t_0)(\epsilon - t_0 - U) = t_{\mathbf{k}}[\epsilon - t_0 - U(1 - \tfrac{1}{2}n)]. \tag{8.2.11}$$

This indicates that the single band $\epsilon_{\mathbf{k}}$ (for $U = 0$) is split into two sub-bands, one centered around energy t_0, the other around $t_0 + U$; the Coulomb-generated energy gap between the sub-bands increases as U increases relative to the bandwidth $\Delta = t_{\max}$. However, the above approximate solution contains many deficiencies (for example, the band-gap is always non-zero for $U \neq 0$, so that in this approximation there is in fact no Mott transition between a band-like and an insulating state as U increases), and the subject is still under active theoretical study. [See the reviews by Adler (1968), Mott (1969), and Doniach (1969).]

8.3. THE INSULATING MAGNET

In the atomic limit, with $n = 1$, each atom is occupied by one electron, and the only remaining variable in the problem is the electron spin **S**.

(For multiorbital atoms this is a multielectron spin variable determined by Hund's rule.) As the overlap operator t_{ij} is switched on slowly the system remains insulating, but the spins become coupled together. The simplest expression for this coupling is the Heisenberg hamiltonian which, in the absence of any applied magnetic field, may be written

$$H = -\tfrac{1}{2}J \sum_{ij} \gamma_{ij} \mathbf{S}_i \cdot \mathbf{S}_j, \tag{8.3.1}$$

where $\mathbf{S}_i$ is the vector spin operator for the atom on site i. We assume nearest-neighbor interactions, so that $\gamma_{ij} = 1$ when i and j are nearest neighbors and is zero otherwise. In fact the use of the Hubbard hamiltonian allows for a derivation of (8.3.1) which is essentially equivalent to a highly simplified version of Anderson's (1950) theory of superexchange. In the simple model J comes out to be negative (antiferromagnetic) and with order of magnitude

$$J = -t^2/U. \tag{8.3.2}$$

We will not discuss further here the foundations of the Heisenberg model [see the review by Anderson (1963)], but go on to a discussion of the excitation spectrum and phase transition in (8.3.1). For simplicity we will confine ourselves to positive J, which leads to a ferromagnetic state.

We shall study the properties of this spin system by a method due to Bogoliubov and Tyablikov (1959) which uses approximate equations of motion. This method emphasizes the close analogy between the present problem and the system of coupled oscillators discussed in Chaps. 1–3. We suppose that S is the total spin of each atom, and classify the spin state of the ith atom by the eigenstates of S_i^z, the z-component of the spin. Thus

$$S_i^z|m\rangle_i = m|m\rangle_i, \qquad -S \leqslant m \leqslant S, \tag{8.3.3}$$

for a state with spin component m in the z-direction. It may be shown that, for $J > 0$, the ground state of the hamiltonian (8.3.1) is the totally aligned state

$$|\Psi_G\rangle = |S\rangle_1|S\rangle_2 \ldots |S\rangle_N, \tag{8.3.4}$$

in which S^z has its maximum value S at each lattice site.

It is easy to show that this state is an exact eigenstate of H. We define the *spin deviation operators* [analogous to $\sigma^{\pm}$ defined by (7.2.12)]

$$S_i^+ = S_i^x + iS_i^y, \qquad S_i^- = S_i^x - iS_i^y. \tag{8.3.5}$$

These are such that, for any lattice site,

$$S^+|m\rangle = |m+1\rangle, \qquad S^-|m\rangle = |m-1\rangle, \tag{8.3.6}$$

together with

$$S^+|S\rangle = 0, \qquad S^-|-S\rangle = 0, \tag{8.3.7}$$

and the hamiltonian (8.3.1) may be written

$$H = -\tfrac{1}{2}J \sum_{ij} \gamma_{ij}\{S_i^z S_j^z + \tfrac{1}{2}(S_i^+ S_j^- + S_i^- S_j^+)\}. \tag{8.3.8}$$

Clearly, when H operates on $|\Psi_G\rangle$, only the z-components of spin contribute, and

$$H|\Psi_G\rangle = -\tfrac{1}{2}J \sum_{ij} \gamma_{ij} S^2 |\Psi_G\rangle; \tag{8.3.9}$$

the ordered state (8.3.4) is thus an eigenstate of the hamiltonian. Other spin configurations can be shown to have higher energies by evaluating matrix elements of H.

Since the Heisenberg interaction (8.3.1) is isotropic, the direction of the spin alignment in the ground state is arbitrary, and the state (8.3.4) is in fact $(N+1)$-fold degenerate. As a result of this degeneracy motions of the spins exist at $T = 0$, corresponding to a rotation of the system as a whole, which require zero excitation energy. These motions can be regarded as a spin-wave excitation of infinite wavelength; thus, as will be verified below, the spin-wave excitation spectrum necessarily has $\omega(\mathbf{q}) = 0$ at $\mathbf{q} = 0$. This is again a case of the Goldstone theorem as in Sec. 7.7. To define a particular direction of magnetization we "break the symmetry" by considering an infinitesimal magnetic field to be present; we assume that the z-axis has been defined in this way.

Our aim is to study the temperature variation of the magnetization

$$M = \left\langle \sum_i S_i^z \right\rangle. \tag{8.3.10}$$

The brackets here denote a thermal average at temperature T, as in Chap. 2:

$$\langle A \rangle = \frac{\mathrm{Tr}\,(e^{-\beta H} A)}{\mathrm{Tr}\,(e^{-\beta H})} \qquad (\beta = 1/k_B T). \tag{8.3.11}$$

We introduce the retarded spin susceptibility Green's function (compare Sec. 7.2)

$$R_{ij}(t) = -i\theta(t)\langle[\tilde{S}_i^-(t), \tilde{S}_j^+(0)]\rangle, \tag{8.3.12}$$

with Heisenberg operators

$$\tilde{S}(t) = e^{iHt} S\, e^{-iHt}. \tag{8.3.13}$$

The commutation rules for the spin operators

$$[S_i^x, S_j^y] = i\delta_{ij} S_i^z \qquad (+\text{ cyclic permutations}) \tag{8.3.14}$$

give

$$[S_i^z, S_j^-] = -\delta_{ij} S_i^-, \qquad [S_i^z, S_j^+] = \delta_{ij} S_i^+,$$

and

$$[S_i^-, S_j^+] = -2\delta_{ij} S_i^z. \tag{8.3.15}$$

Hence

$$\lim_{t\to 0+} \sum_{ij} R_{ij}(t) = -i \sum_{ij} \langle[S_i^-, S_j^+]\rangle = 2i\left\langle \sum_i S_i^z \right\rangle = 2iM, \tag{8.3.16}$$

giving a direct relation between the susceptibility response and the magnetization.

The equation of motion for $R_{ij}(t)$ is

$$i\frac{\partial R_{ij}(t)}{\partial t} = \delta(t)\langle[S_i^-, S_j^+]\rangle - i\theta(t)\langle[[\tilde{S}_i^-(t), H], \tilde{S}_j^+(0)]\rangle. \tag{8.3.17}$$

Evaluating the commutator $[S_i^-, H]$ (remembering that $\gamma_{ii} = 0$, $\gamma_{ij} = \gamma_{ji}$) we obtain

$$\begin{aligned} i\frac{\partial R_{ij}(t)}{\partial t} &= -2\delta(t)\delta_{ij}\langle S_i^z\rangle \\ &\quad - i\theta(t)\,\langle[J \sum_{i'} \gamma_{i'i}\{\tilde{S}_{i'}^-(t)\tilde{S}_i^z(t) - \tilde{S}_{i'}^z(t)\tilde{S}_i^-(t)\}, \\ &\quad \tilde{S}_j^+(0)]\rangle. \end{aligned} \tag{8.3.18}$$

We thus have a higher order Green's function on the right-hand side, and the system of equations of motion does not close up. We approximate, in the sense of a molecular field approximation, by replacing S_i^z everywhere by its average $\langle S^z\rangle$ which, because of translational symmetry, will be independent of the lattice site. This amounts to a neglect of fluctuations in the system. The hierarchy of equations of motion then decouples, and we have

$$i\,\frac{\partial R_{ij}(t)}{\partial t} = -2\delta(t)\delta_{ij}\langle S^z\rangle + J\langle S^z\rangle \sum_{i'} \gamma_{i'i}\{R_{i'j}(t) - R_{ij}(t)\}. \tag{8.3.19}$$

This is analogous to the equation of motion (1.4.14) for a set of coupled oscillators, except that a first derivative $\partial/\partial t$ appears on the left-hand side instead of $\partial^2/\partial t^2$.

We can now obtain a solution by diagrammatic expansion, leading to diagrams similar to those of Chap. 1 which correspond to an excited state propagating through the crystal. This is the *spin-wave* solution, which is exact in the limit $T \to 0$ when the averages are ground state averages and $S_i^z = S$ exactly. Because of the translational invariance of H [with $\gamma_{ij} = \gamma(|i-j|)$], we can get a closed *normal mode* solution as in the lattice case. Fourier-transforming to **k**-space the equation of motion (8.3.19) becomes

$$i\,\frac{\partial R_{\mathbf{k}}(t)}{\partial t} = -2\delta(t)\langle S^z\rangle + J\langle S^z\rangle(\gamma_{\mathbf{k}} - \gamma_0)R_{\mathbf{k}}(t), \tag{8.3.20}$$

where

$$\gamma_{\mathbf{k}} = \sum_i \gamma_{ij}\, e^{-i\mathbf{k}\cdot(\mathbf{X}_i - \mathbf{X}_j)} \tag{8.3.21}$$

(and $\gamma_0 = \gamma_{\mathbf{k}=0}$). (The $\mathbf{X}_i$ are lattice vectors.)

Consider first $T = 0$. Here $\langle S^z\rangle = S$, a given constant, and we write

$$i\,\frac{\partial R_{\mathbf{k}}}{\partial t} = -2\delta(t)S - \omega_{\mathbf{k}}R_{\mathbf{k}}, \tag{8.3.22}$$

where

$$\omega_{\mathbf{k}} = JS(\gamma_0 - \gamma_{\mathbf{k}}) \tag{8.3.23}$$

represents the frequencies of the spin-wave excitations. We can solve the equation of motion as usual by Fourier transforms. With

$$R_{\mathbf{k}}(\omega) = \int_{-\infty}^{\infty} R_{\mathbf{k}}(t)\, e^{i\omega t}\, dt \tag{8.3.24}$$

we obtain

$$(\omega + \omega_{\mathbf{k}})R_{\mathbf{k}}(\omega) = -2S,$$

and thus

$$R_{\mathbf{k}}(\omega) = -\frac{2S}{\omega + \omega_{\mathbf{k}} + i\eta}, \tag{8.3.25}$$

where we have introduced $\eta = 0+$ to give

$$R_{\mathbf{k}}(t) = 2iS\, e^{i\omega_{\mathbf{k}}t} \tag{8.3.26}$$

for $t > 0$.

The frequencies $\omega_{\mathbf{k}}$ are easily evaluated from (8.3.21) and (8.3.23). For nearest-neighbor interactions we have

$$\gamma_{\mathbf{k}} = \sum_{\mathbf{a}} e^{-i\mathbf{k}\cdot\mathbf{a}} = \sum_{\mathbf{a}} \cos \mathbf{k}\cdot\mathbf{a}, \tag{8.3.27}$$

where **a** are the vectors from the origin to the nearest-neighbor sites, which occur in pairs $\pm\mathbf{a}$. Hence

$$\omega_{\mathbf{k}} = JS \sum_{\mathbf{a}} (1 - \cos \mathbf{k}\cdot\mathbf{a}), \tag{8.3.28}$$

and for small $|\mathbf{k}|$ (long waves)

$$\omega_{\mathbf{k}} \simeq \tfrac{1}{2}JS \sum_{\mathbf{a}} (\mathbf{k}\cdot\mathbf{a})^2. \tag{8.3.29}$$

For a simple cubic lattice this reduces to

$$\omega_{\mathbf{k}} \simeq JSa^2|\mathbf{k}|^2, \tag{8.3.30}$$

where a is the lattice constant, with corresponding results for other lattices.

At *finite temperatures* T we replace S everywhere by the thermal average $\langle S^z\rangle$, which is an unknown function of T. Then

$$R_{\mathbf{k}}(\omega) = -\frac{2\langle S^z\rangle}{\omega + \omega_{\mathbf{k}}(T) + i\eta}, \qquad \text{with } \omega_{\mathbf{k}}(T) = J\langle S^z\rangle(\gamma_0 - \gamma_{\mathbf{k}}), \tag{8.3.31}$$

and we still have an approximate description in terms of spin waves. However, (8.3.31) is just a relation between the two unknown quantities $R_{\mathbf{k}}$ and $\langle S^z\rangle$. To obtain these functions separately, we note that we can in turn express $\langle S^z \rangle$ in terms of $R_{\mathbf{k}}$ by means of the fluctuation-dissipation theorem. If $J_{\mathbf{k}}(\omega)$ is the spectral density associated with $R_{\mathbf{k}}(\omega)$, Eq. (A.2.9) (Appendix 2) gives

$$-\tfrac{1}{2}(1 - e^{-\beta\omega})J_{\mathbf{k}}(\omega) = \text{Im } R_{\mathbf{k}}(\omega) = 2\pi\langle S^z \rangle \delta(\omega + \omega_{\mathbf{k}}).$$

The spectral density $J_{ij}(\omega)$ associated with $R_{ij}(\omega)$ is therefore

$$\begin{aligned} J_{ij}(\omega) &= \frac{1}{N} \sum_{\mathbf{k}} e^{i\mathbf{k}\cdot(\mathbf{X}_i - \mathbf{X}_j)} J_{\mathbf{k}}(\omega) \\ &= \frac{4\pi}{N} \langle S^z \rangle \sum_{\mathbf{k}} \frac{e^{i\mathbf{k}\cdot(\mathbf{X}_i - \mathbf{X}_j)}\delta(\omega + \omega_{\mathbf{k}})}{e^{\beta\omega_{\mathbf{k}}} - 1}, \end{aligned} \tag{8.3.32}$$

and the associated time correlation function is

$$\langle \tilde{S}_i^-(t)\tilde{S}_j^+(0) \rangle = \frac{1}{2\pi} \int_{-\infty}^{\infty} J_{ij}(\omega)\, e^{-i\omega t}\, d\omega. \tag{8.3.33}$$

Hence, putting $t = 0$ and $i = j$, we have from (8.3.32) and (8.3.33)

$$\begin{aligned} \langle S_i^- S_i^+ \rangle &= \frac{1}{2\pi} \int_{-\infty}^{\infty} J_{ii}(\omega)\, d\omega \\ &= \frac{2}{N} \langle S^z \rangle \sum_{\mathbf{k}} \frac{1}{e^{\beta\omega_{\mathbf{k}}} - 1}. \end{aligned} \tag{8.3.34}$$

Finally, we write on the left-hand side

$$\begin{aligned} S_i^- S_i^+ &= (S_i^x - iS_i^y)(S_i^x + iS_i^y) = (S_i^x)^2 + (S_i^y)^2 + i[S_i^x, S_i^y] \\ &= \mathbf{S}^2 - (S_i^z)^2 - S_i^z = S(S+1) - S_i^z(S_i^z + 1), \end{aligned}$$

and, with the approximation $\langle (S_i^z)^2 \rangle = \langle S_i^z \rangle^2$, which is consistent with our earlier approximations, we have

$$\langle S_i^- S_i^+ \rangle \simeq S(S+1) - \langle S^z \rangle \{\langle S^z \rangle + 1\}. \tag{8.3.35}$$

Thus our final equation is

$$S(S+1) - \langle S^z \rangle \{\langle S^z \rangle + 1\} = \frac{2}{N} \langle S^z \rangle \sum_{\mathbf{k}} \frac{1}{e^{\beta\omega_{\mathbf{k}}} - 1}, \tag{8.3.36}$$

with

$$\omega_{\mathbf{k}} = J\langle S^z \rangle(\gamma_0 - \gamma_{\mathbf{k}}). \tag{8.3.36}$$

This is an implicit equation for the magnetization $\langle S^z \rangle$ as a function of T. It is seen that, in this approximation, the excitations form a set of bosons described by a Planck distribution function.

For $J > 0$, (8.3.36) has a solution $\langle S^z \rangle$ which decreases steadily from $\langle S^z \rangle = S$ at $T = 0$ to $\langle S^z \rangle = 0$ at the *Curie temperature* T_c. To find an expression for T_c we have to take the limit $\langle S^z \rangle \to 0$, $\omega_{\mathbf{k}} \to 0$ on the right-hand side of (8.3.36), and obtain

$$S(S+1) = \frac{2}{N} \langle S^z \rangle \sum_{\mathbf{k}} \frac{1}{\beta_c \omega_{\mathbf{k}}} = \frac{2}{N} \sum_{\mathbf{k}} \frac{k_B T_c}{J(\gamma_0 - \gamma_{\mathbf{k}})},$$

where k_B is Boltzmann's constant. Hence, using (8.3.27),

$$k_B T_c = \frac{\frac{1}{2} NJS(S+1)}{\sum_{\mathbf{k}} \frac{1}{\gamma_0 - \gamma_{\mathbf{k}}}} = \frac{\frac{1}{2} NJS(S+1)}{\sum_{\mathbf{k}} \frac{1}{\sum_{\mathbf{a}} (1 - \cos \mathbf{k} \cdot \mathbf{a})}}. \tag{8.3.37}$$

In particular, if we neglect the $\cos \mathbf{k} \cdot \mathbf{a}$ term, we have the case of the *Ising model.* In this case the denominator reduces to N/z, where z is the number of nearest neighbors, and thus

$$kT_c = \tfrac{1}{2} JzS(S+1). \tag{8.3.38}$$

This agrees with the expression for the Curie temperature of the Ising model given by molecular field theory. A smaller (and more nearly correct) value of T_c is obtained when the **k**-dependence is included.

Because of the neglect of correlations between spins, the details of the temperature variation of the magnetization are not given correctly by (8.3.36). In a more exact theory, interactions between the spin waves must be taken into account; this was first done by Dyson (1956a). Corrections to the present theory can be calculated by decoupling the Green's function equations of motion in higher order, but the approximations involved are difficult to control.

Another method is due to Holstein and Primakoff (1940) and is useful for low temperatures. This is the method which was later improved by Dyson. In this approach the spin deviation operators acting on the ferromagnetic ground state are replaced by equivalent boson operators through the correspondence

$$\begin{aligned} S^+ &= \sqrt{(2S)}a^\dagger, \\ S^- &= \sqrt{(2S)}\left(1 - \frac{n}{2S}\right)a, \\ n &= a^\dagger a, \\ S^z &= -S + n. \end{aligned} \tag{8.3.39}$$

The spin hamiltonian then becomes a boson hamiltonian with fourth order boson interactions (the "Dyson hamiltonian"), of the form

$$H = 2SJ \sum \gamma_{ij} a_i^\dagger a_j + \tfrac{1}{2}J \sum \gamma_{ij} \{a_i^\dagger a_j^\dagger a_j a_j - a_i^\dagger a_j^\dagger a_i a_j\}. \tag{8.3.40}$$

This turns out to be non-hermitian as (8.3.39) is actually a similarity transformation rather than a unitary transformation. However, it may be shown that it gives the correct thermodynamics for temperatures well below T_c [Dyson (1956b), Wortis (1965)]. This model has the advantage that one can do straightforward Feynman–Dyson perturbation theory and develop a theory of renormalized spin waves which turns out to be remarkably effective over a wide range of temperatures below T_c.

Chapter 9

Transient Response of the Fermi Gas—the X-ray and Kondo Problems

As a result of the Pauli principle, the non-interacting fermi gas is a stable "liquid" at zero temperature. This stability is believed to persist in the presence of relatively weak repulsive forces (such as the screened Coulomb potential of Chap. 6). The stability is described in a general way by the Landau phenomenological theory of fermi liquids. The dominant physical assumption of this theory is that the Pauli principle quenches out particle–particle scattering between pairs of quasiparticle states with momentum p near the fermi momentum p_f which undergo energy transfers very small compared to the fermi energy ϵ_f.

Paradoxically, however, the dominance of the Pauli principle for low-lying fermion excitations can also lead to very singular effects. The most dramatic of these is the occurrence of superconductivity in the presence of weak attractive forces. This will be discussed in Chap. 10. A some-

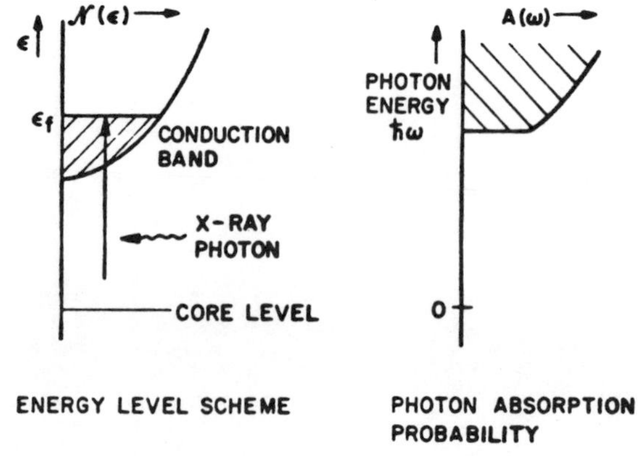

Fig. 9.1. Soft x-ray absorption in a metal.

what less dramatic, but still interesting, effect occurs in the response of the fermi gas to a time-dependent localized perturbation.

One of the places where such an effect arises physically is in the process of soft x-ray absorption in a metal. In this process a core electron (e.g., the 1*s* electron in Li metal) absorbs an x-ray photon and is excited to the fermi level or above (Fig. 9.1).

As a result of this one-electron excitation there is a localized change of the screened Coulomb potential due to the ions seen by the conduction electrons of the metal, as one of the metal atoms has now become a positive metal ion. This sudden change of potential leads to a transient perturbation of the electron gas which turns out to be very singular in

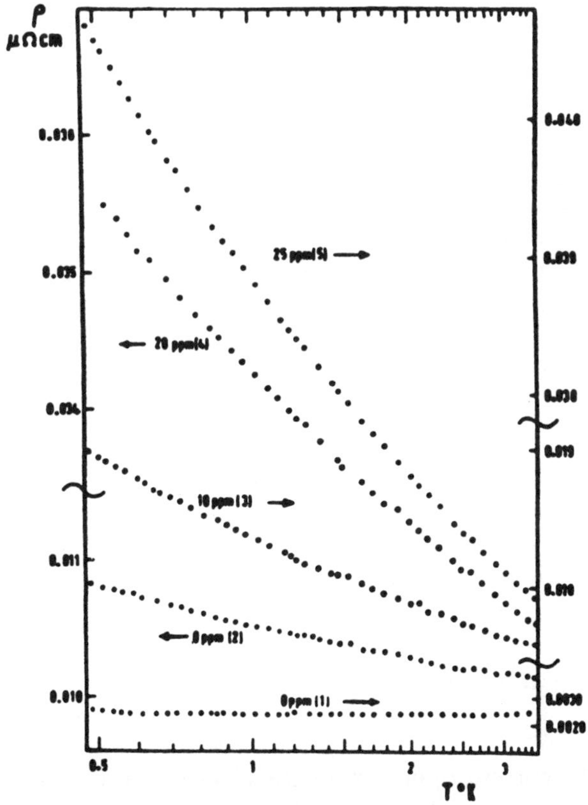

Fig. 9.2. The low-temperature resistivity of dilute AuFe alloys as a function of log *T*. [After Loram, Whall and Ford (1970).]

character as a result of the Pauli principle (i.e., fermi statistics governing the low-lying electron gas excitations). The singularity shows up in the energy dependence of the cross-section for ejecting photoelectrons from the core state, in the total absorption cross-section for the soft x-ray photons at the threshold of the core electron → fermi level transition, and also in the soft x-ray emission spectrum of metals.

A closely related singular phenomenon is found in certain dilute alloys in which a magnetic transition metal (typically iron) is dissolved in a non-magnetic host such as gold. When alloys of this kind (with one per cent of Fe or so) are cooled to a few degrees Kelvin, the electrical resistivity is found to show low-temperature anomalies of the approximate form $\rho \simeq \rho_0 - \rho_1 \log (T/T_f)$, where T_f is some reference temperature of the order of the fermi temperature ($\epsilon_f = k_B T_f$) (Fig. 9.2). A theoretical model of this anomaly was first proposed by Kondo (1964). We now know that the effect is related to the singular transient response of the fermi gas, and this will be discussed in the second half of this chapter.

9.1. THE X-RAY SINGULARITY SEEN IN PHOTOEMISSION

The simplest experiment in which to understand the singular response of the fermi gas is that of x-ray photoemission. In practice this effect is somewhat complicated by the scattering of the ejected photoelectrons which have relatively short mean free paths (10 Å or so) and are therefore strongly influenced by conditions at the metal surface.

In theory, however, the experiment provides a direct way of seeing the spectrum of excitations associated with the ejection of the core electrons by an atom, or creation of a "deep hole." What is being measured is a production cross-section $d^2\sigma/d\omega\, d\epsilon$ for absorption of a photon of energy ω and subsequent emission of a fast electron of energy ϵ. To calculate this cross-section we introduce the coupling to the x-ray photon

$$H_{\text{x-ray}} = -\int d^3x\mathbf{j}(\mathbf{x}) \cdot \mathbf{A}(\mathbf{x}), \tag{9.1.1}$$

where $\mathbf{j}(\mathbf{x})$ is the current operator for the electrons in the metal and $\mathbf{A}(\mathbf{x})$ is the vector potential operator describing the x-ray electromagnetic field [see any standard book on quantum electrodynamics, e.g., Akhiezer and Berestetskii (1965), p. 272]. Then the total absorption cross-section for a photon is given by

$$\frac{d\sigma}{d\omega} = \sum_f |\langle f, 0|H_{\text{x-ray}}|i, \mathbf{q}\rangle|^2 \delta(E_i + \omega_\mathbf{q} - E_f), \tag{9.1.2}$$

where $|i, \mathbf{q}\rangle$ represents the initial state of metal + x-ray photon of wave vector $\mathbf{q}$, $|f, 0\rangle$ is a final (excited) state in which the photon is no longer present and $\omega_\mathbf{q} = c|\mathbf{q}|$ is the x-ray photon energy.

We can now classify final states by expressing the current operator $\mathbf{j}(\mathbf{x})$ in second-quantized form with respect to the electrons in the metal

$$\mathbf{j}(\mathbf{x}) = \sum_{m,n} a_n^\dagger a_m \langle n|\mathbf{j}(\mathbf{x})|m\rangle, \tag{9.1.3}$$

where $|m\rangle$ and $|n\rangle$ are a basis set of one-electron states (see Appendix 1). Now, for photoemission from a core state, we are interested in transitions where one of the electrons $|m\rangle$ is initially in a core state

$$\langle \mathbf{x}|m\rangle = \varphi_{\text{core}}(\mathbf{x}), \tag{9.1.4}$$

and the final electron $|n\rangle$ is in a high-energy plane wave state $\langle \mathbf{x}|n\rangle = e^{i\mathbf{p}_n\cdot\mathbf{x}}$. This is energetically possible since we can make the photon energy ω much larger than the binding energy $|\epsilon_{\text{core}}|$ of the core electron. The part of the total cross-section for such transitions then becomes

$$\left.\frac{d\sigma}{d\omega}\right|_{\text{photoel}} = \sum_f \sum_\mathbf{p} \left| \langle f|a_\mathbf{p}^\dagger d|i\rangle \int d^3x\, e^{i\mathbf{q}\cdot\mathbf{x}} \right.$$

$$\left. \times \langle \mathbf{p}|\mathbf{j}(\mathbf{x})|\text{core}\rangle \cdot \boldsymbol{\epsilon}_\mathbf{q} \right|^2 \delta(E_i + \omega_\mathbf{q} - E_f), \tag{9.1.5}$$

where d is a core state annihilation operator and $\boldsymbol{\epsilon}_\mathbf{q}$ is an electromagnetic polarization vector.

Since the final electron momentum $\mathbf{p}$ is assumed very large we can simplify the analysis by projecting onto final states

$$|f\rangle = |\tilde{f}\rangle|\mathbf{k}\rangle, \tag{9.1.6}$$

where $|\tilde{f}\rangle$ is the excited state of the metal (with one electron missing), and $|\mathbf{k}\rangle$ is a particular plane wave final state selected by the photoelectron spectrometer. For such states, neglecting interaction of the outgoing electron with the electrons in the metal, we have

$$E_f = E_{\tilde{f}} + \epsilon_\mathbf{k}, \tag{9.1.7}$$

where $\epsilon_{\mathbf{k}}$ is the photoelectron kinetic energy. Thus

$$\frac{d^2\sigma}{d\omega d\epsilon_{\mathbf{k}}} = \sum_{\tilde{f}} \left| m_{\mathbf{k},\text{core}} \langle \tilde{f} | d | i \rangle \right|^2 \delta(E_i + \omega_{\mathbf{q}} - E_{\tilde{f}} - \epsilon_{\mathbf{k}}), \tag{9.1.8}$$

where $m_{\mathbf{k},\text{core}}$ is the current matrix element of Eq. (9.1.5).

Finally, (9.1.8) can be expressed in terms of a time-dependent correlation function as in the derivation of the van Hove–Placzek formula [Chap. 1, Eq. (1.3.10)] by inserting the hamiltonian H of the metal electrons which satisfies

$$H|\tilde{f}\rangle = E_{\tilde{f}}|\tilde{f}\rangle, \qquad H|i\rangle = E_i|i\rangle, \tag{9.1.9}$$

to give

$$\frac{d^2\sigma}{d\Omega d\epsilon_{\mathbf{k}}} = |m|^2 \frac{1}{2\pi} \int_{-\infty}^{\infty} dt\, g(t)\, e^{i\Omega t}, \tag{9.1.10}$$

where $g(t)$ is a "deep-hole" correlation function

$$g(t) = \langle i | \tilde{d}^\dagger(t)\, \tilde{d}(0) | i \rangle. \tag{9.1.11}$$

Here $\tilde{d}(t) = e^{iHt} d\, e^{-iHt}$, and $g(t)$ is obtained from the deep-hole Green's function

$$\mathscr{G}(t) = \langle i | T[\tilde{d}^\dagger(t) \tilde{d}(0)] | i \rangle, \tag{9.1.12}$$

which is identical with $g(t)$ for $t > 0$. [Note that $\mathscr{G}(t)$ is zero for $t < 0$: "propagation" of the deep hole is only in one time direction.] In (9.1.10) Ω is now the energy difference between the initial x-ray $\omega_{\mathbf{q}}$ and the final photoelectron $\epsilon_{\mathbf{k}}$:

$$\Omega = \omega_{\mathbf{q}} - \epsilon_{\mathbf{k}}, \tag{9.1.13}$$

i.e., Ω is the energy transferred to the solid.

The x-ray singularity arises from the nature of the interaction of the conduction electrons in the metal with the core hole. As explained in the introduction to this chapter, the core state operator d acts on the electrons in the metal by introducing a localized additional screened Coulomb potential V at an atomic site in the metal. Thus H may be written as

$$H = \sum_{\mathbf{p}} \epsilon_{\mathbf{p}} a_{\mathbf{p}}^\dagger a_{\mathbf{p}} + \epsilon_{\text{core}} d^\dagger d + \frac{V}{N} \sum_{\mathbf{p}\mathbf{p}'} a_{\mathbf{p}}^\dagger a_{\mathbf{p}'} d d^\dagger. \tag{9.1.14}$$

In this hamiltonian, electron–electron interactions have all been neglected except those due to the presence of the hole ($dd^\dagger$ is zero unless it acts on a state in which the core hole is present). V is an effective interaction strength confined to one atom in the metal. Thus the model is essentially that of a local impurity in a metal as in Chap. 4, except that it switches on and off as a result of the hole-number operator $dd^\dagger$. We thus have in effect a one-body scattering problem with a time-dependent scattering potential.

9.2. THE NATURE OF THE X-RAY SINGULARITY

What is it about the model hamiltonian (9.1.14) for the heavy hole which leads to a singular time dependence for the heavy-hole propagator (9.1.12)? It is essentially the fact that the heavy hole can transmit momentum to electron-hole pairs (charge density perturbations) in the fermi gas without itself taking up recoil energy. Now a fermi gas has the property that an electron-hole pair can have large momentum but very small total energy. This is because an electron can be moved from just below the fermi surface to just above (Fig. 9.3). Thus electron-hole pairs can be excited without expenditure of energy.

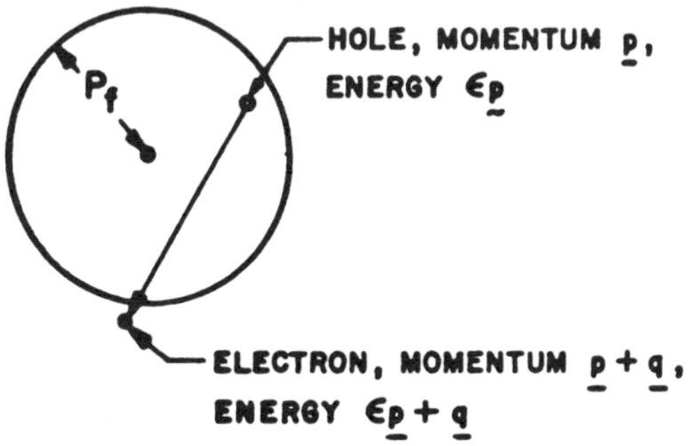

Fig. 9.3. Total electron-hole pair energy $\Delta\epsilon = \epsilon_{\mathbf{p+q}} - \epsilon_{\mathbf{p}} \to 0$ as $\mathbf{p} \to p_f -$ while $|\mathbf{p} + \mathbf{q}| \to p_f +$.

If we examine the perturbation expansion of $\mathcal{G}(t)$, this zero-energy denominator shows up as a logarithmic behavior for large positive t. To see this we go to the limit that the binding energy $|\epsilon_{\text{core}}|$ of the hole becomes very large. In the interaction picture (see Sec. 3.1), $\mathscr{G}(t)$ may be written (for $t > 0$)

$$\mathscr{G}(t) = e^{iE_i^0 t}\langle i|\, T[\exp\{-i \int_{-\infty}^{\infty} H_{\text{int}}(\tau)d\tau\} d^\dagger(t)d(0)]\,|i\rangle, \tag{9.2.1}$$

where H has been written as a sum of H_0 and H_{int}. Now, because of the limit $|\epsilon_{\text{core}}| \gg \epsilon_f$, all vacuum fluctuations involving the virtual excitation of a deep electron will have very small amplitude (they will involve energy denominators of order $1/|\epsilon_{\text{core}}|$). So the only terms of importance are those for which the deep-hole creation operator d has already acted. Thus, as $|\epsilon_{\text{core}}|/\epsilon_f \to \infty$,

$$\mathscr{G}(t) \to e^{iE_i^0 t}\langle i|\, T[d^\dagger(t) \exp\{-i \int_0^t H_{\text{int}}(\tau)d\tau\} d(0)]\,|i\rangle. \tag{9.2.2}$$

In this heavy-hole limit the effect of H_{int} is only felt for states in which the deep hole is present, i.e., for which $\langle dd^\dagger\rangle = 1$. So we can remove the d operators entirely from $\mathscr{G}(t)$ and rewrite it as

$$\mathscr{G}(t) = e^{iE_i^0 t}\langle i|\, T[\exp\{-i \int_0^t H_V(\tau)d\tau\}]\,|i\rangle, \tag{9.2.3}$$

where $H_V = (V/N) \sum_{\mathbf{pp}'} a_{\mathbf{p}}^\dagger a_{\mathbf{p}'}$. Thus we now have a transient problem in which the hole potential is switched on at time 0 and switched off again at time t.

It turns out that this problem is exactly soluble, as first shown by Nozières and DeDominicis (1969). However, it is instructive to look at it first using perturbation theory [this was originally the approach of Mahan (1967)]. Expanding (9.2.3) in powers of V we have as the lowest non-trivial term in second order

$$\begin{aligned}\mathscr{G}^{(2)}(t) &= \langle i|\, T\,[\tfrac{1}{2}i^2 \int_0^t d\tau_1 \int_0^t d\tau_2 H_{\text{int}}(\tau_1)H_{\text{int}}(\tau_2)]\,|i\rangle \\ &= -C(t), \text{ say,}\end{aligned} \tag{9.2.4}$$

where $|i\rangle$ is the fermi sea and we have removed the phase factor $e^{iE_i^0 t}$. This expression is a simple density response function of the non-interacting fermi gas

$$C(t) = \tfrac{1}{2}V^2 \int_0^t d\tau_1 \int_0^t d\tau_2 \sum_{\mathbf{pp}'} e^{i(\epsilon_{\mathbf{p}}-\epsilon_{\mathbf{p}'})|\tau_1-\tau_2|} f(\epsilon_{\mathbf{p}})\{1-f(\epsilon_{\mathbf{p}'})\}. \tag{9.2.5}$$

Replacing the sums over **p** by energy integrals, neglecting the energy dependence of the free-electron density of states $\mathcal{N}(\epsilon_b)$, and introducing a cut-off $\pm\epsilon_b$ of the order of the conduction bandwidth, we have

$$C(t) = \bar{V}^2 \int_0^t d\tau_1 \int_0^{\tau_1} d\tau_2 \int_{-\epsilon_b}^{0} d\epsilon \int_0^{\epsilon_b} d\epsilon' \, e^{i(\epsilon-\epsilon')(\tau_1-\tau_2)}$$

$$= \bar{V}^2 \int_0^t d\tau_1 \int_0^{\tau_1} d\tau_2 \left\{ \frac{1-e^{-i\epsilon_b(\tau_1-\tau_2)}}{i(\tau_1-\tau_2)} \right\}^2, \qquad (9.2.6)$$

where $\bar{V} = \mathcal{N}(0)V$. (The fermi level is now taken as the energy zero.)

For large times t, the dominant contribution comes from the singularity $1/(\tau_1-\tau_2)^2$. The integral over τ_1 and τ_2 can be approximately evaluated by replacing the oscillating term in (9.2.6) by a short time cut-off at $\tau_1 = \tau_2 = i/\epsilon_b$ (the cut-off must be pure imaginary in order to give a real spectral density for the response function). Then

$$\int_0^{\tau_1} d\tau_2 \left\{ \frac{1-e^{-i\epsilon_b(\tau_1-\tau_2)}}{i(\tau_1-\tau_2)} \right\}^2 \sim \int^{i/\epsilon_b} \frac{d\tau_2}{(\tau_1-\tau_2)^2}$$

$$= 1 \Big/ \left(\tau_1 - \frac{i}{\epsilon_b}\right),$$

giving

$$C(t) \simeq \bar{V}^2 \log(i\epsilon_b t), \qquad (9.2.7)$$

where ϵ_b is the bandwidth parameter. Thus the perturbation slowly diverges as $t \to \infty$.

This divergence occurs in each term of perturbation theory for $\mathcal{G}(t)$. Notice that this is in a sense a weak divergence: any shift $\delta E_i{}^0$ in the threshold energy will show up in a power series expansion in $\delta E_i{}^0$ as a perturbation term which grows as $(\delta E_i{}^0)^n (it)^n/n!$. Thus what is actually happening is a weakly singular modification of the photoelectric cross-section from the singular δ-function of energy which occurs in the simple one-electron theory to a skew, weaker singularity which turns out to be of the form $\Omega^{-\alpha}$, where Ω is the photoelectron energy measured relative to the threshold. However, this kind of singular response of the fermi gas to a transient perturbation has very strong ramifications when it comes to superconductivity. In this case the interaction is between all the electrons in the system. Thus it becomes extensive in effect and leads to the superconducting instability of the fermi ground state even for extremely weak electron–electron attractive forces.

On account of the divergence of (9.2.7) for large t (which leads to a divergence of the Fourier transform of $\mathcal{G}(t)$ for small Ω), it is not a good approximation to work only to second order in V even for $V \ll 1$. What one can do, however, is to work to leading *logarithmic* accuracy in the higher-order terms. One then finds that the term of order V^{2n} diverges as $(\log t)^n$ + corrections which diverge more slowly. There are various ways to establish the nature of the resulting series sum, of which perhaps the most physically appealing is that pointed out by Schotte and Schotte (1969). They showed, using an argument originally established by Tomonaga for a one-dimensional fermi gas that the local density perturbation operator H_{int} due to the core hole behaves very much like a sum of displacement operators for a set of harmonic oscillators representing the low-lying excitation spectrum of the fermi gas. We will not establish this here, but simply assume this property in order to get the required result. One uses the fact that, if two hermitian operators A and B satisfy the relation

$$[A, B] = c, \tag{9.2.8}$$

where c is a c-number (commuting with both A and B), then

$$e^{A+B} = e^A\, e^B\, e^{-(1/2)[A,B]}. \tag{9.2.9}$$

[This is a special case of the Baker–Hausdorff theorem (see Magnus, 1954).] Now, if ρ is an oscillator displacement operator, then in terms of boson creation and annihilation operators ρ may be written

$$\rho = \frac{1}{\sqrt{(2\Omega)}}\,(B + B^{\dagger}), \tag{9.2.10}$$

where $[B, B^{\dagger}] = 1$ (see Sec. 1.5). Hence one has the general result that the ground state expectation value of e^{ρ} is given by

$$\langle e^{\rho}\rangle_0 = \langle e^{B^{\dagger}/\sqrt{(2\Omega)}}\, e^{B/\sqrt{(2\Omega)}}\rangle\, e^{-1/4\Omega} = e^{-(1/2)\langle\rho^2\rangle_0}, \tag{9.2.11}$$

since $(B)^n|0\rangle = 0$ for all n. Now, going back to (9.2.3), we can extract the time dependence of $H_V(\tau)$ to give

$$\mathcal{G}(t) = \langle i|\,T[\exp\{-i\int_0^t V\rho_0(\tau)d\tau\}]\,|i\rangle, \tag{9.2.12}$$

where

$$\rho_0(\tau) = \frac{1}{N} \sum_{\mathbf{pp}'} e^{i(\epsilon_{\mathbf{p}} - \epsilon_{\mathbf{p}'})\tau} a_{\mathbf{p}}^{\dagger} a_{\mathbf{p}'},$$

and we have again removed the phase factor exp $(iE_i^0 t)$.

Assuming that for small V it is a good approximation to treat the operator products $a_{\mathbf{p}}^{\dagger} a_{\mathbf{p}'}$ as though they were oscillators, Eq. (9.2.11) is then equivalent to the statement that $C(t)$ in the expression

$$\mathscr{G}(t) = e^{-C(t)} \tag{9.2.13}$$

is quadratic in V, i.e.,

$$C(t) = -\tfrac{1}{2} V^2 \left. \frac{\partial^2}{\partial V^2} \mathscr{G}(t) \right|_{V=0} . \tag{9.2.14}$$

Differentiating (9.2.12) we then have

$$C(t) = \tfrac{1}{2} V^2 \int_0^t d\tau \int_0^t d\tau' \langle T[\rho_0(\tau)\rho_0(\tau')]\rangle, \tag{9.2.15}$$

which is identical with the second-order expression (9.2.4) with the asymptotic form (9.2.7). Hence the oscillator limit (valid for $V \ll 1$) gives, for large t,

$$g(t) = \mathscr{G}(t) = e^{-C(t)} = e^{-\bar{V}^2 \log(i\epsilon_b t)} = \frac{1}{(i\epsilon_b t)^{\alpha}}, \tag{9.2.16}$$

where the exponent α is given by

$$\alpha = \bar{V}^2 = \{V \mathscr{N}(0)\}^2. \tag{9.2.17}$$

To see the effect of this singularity on the resulting spectrum we need the Fourier transform $g(\Omega)$ of the time correlation function $g(t)$, which determines the photoelectric cross-section according to Eq. (9.1.10). This clearly must have a threshold energy ($\Omega = 0$) corresponding to the maximum energy $\epsilon_{\mathbf{k}}$(max) of the emitted photoelectron: this occurs when all the initial photon energy is transferred to the photoelectron, leaving the hole + metal in the ground state. Thus the resulting transform is a one-sided function: $g(\Omega) = 0$ for $\Omega < 0$ (adjusted to the above threshold) and, for $\Omega > 0$,

$$g(\Omega) \simeq \int_0^{\infty} dt\, e^{i\Omega t} \frac{1}{(i\epsilon_b t)^{\alpha}} . \tag{9.2.18}$$

[Note that the integrand is only a good approximation for $\epsilon_b t \gg 1$. This is sufficient to determine the form of $g(\Omega)$ near threshold.] Hence, for small $\Omega > 0$,

$$g(\Omega) \sim \frac{1}{\Omega^{1-\alpha}} \int_0^\infty dx\, e^{ix} \frac{1}{(i\epsilon_b x)^\alpha}, \tag{9.2.19}$$

from which

$$\text{Im } g(\Omega) \sim \frac{\sin(\frac{1}{2}\pi\alpha)}{\Omega^{1-\alpha}}, \tag{9.2.20}$$

[for $\alpha = 0$, Im g is just a δ-function $\delta(\Omega)$].

Thus the photoemission δ-function is broadened out into an asymmetric peak. The singularity follows a power law with an exponent determined by the interaction strength. In real life the hole state has a finite lifetime (due to Auger processes and radiative decay) which smears the function out [see Doniach and Šunjić (1970)]. The low-energy tail of the photoelectron line corresponds to final states in which many low-energy electron-hole pairs are excited in the metal in addition to the final fast photoelectron. The singular cross-section means that this process becomes more and more probable as the energy loss involved [Ω in Eq. (9.2.20)] tends to zero.

Experimentally, skew line shapes for emitted photoelectrons have been observed by Siegbahn and collaborators [Nordling *et al.* (1958)], but experiments with higher resolution are required for a detailed test of the theory.

9.3. THE EDGE SINGULARITY IN SOFT X-RAY ABSORPTION AND EMISSION EXPERIMENTS

The case of most practical interest in x-ray experiments is the straightforward emission and absorption of x-rays by core level to fermi surface transitions. The theory of these processes is slightly more complicated than in the photoemission case as the final electron does not escape from the metal, but at energies near threshold stays in the vicinity of the heavy hole and scatters from it in a singular fashion.

The absorption (or emission) cross-section is given by the straightforward van Hove formula which, using the notation of Eq. (9.1.10), may be written as

$$\frac{d\sigma}{d\omega} = |m|^2 \frac{1}{2\pi} \int_{-\infty}^{\infty} dt\, f(t)\, e^{i\omega t}, \tag{9.3.1}$$

where $f(t)$ is now, for $t > 0$, identical with a deep-hole-conduction electron pair propagator:

$$F(t) = \langle i|\, T[\tilde{d}^\dagger(t)\tilde{a}_0(t)\tilde{a}_0^\dagger(0)\tilde{d}(0)]\,|i\rangle, \tag{9.3.2}$$

and $a_0{}^\dagger$ is a localized (or Wannier) electron creation operator [compare Eq. (7.1.5)]:

$$a_0{}^\dagger = N^{-1/2} \sum_{\mathbf{p}} a_{\mathbf{p}}{}^\dagger. \tag{9.3.3}$$

If the state $|i\rangle$ is the ground state of the fermi sea, then F measures the absorption probability. To simulate the emission case, Nozières and DeDominicis (1969) assumed that $|i\rangle$ could be replaced by a state in which a deep hole was already present in the initial state. Strictly speaking this is an unstable state, so this procedure is not really allowed, but in our model hamiltonian (9.1.14) it is in fact stable (since the Auger and radiative interactions are not included), so that the procedure is consistent with the model. For the emission case the order of operation of $d^\dagger a$ and $a^\dagger d$ in (9.3.2) is to be reversed of course. $F(t)$ can now be studied in the same deep-hole limit which we used for the hole propagator $\mathcal{G}(t)$ in (9.2.2). Again the potential acts only between times 0 and t when the hole is present. So $F(t)$ may be rewritten as

$$F(t) = e^{iE_i^0 t}\langle i|\, T\big[a_0(t)\exp\{-i\int_0^t H_V(\tau)d\tau\}\, a_0{}^\dagger(0)\big]\,|i\rangle. \tag{9.3.4}$$

This is identical with (9.2.3) except for the presence of the conduction electron operators and may again be evaluated exactly (for large t), as shown by Nozières and DeDominicis (1969). To see the physics of the result we again use the Schotte and Schotte approximation. They showed that the important effect of the a_0 operators appearing in (9.3.4) was to introduce an *additional* perturbation of the electron gas "oscillators" as a result of the appearance of an additional electron at the site ($\mathbf{X} = 0$) of the deep hole. This may be thought of as a transient change of the fermi level at this site and hence a local change in conduction electron density.

To see this within the oscillator approximation for the electron gas, we introduct a set of operators which are approximately conjugate to the Fourier components of the density operator [compare Eq. (4.2.14)]

$$\rho_{\mathbf{q}} = \frac{1}{N} \sum_{\mathbf{p}} a^\dagger_{\mathbf{p}+\mathbf{q}} a_{\mathbf{p}} \tag{9.3.5}$$

(N is the number of atoms in the system), by means of the definition

$$\Pi_{\mathbf{q}} = \frac{1}{\mathcal{N}(0)} \sum_{\mathbf{p}} \frac{a^\dagger_{\mathbf{p}-\mathbf{q}} a_{\mathbf{p}}}{\epsilon_{\mathbf{p}} - \epsilon_{\mathbf{p}-\mathbf{q}}}, \tag{9.3.6}$$

where $\mathcal{N}(0)$ is the density of states at the fermi level. Then we have the commutator

$$[\Pi_{\mathbf{q}}, \rho_{\mathbf{q}}] = \frac{1}{N\mathcal{N}(0)} \sum_{\mathbf{p}} \frac{n_{\mathbf{p}-\mathbf{q}} - n_{\mathbf{p}}}{\epsilon_{\mathbf{p}} - \epsilon_{\mathbf{p}-\mathbf{q}}}. \tag{9.3.7}$$

This has the property that its average value is unity for small q [compare Eq. (7.4.6)]. Now the localized fermion creation operator (9.3.3) has the property that

$$[\rho_0, a_0{}^\dagger] = a_0{}^\dagger, \tag{9.3.8}$$

where $\rho_0 = \sum_{\mathbf{q}} \rho_{\mathbf{q}}$ is the local density operator (note that $H_V = V\rho_0$).

Hence, if we make the representation

$$a_0{}^\dagger \to e^{-\Pi_0}, \quad \text{with } \Pi_0 = \frac{1}{N} \sum_{\mathbf{q}} \Pi_{\mathbf{q}}, \tag{9.3.9}$$

then we find, working in a small-q approximation, that

$$e^{\Pi_0} \rho_0 \, e^{-\Pi_0} = \rho_0 + [\Pi_0, \rho_0] + \cdots \simeq \rho_0 + 1.$$

Thus

$$[\rho_0, e^{-\Pi_0}] \simeq e^{-\Pi_0}, \tag{9.3.10}$$

i.e., $e^{-\Pi_0}$ simulates the effect of the creation operator $a_0{}^\dagger$ on the electron oscillators.

Now we can go back to the calculation of $F(t)$:

$$F(t) = e^{iE_i{}^0 t} \langle i | T[e^{iH_0 t} a_0 \, e^{-iH_0 t} \exp\{-i \int_0^t V\rho_0(\tau) d\tau\} \, a_0{}^\dagger] | i \rangle, \tag{9.3.11}$$

and, substituting our approximate representation of a_0, we have

$$F(t) = e^{iE_i^0 t}\langle i|\,T[e^{iH_0 t}\, e^{\Pi_0} \exp\{-i \int_0^t (H_0 + V\rho_0(\tau))d\tau\}\, e^{-\Pi_0}]\,|i\rangle. \tag{9.3.12}$$

Now the e^{Π_0} operators have the effect of a unitary transformation:

$$e^{\Pi_0} H_0\, e^{-\Pi_0} = H_0 + [\Pi_0, H_0] + \cdots \simeq H_0 + \mathcal{N}(0)^{-1}\rho_0, \tag{9.3.13}$$

using the definition (9.3.6), (9.3.9) of Π_0. Hence, neglecting a constant frequency shift term,

$$F(t) = e^{iE_i^0 t}\langle i|\,T[\exp\{-i \int_0^t (V + \mathcal{N}(0)^{-1})\rho_0(\tau)d\tau\}]\,|i\rangle. \tag{9.3.14}$$

We have thus reduced F to an identical form to that [Eq. (9.2.12)] which occurred in the calculation of the deep-hole propagator. So we can use exactly the same procedure to derive, for large t,

$$F(t) \simeq e^{-\{1+V\mathcal{N}(0)\}^2 \log t} = \frac{1}{t^{(1+\bar{V})^2}}. \tag{9.3.15}$$

On making the Fourier transform to energy space we have

$$\frac{d\sigma}{d\omega} \propto \text{Im } F(\omega) = \int_0^\infty dt\, \frac{e^{i\omega t}}{t^{(1+\bar{V})^2}} \sim \omega^{\bar{V}^2 + 2\bar{V}}. \tag{9.3.16}$$

In terms of the phase shift for Born approximation

$$\delta_B/\pi = -\bar{V} \tag{9.3.17}$$

this becomes

$$\frac{d\sigma}{d\omega} \propto \frac{1}{\omega^{2\delta/\pi - (\delta/\pi)^2}}. \tag{9.3.18}$$

Thus, for $\delta > 0$ (an attractive potential) the x-ray absorption and emission cross-sections become singular at threshold. The above result has been generalized to strong potentials by Nozières and DeDominicis (1969) who find

$$\frac{d\sigma_l}{d\omega} \propto \frac{1}{\omega^{\alpha_l}}, \tag{9.3.19}$$

where

$$\alpha_l = \frac{2\delta_l}{\pi} - \sum_{l'=0}^{\infty} (2l'+1)\left(\frac{\delta_{l'}}{\pi}\right)^2 \tag{9.3.20}$$

In this equation the angular momentum quantum number l is dictated by the quantum number of the core state involved in a particular x-ray absorption or emission event (as determined by the photon energy) and the dipole selection rule determining the matrix element for transition to the conduction band.

This is an important consideration in practice [Ausman and Glick (1969)]. For instance, the K-edge in Li metal involves a transition from a 1s core state. The conduction electrons must therefore be in a p-wave state relative to the Li atom which absorbed the x-ray. Now a p-wave at the fermi surface will not scatter too strongly from the Li ion as the ion will be screened and the centrifugal barrier will reduce the phase shift. On the other hand, the hole propagator contribution in (9.3.20) [the $\Sigma\,(2l+1)(\delta_l/\pi)^2$ term] will contain an s-wave term which may be expected to be large since the Li^+ ion must attract a unit of charge, by the Friedel sum rule (i.e., $\delta_{s\text{-wave}} \simeq \frac{1}{2}\pi$). Hence the second term in (9.3.20) will dominate and lead to a suppression of the edge. On the other hand, the Na L_2 edge will involve a 2p core electron. In this case the final state is a conduction electron s-wave. Hence in this case the $2\delta_l/\pi$ term will dominate for α_l and the edge will have a singular peak (Fig. 9.4).

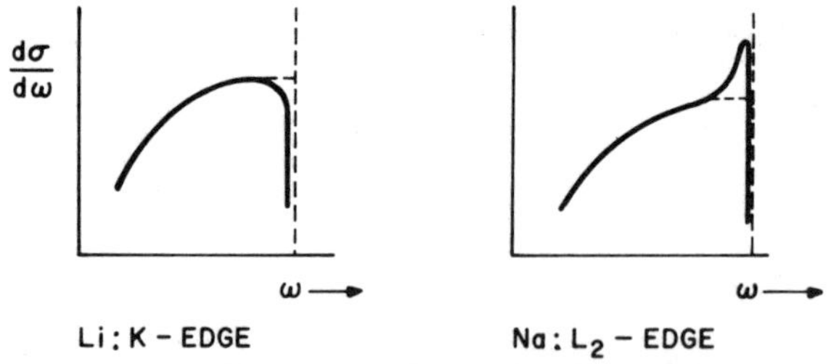

Fig. 9.4. Emission spectra of alkali metals.

9.4. ATOMIC LIMIT FOR LOCALIZED MAGNETIC MOMENTS IN METALS—THE KONDO EFFECT

When the local Coulomb interaction for a magnetic atom in a metal becomes large relative to the width of the Friedel–Anderson virtual state, as discussed in Sec. 7.8, the atom becomes strongly magnetized. However, since the conduction electrons in the bulk of the metal are not assumed strongly interacting, the fermi surface still remains and a new effect comes into play as a result of the degeneracy of the spin states of the local impurity atom. This is a somewhat analogous situation to that of the insulating magnet in that in this regime it is a reasonable approximation to treat the impurity atom in the atomic limit and expand the Anderson hamiltonian [Eqs. (7.8.1), (7.8.2)] in powers of the coupling $V_{\mathbf{k}d}$ to the conduction electrons.

The simplest form of the resulting interaction is a phenomenological hamiltonian of the Heisenberg type (studied in the 1950's by Kasuya and Yosida, but often referred to as the Kondo hamiltonian)

$$\left.\begin{aligned} H_0 &= \sum_{\mathbf{k}} \epsilon_{\mathbf{k}} a_{\mathbf{k}}{}^{\dagger} a_{\mathbf{k}}, \\ H_1 &= -J\mathbf{S} \cdot \boldsymbol{\sigma}(0), \end{aligned}\right\} \tag{9.4.1}$$

where $\boldsymbol{\sigma}(0)$ is the spin density operator of the conduction electrons at the position of the impurity atom:

$$\sigma_\alpha(0) = \sum_{\mu\nu} a_{0\mu}^{\dagger} \sigma_{\mu\nu}^{\alpha} a_{0\nu}, \tag{9.4.2}$$

and $a_{0\mu}^{\dagger}$ is the Wannier creation operator for an electron at the impurity lattice site, μ a spin suffix and $\sigma_{\mu\nu}^{\alpha}$ a Pauli matrix. Eq. (9.4.1) may be derived from the Anderson hamiltonian by means of a unitary transformation which expands the effective conduction electron-d-state interaction to second order in $V_{\mathbf{k}d}$. This was done by Schrieffer and Wolff (1966).

In this section we will describe briefly the singular properties of (9.4.1) discovered by Kondo. It turns out that the local spin interaction operator (9.4.1) acts as a time-dependent localized perturbation of the electron gas density in a rather similar way to the action of the transient hole potential in the x-ray problem. The present, Kondo, problem is more complicated in that the perturbing potential has its own degrees of freedom represented by the set of $2S + 1$ spin states $|S, m\rangle$ which are degenerate in zero applied magnetic field.

We first demonstrate the relationship between the Kondo problem and the x-ray problem, following the approach of Yuval and Anderson (1970). We then go on in the next section to show how the local spin interaction affects the scattering of conduction electrons from the magnetic impurity, leading to an unusual temperature-dependent electrical resistivity at temperatures of a few degrees K.

The way Yuval and Anderson approach the problem is to separate the interaction (9.4.1) with the localized spin into an Ising-like part

$$H_{\parallel} = -JS^z\sigma_z(0) \tag{9.4.3}$$

and a spin-flip part

$$H_{\perp} = -\tfrac{1}{2}J\{S^+\sigma^-(0) + S^-\sigma^+(0)\}. \tag{9.4.4}$$

For a given spin state of the impurity (for simplicity we will take $S = \frac{1}{2}$) the hamiltonian $H_0 + H_{\parallel}$ for the conduction electrons + impurity can be diagonalized in terms of scattering states $|\downarrow\rangle$ and $|\uparrow\rangle$ for the local potential due to S^z:

$$\begin{aligned} H_{\downarrow} &= \sum_{\mathbf{pp}'} \left\{\epsilon_{\mathbf{p}}\delta_{\mathbf{pp}'} + \frac{J}{N}\right\} a_{\mathbf{p}}^{\dagger}a_{\mathbf{p}'}, \\ H_{\uparrow} &= \sum_{\mathbf{pp}'} \left\{\epsilon_{\mathbf{p}}\delta_{\mathbf{pp}'} - \frac{J}{N}\right\} a_{\mathbf{p}}^{\dagger}a_{\mathbf{p}'}. \end{aligned} \tag{9.4.5}$$

Suppose we now look at the Green's function for the localized spin operator:

$$D_{\perp}(t) = \langle\downarrow| T[\tilde{S}^-(t)\tilde{S}^+(0)] |\downarrow\rangle. \tag{9.4.6}$$

A Green's function of this type would enter, for instance, in the calculation of the radio-frequency response in an electron spin resonance experiment.

We can now formally expand with respect to the spin-flip part of the interaction, $H_{\perp}$. To zero order in $H_{\perp}$, but to all orders in $H_{\parallel}$, the spin Green's function becomes (for $t > 0$)

$$\begin{aligned} D_{\perp}^{(\text{Ising})}(t) &= \langle\downarrow| e^{iH_{\downarrow}t} S^- e^{-iH_{\uparrow}t} S^+ |\downarrow\rangle \\ &= e^{iE_{\downarrow}t}\langle N| e^{i(H_{\downarrow}-H_{\uparrow})t} |N\rangle. \end{aligned} \tag{9.4.7}$$

This is precisely the deep-hole Green's function of Sec. 9.2 in which the strength of the local potential is now the exchange constant J. Yuval and Anderson showed that the effect of the spin-flip part of the interaction is to introduce a series of overlapping x-ray singularities at intermediate times which complicate the nature of the singular response in (9.4.7). We will not go into this here, but consider instead how the x-ray singularity of (9.4.7) shows up in the electron scattering cross-section from the magnetic impurity.

9.5. TEMPERATURE-DEPENDENT ELECTRICAL RESISTIVITY DUE TO A MAGNETIC IMPURITY

In order to demonstrate this singular scattering we expand the T matrix for electron-impurity scattering in powers of J (i.e., powers of both $H_\parallel$ and $H_\perp$) by Feynman–Dyson perturbation theory. To perform the expansion we require a Wick's theorem for the localized impurity spin operator S. This is a non-trivial problem because of the form of the spin commutation rules. Various ways round the difficulty have been proposed, of which we adopt one which works straightforwardly at zero temperature [Mattis (1965)]. This is a so-called drone fermion representation for the case of a spin $\frac{1}{2}$ localized impurity. We define

$$\begin{aligned} S^+ &= c^\dagger\varphi, \\ S^z &= c^\dagger c - \tfrac{1}{2}, \\ \varphi &= d + d^\dagger, \end{aligned} \tag{9.5.1}$$

in which both the spin deviation "c-on" operators and the drone "d-on" operators are mutually anticommuting fermion operators, with $\{c, c^\dagger\} = 1$, $\{d, d^\dagger\} = 1$ and all other pairs of operators anticommuting.

It is easily verified, using $c^\dagger c^\dagger = 0$, $\varphi^\dagger = \varphi$ and $\varphi^2 = 1$, that the operators (9.5.1) satisfy the spin commutation rules

$$[S^z, S^+] = S^+, \qquad [S^+, S^-] = 2S^z.$$

In fact (9.5.1) provides a double representation of the spin $\frac{1}{2}$ commutation rules based on the orthogonal pair of spin-down states $|\downarrow\rangle = |0\rangle$ and $|\downarrow\rangle = d^\dagger|0\rangle$; however, thermodynamic averages may be calculated by taking a trace over both sets of states without various complications which arise with other possible fermion representations. Using this representation, Wick's theorem for fermions applies in its usual form to time- or temperature-ordered products of spin $\frac{1}{2}$ operators [Spencer (1968)].

We can now proceed with a perturbation expansion of the one-electron Green's function in powers of H_1. We then find diagrams in which propagators for the drone fermions appear: these are the step functions

$$\begin{aligned} C^0(t) &= -i\langle T[c(t)c^\dagger(0)]\rangle \\ &= -i\{-\tfrac{1}{2} + \theta(t)\}, \end{aligned} \tag{9.5.2}$$

$$\begin{aligned} D^0(t) &= -i\langle T[\varphi(t)\varphi(0)]\rangle \\ &= -i\{2\theta(t) - 1\}, \end{aligned} \tag{9.5.3}$$

where $\theta(t)$ is the unit step function. The singularity in the T matrix then appears first in *third* order with a diagram of the form of Fig. 9.5, where

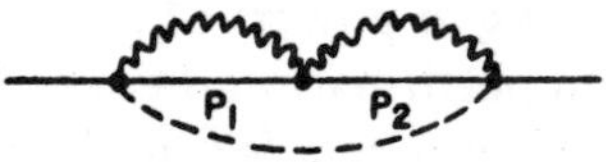

Fig. 9.5. Third-order T matrix diagram leading to singularity.

the wavy line represents a c-on and the dotted line a d-on. The mathematical reason why this leads to a singularity can be seen by considering the propagator

$$\tilde{G}^0(\mathbf{p}, t) = G^0(\mathbf{p}, t)C(t), \tag{9.5.4}$$

which represents the joint propagation of a c-on and an electron in Fig. 9.5. The effect of the $C(t)$ is to change the sign of $G(t)$ for $t < 0$, thus removing the usual antisymmetry of fermion operators.

We then have for the Fourier transform

$$\tilde{G}^0(\mathbf{p}, \epsilon) = -\tfrac{1}{2}i \left\{ \frac{f_{\mathbf{p}}^+}{\epsilon - \epsilon_{\mathbf{p}} + i\eta} - \frac{f_{\mathbf{p}}^-}{\epsilon - \epsilon_{\mathbf{p}} - i\eta} \right\}, \tag{9.5.5}$$

instead of the expression with a plus sign which appears for a usual electron propagator [see Eq. (4.6.6)]. Note that this would not enter in second order (Fig. 9.6), as the d-on propagator then combines with the c-on to give a single sign for all t.

Fig. 9.6. Second-order T matrix diagram (no singularity).

The contribution from Fig. 9.5 to the T matrix then comes out to be

$$T^{(3)}(\epsilon) = \tfrac{1}{4}iJ^3 \int_{-\infty}^{\infty} d\omega D^0(\omega)\{\tilde{G}_0{}^0(\epsilon + \omega)\}^2, \tag{9.5.6}$$

where

$$\tilde{G}_0{}^0(\epsilon) = \frac{1}{N}\sum_{\mathbf{p}} \tilde{G}^0(\mathbf{p}, \epsilon) \tag{9.5.7}$$

comes in, since, as in the potential scattering case (Chap. 4), H_1 is a localized interaction so that the electron lines are summed over $\mathbf{p}$ in the intermediate states. $D^0(\omega)$ is the Fourier transform of (9.5.3) which is

$$D^0(\omega) = \frac{1}{\omega - i\eta} + \frac{1}{\omega + i\eta} = 2\mathscr{P}\frac{1}{\omega}, \tag{9.5.8}$$

where $\mathscr{P}$ is the principal value operator. The singularity arises from the fact that, as in the x-ray case, the real part of $\tilde{G}_0{}^0$ now diverges logarithmically:

$$\begin{aligned} \mathrm{Re}\,\tilde{G}_0{}^0(\epsilon) &= \frac{1}{N}\sum_{\mathbf{p}} \mathscr{P}\frac{f_{\mathbf{p}}^+ - f_{\mathbf{p}}^-}{\epsilon - \epsilon_{\mathbf{p}}} \\ &= \int d\epsilon' \, \mathscr{N}(\epsilon') \frac{\{1 - 2f^-(\epsilon')\}}{\epsilon - \epsilon'}, \end{aligned} \tag{9.5.9}$$

of which the divergent part is

$$-2\int_{-\epsilon_0}^{\epsilon_f} \frac{d\epsilon'}{\epsilon - \epsilon'} \simeq -2\log\left(\frac{\epsilon - \epsilon_f}{\epsilon_0}\right), \tag{9.5.10}$$

where ϵ_0 is the band edge. This is closely analogous to the second-order contribution (9.2.7) to the deep-hole propagator and clearly breaks down at sufficiently small $(\epsilon - \epsilon_f)$, corresponding to large t in the Fourier transform.

Going back to (9.5.6) we can insert (9.5.8) and (9.5.5) and do the integration over ω by a contour in the upper or lower half-plane, to pick up the residue at the zero-frequency poles of $D^0(\omega)$. The divergent term turns out to be a cross term of the form

$$\begin{aligned} T^{(3)}(\epsilon) &\simeq \tfrac{1}{4} i J^3 \cdot 2 \sum_{\mathbf{p}\mathbf{p}'} f_{\mathbf{p}}^+ f_{\mathbf{p}'}^- \frac{1}{\epsilon_{\mathbf{p}'} - \epsilon_{\mathbf{p}} + i\eta} \\ &\quad \times \left\{\frac{1}{\epsilon_{\mathbf{p}'} - \epsilon + i\eta} + \frac{1}{\epsilon_{\mathbf{p}} - \epsilon - i\eta}\right\}. \end{aligned} \tag{9.5.11}$$

On taking the imaginary part of the second and third denominators this leads to

$$T^{(3)}(\epsilon) \simeq \tfrac{1}{2}\pi J^3 \sum_{\mathbf{pp}'} f_{\mathbf{p}}^{+} f_{\mathbf{p}'}^{-} \left\{ \frac{f^{-}(\epsilon)}{\epsilon - \epsilon_{\mathbf{p}} + i\eta} - \frac{f^{+}(\epsilon)}{\epsilon_{\mathbf{p}'} - \epsilon + i\eta} \right\}$$

$$= \tfrac{1}{2}\pi J^3 \{\mathcal{N}(\epsilon_f)\}^2 \{-n f^{-}(\epsilon) \log(\epsilon_f - \epsilon) - (1-n) f^{+}(\epsilon) \times \log(\epsilon - \epsilon_f)\}. \quad (9.5.12)$$

Thus the third-order contribution to the T matrix diverges logarithmically, for ϵ both above and below ϵ_f. As may be seen from the form of (9.5.9), higher-order diagrams of the form in Fig. 9.7 will contain con-

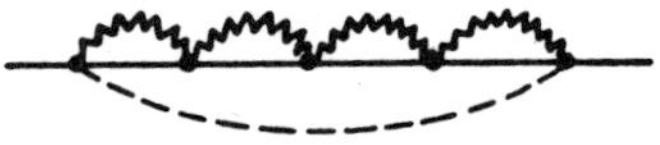

Fig. 9. 7. **Higher order diagrams.**

tributions of the form $\{\tilde{G}_0{}^0(t)\}^n$ and hence will diverge as

$$J^{n+2}\{\log(\epsilon - \epsilon_f)\}^n.$$

Thus the perturbation series is divergent in every order, as in the x-ray case.

The effect of this strong energy dependence of the T matrix, when treated correctly at finite temperatures, is to lead to an electron scattering rate which tends to diverge logarithmically as T is decreased, leading to a logarithmically increasing resistivity as the temperature is decreased. This model can be used to explain the resistance minimum phenomenon observed in many dilute alloys containing magnetic transition metal impurities: at high T the resistivity increases with T due to phonon scattering, while at low T it increases as T decreases due to spin scattering, hence the minimum.

The series of diagrams of the type in Fig. 9.7 can be summed and leads to a T matrix of the form

$$T(\epsilon) \simeq \frac{J^2}{1 + \mathcal{N}(\epsilon_f) J \log[(\epsilon - \epsilon_f)/\epsilon_f]}, \quad (9.5.13)$$

which has an unphysical pole at an energy (measured from the fermi level)

$$\epsilon_K = \epsilon_f\, e^{-1/J\mathcal{N}(\epsilon_f)}. \quad (9.5.14)$$

This pole signals the need for a self-consistently renormalized solution for the T matrix. Solutions of this type have been studied by approximate equation-of-motion methods by Nagaoka (1965) and others, and by other methods of approximation. They lead to the appearance of a characteristic temperature, the *Kondo temperature* $k_B T_K = \epsilon_K$, below which the scattering starts to saturate; this appears to be the case experimentally for dilute copper-iron alloys (Fig. 9.8).

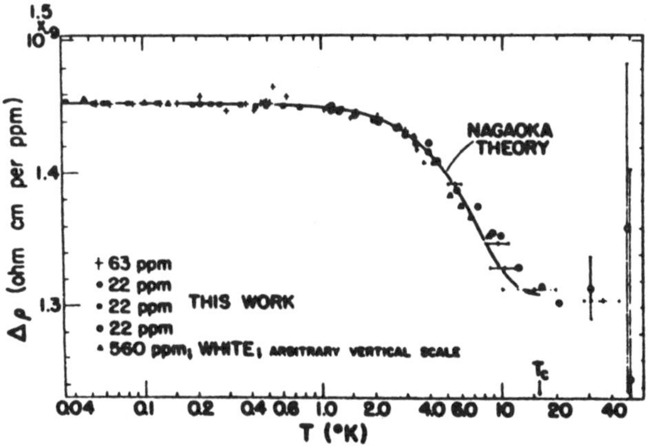

Fig. 9.8. Low-temperature resistivity of CuFe. [After Daybell and Steyert (1967).]

Chapter 10

Superconductivity

10.1. THE COOPER PAIR INSTABILITY OF THE FERMI GAS

In 1956 Leon Cooper suggested a very simple idea whereby the instability of the fermi gas under attractive electron interactions could be understood in terms of a bound state between a pair of fermi particles, interacting with each other but not with the other electrons in the fermi sea, and restricted by the Pauli principle to lie outside the fermi surface. To understand this effect let us write down the wave function for such a pair of electrons, using Cooper's original argument. We suppose that the total momentum of the pair is zero, so that we are in the center-of-mass frame for the pair of electrons. In order to apply the Pauli principle in the intermediate states we must work in momentum space. Let $\psi(\mathbf{p})$ be the wave function of the pair in momentum representation as a function of the relative momentum of the pair. Then $\psi(\mathbf{p})$ obeys the Schrödinger equation

$$\frac{\mathbf{p}^2}{2m}\,\psi(\mathbf{p}) + \int d^3p' V(\mathbf{p}-\mathbf{p}')\psi(\mathbf{p}') = E\psi(\mathbf{p}) \qquad \text{for } p > p_f,$$

$$\psi(\mathbf{p}) = 0 \qquad \text{for } p < p_f, \tag{10.1.1}$$

where we impose the restriction that no components of $\psi(\mathbf{p})$ are allowed for momentum states below the fermi surface.

The origin of the attractive force between the electrons is the attraction resulting from the distortion of the lattice ionic positions due to the electron–phonon interaction. Pictorially speaking, an electron which moves through the lattice can cause the spontaneous creation of a wave packet of lattice distortion. This will live for a time determined by the maximum lattice frequency,

$$\tau \simeq 1/\omega_{\max} \simeq \hbar/k_B\Theta_D,$$

where Θ_D is the Debye temperature of the lattice. During this time a second electron can come into the influence of the lattice distortion which has occurred. This leads to an effective electron–electron attraction, and means that the maximum energy transfer occurring during the interaction between the electrons will be of order $k_B\Theta_D$, so that the matrix elements $V(\mathbf{p}-\mathbf{p}')$ can be represented as

$$\begin{aligned} V(\mathbf{p}-\mathbf{p}') &= V \qquad \text{for } \epsilon_{\mathbf{p}}-\epsilon_f, \quad \epsilon_{\mathbf{p}'}-\epsilon_f < k_B\Theta_D, \\ &= 0 \qquad \text{otherwise.} \end{aligned} \tag{10.1.2}$$

With this model the integral equation (10.1.1) becomes separable. We simplify it by replacing the integral over $\mathbf{p}'$ by an integral over the electron energy $\epsilon_{\mathbf{p}'}$, and assume that $k_B\Theta_D$ is much less than ϵ_f so that the density of states $\mathcal{N}(\epsilon)$ of the electron gas is a slowly varying function which can be kept approximately constant in the integration. Eq. (10.1.1) then becomes

$$\epsilon_{\mathbf{p}}\psi(\mathbf{p}) + V\mathcal{N}(\epsilon_f) \int_{\epsilon f}^{\epsilon f + kB\Theta D} d\epsilon_{\mathbf{p}'}\,\psi(\mathbf{p}') = E\psi(\mathbf{p}). \tag{10.1.3}$$

Writing the (constant) integral as

$$\Lambda = \int_{\epsilon f}^{\epsilon f + kB\Theta D} d\epsilon_{\mathbf{p}'}\,\psi(\mathbf{p}'), \tag{10.1.4}$$

we can solve (10.1.3) to give

$$\psi(\mathbf{p}) = -\frac{V\mathcal{N}(\epsilon_f)\Lambda}{\epsilon_{\mathbf{p}} - E}, \tag{10.1.5}$$

and substituting this back into (10.1.4) the condition for the eigenvalue E becomes

$$1 = -V\mathcal{N}(\epsilon_f) \int_{\epsilon f}^{\epsilon f + kB\Theta D} \frac{d\epsilon_{\mathbf{p}'}}{\epsilon_{\mathbf{p}'} - E}. \tag{10.1.6}$$

We can now look for bound state solutions of this equation by assuming that E is lower than the lowest value of $\epsilon_{\mathbf{p}'}$ occurring in the integral, i.e., that E is below the fermi level ϵ_f. Then (10.1.6) can be integrated to give

$$\log\left(\frac{\epsilon_f + k_B\Theta_D - E}{\epsilon_f - E}\right) = -\frac{1}{V\mathcal{N}(\epsilon_f)}, \tag{10.1.7}$$

which has a solution, provided V is negative (attractive), giving a binding energy $W = \epsilon_f - E$ of

$$W \simeq k_B\Theta_D\ e^{-1/|V|\mathcal{N}(\epsilon_f)}. \tag{10.1.8}$$

Thus, by forming the Cooper pair, the pair of electrons can lower the ground state energy of the system below the original fermi energy. This is obviously an unstable situation and every pair of electrons in the system will try to form Cooper pairs. In the process the fermi distribution which has been introduced in the above argument will become distorted, so that the whole problem must be treated self-consistently in order to arrive at the new ground state. This is what the wave function of Bardeen, Cooper and Schrieffer was designed to achieve.

However, the importance of the above argument is to show that, whatever the strength of the interaction V, it will lead to an instability provided the interaction is attractive. This will happen even if the potential is extremely weak compared to the fermi energy of the system. The argument also suggests that the binding energy of the new state cannot be obtained as a power series expansion in terms of the interaction strength, thus implying that perturbation theory with respect to the normal ground state of the fermi gas cannot possibly converge in the superconducting state.

10.2. THE SUPERCONDUCTING INSTABILITY AT FINITE TEMPERATURES

In order to demonstrate the detailed nature of the instability of the electron gas, which we did heuristically in Sec. 10.1 using the Cooper argument, we need to follow the example of the magnetism case and look at a response function which diverges as the instability point is approached. This will happen at finite temperatures as T decreases, thus determining the superconducting critical temperature T_c.

As we saw in Sec. 10.1, the instability results from a tendency for electron pairs to bind together. This suggests that the instability will be seen by examining the propagator for two electrons inserted at time 0 and removed from the system at time t. Unfortunately the process described by this function is rather hard to observe directly in the laboratory. (Something like it can be seen in a tunneling experiment in which electrons are inserted at one side of the sample through a thin oxide film; however, such measurements tend to be dominated by single-particle injection.) The resulting two-electron propagator is therefore more in the nature of a mathematical construct rather than an easily measured response function.

Since we need to discuss the temperature dependence of the electron pairing phenomenon rather carefully we work with temperature Green's functions, as introduced in Secs. 2.2 and 3.3, instead of zero-temperature time-dependent Green's functions. We consider the two-electron propagator, describing correlations between particles of opposite spin,

$$F(\mathbf{p}, \mathbf{p}', \mathbf{q}; \sigma) = \langle T[\tilde{a}_{\mathbf{p}+\mathbf{q}\uparrow}(\sigma)\tilde{a}_{-\mathbf{p}\downarrow}(\sigma)\tilde{a}^{\dagger}_{\mathbf{p}'+\mathbf{q}\uparrow}(0)\tilde{a}^{\dagger}_{-\mathbf{p}'\downarrow}(0)]\rangle\,, \qquad (10.2.1)$$

where σ is a temperature variable ($-\beta \leqslant \sigma \leqslant \beta$, $\beta = 1/k_B T$), $\mathbf{q}$ is measured in the center-of-mass frame of the pair, and the brackets denote a thermodynamic average with respect to the interacting electron B.C.S. hamiltonian

$$H = H_0 + H_1,$$

$$H_1 = -\tfrac{1}{2}V \sum_{\substack{\{\mathbf{p},\mathbf{p}',\mathbf{q}\}\\ \sigma,\sigma'}} a^{\dagger}_{\mathbf{p}+\mathbf{q},\sigma}a^{\dagger}_{\mathbf{p}'-\mathbf{q},\sigma'}a_{\mathbf{p}'\sigma'}a_{\mathbf{p}\sigma}\,. \qquad (10.2.2)$$

(Note that $V > 0$ now corresponds to an attractive potential. σ, σ' are spin suffixes.) As discussed above, V arises from electron–phonon coupling in which a single phonon is exchanged between a pair of electrons. This introduces an energy dependence of the effective interaction which is is simulated in the B.C.S. model by the procedure of cutting off the momentum sums which we used in Sec. 10.1 on each side of the fermi level. However, we are no longer restricting to $\mathbf{q} = 0$, as we will show that this limitation arises naturally in the formation of the instability.

We now expand (10.2.1) by a finite-temperature Feynman–Dyson expansion, as discussed in Sec. 3.3. In the resulting diagrams the unperturbed electron lines have the propagator (independent of spin)

$$G^0(\mathbf{p}, \sigma) = \langle T[\tilde{a}_{\mathbf{p}}(\sigma)\tilde{a}_{\mathbf{p}}^{\dagger}(0)]\rangle_0, \qquad (10.2.3)$$

where the thermodynamic average is taken with respect to the grand canonical hamiltonian

$$H_0 = \sum_{\mathbf{p}} \epsilon_{\mathbf{p}} a_{\mathbf{p}}^{\dagger} a_{\mathbf{p}}, \tag{10.2.4}$$

in which

$$\epsilon_{\mathbf{p}} = \frac{\mathbf{p}^2}{2m} - \mu \tag{10.2.5}$$

and μ is the chemical potential of the electrons. The unperturbed function (10.2.3) has been evaluated in Sec. 4.7. We found there [see Eq. (4.7.12)] that the Fourier coefficients in the expansion

$$G^0(\mathbf{p}, \sigma) = \sum_{\bar{\nu}} e^{i\bar{\nu}\sigma} G^0(\mathbf{p}, \bar{\nu}) \qquad (-\beta \leqslant \sigma \leqslant \beta) \tag{10.2.6}$$

are given by

$$G^0(\mathbf{p}, \bar{\nu}) = \frac{1/\beta}{i\bar{\nu} + \epsilon_{\mathbf{p}}}, \tag{10.2.7}$$

where

$$\bar{\nu} = \frac{\pi}{\beta}(2\nu + 1) \qquad (\nu = 0, \pm 1, \pm 2, \ldots), \tag{10.2.8}$$

(the restriction to odd integers being a general property of fermions).

Since (10.2.1) is a two-electron propagator, the T-product does not involve a change of sign and the σ-dependence of $F(\mathbf{p}, \mathbf{p}', \mathbf{q}; \sigma)$ has boson symmetry as in the phonon Green's function case of Sec. 2.2, leading to a Fourier expansion

$$F(\mathbf{p}, \mathbf{p}', \mathbf{q}; \sigma) = \sum_{\bar{\mu}} e^{i\bar{\mu}\sigma} F(\mathbf{p}, \mathbf{p}', \mathbf{q}; \bar{\mu}), \tag{10.2.9}$$

with

$$\bar{\mu} = \frac{\pi}{\beta} 2\mu \qquad (\mu = 0, \pm 1, \pm 2, \ldots). \tag{10.2.10}$$

In zero order $F(\mathbf{p}, \mathbf{p}', \mathbf{q}; \sigma)$ can be written as a product of one-particle Green's functions, as discussed in Sec. 5.6. The unperturbed value

$F^0(\mathbf{p}, \mathbf{p}', \mathbf{q}; \bar{\mu})$ of the Fourier coefficient in (10.2.9) is given by

$$F^0(\mathbf{p}, \mathbf{p}', \mathbf{q}; \bar{\mu}) = -\sum_{\bar{\nu}} G^0(\mathbf{p} + \mathbf{q}, \bar{\mu} + \bar{\nu}) G^0(-\mathbf{p}, -\bar{\nu}) \delta_{\mathbf{pp}'}, \qquad (10.2.11)$$

corresponding to a single pair of electron lines. Following the Cooper pair argument, the singular behavior will appear when we consider diagrams in which the electrons in the pair scatter from each other (Fig. 10.1). Because of the momentum independence of V in the B.C.S. model

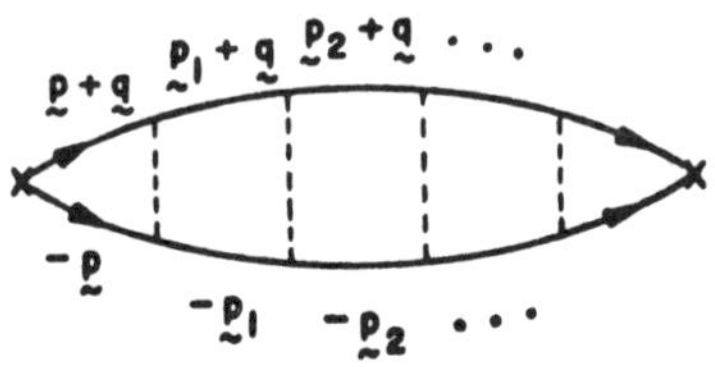

Fig. 10.1. Electron–electron scattering diagrams.

hamiltonian (10.2.2) the resulting sums over $\mathbf{p}_1, \mathbf{p}_2, \ldots$, separate and, with $F(\mathbf{q}) = \sum_{\mathbf{p}} F(\mathbf{p}, \mathbf{p}, \mathbf{q})$, one simply has

$$F(\mathbf{q}, \bar{\mu}) = \frac{F^0(\mathbf{q}, \bar{\mu})}{1 + \beta V F^0(\mathbf{q}, \bar{\mu})}, \qquad (10.2.12)$$

where $F^0(\mathbf{q}, \bar{\mu})$ is now a restricted sum over the internal pair momentum $\mathbf{p}$,

$$F^0(\mathbf{q}, \bar{\mu}) = \sum_{\{\mathbf{p}\}} F^0(\mathbf{p}, \mathbf{p}, \mathbf{q}; \bar{\mu}), \qquad (10.2.13)$$

and $\{\mathbf{p}\}$ is such that $\epsilon_{\mathbf{p}}$ is within $\pm k_B \Theta_D$ of the fermi surface (taken as $\epsilon_{\mathbf{p}} = 0$).

Notice that, since V is attractive, the denominator of (10.2.12) has $+V$ in contrast to the repulsive-potential case of magnetism (Sec. 7.3), so that no instability will occur in (10.2.12) for electron-hole propagation. However, in the electron–electron case, the propagator $F^0(\mathbf{q}, \bar{\mu})$ has a negative sign which makes (10.2.12) unstable at low enough temperatures. We have

$$F^0(\mathbf{q}, \bar{\mu}) = -\frac{1}{\beta^2}\sum_{\{\mathbf{p}\},\bar{\nu}} \frac{1}{\{i(\bar{\nu}+\bar{\mu}) + \epsilon_{\mathbf{p+q}}\}\{-i\bar{\nu} + \epsilon_{-\mathbf{p}}\}}, \tag{10.2.14}$$

from which it may be seen that the absolute value of $F^0(\mathbf{q}, \bar{\mu})$ decreases as $\bar{\mu}$ increases, so that the most unstable term will come about for $\bar{\mu} = 0$.

The sum (10.2.14) is evaluated, as in the case of the phonon Green's function in Sec. 2.2, by representing it as a contour integral

$$F^0(\mathbf{q}, \bar{\mu}) = -\frac{i}{2\pi\beta}\sum_{\{\mathbf{p}\}} \int_C d\omega \frac{1}{(e^{\beta\omega}+1)} \cdot \frac{1}{(\omega + \epsilon_{\mathbf{p+q}} + i\bar{\mu})(-\omega + \epsilon_{\mathbf{p}})},$$

where C is a contour (Fig. 10.2) which surrounds the poles, $\omega = i\bar{\nu}$, of the fermi function $1/(e^{\beta\omega}+1)$ in the anticlockwise sense. The contour

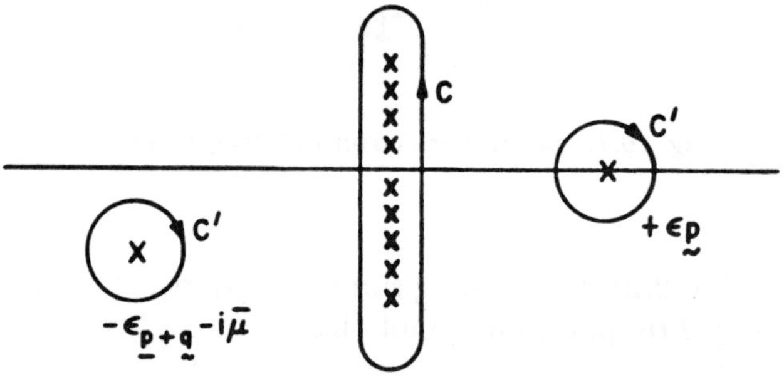

Fig. 10.2. Contours for evaluation of $F^0(\mathbf{q}, \bar{\mu})$.

C may then be deformed to a contour C' surrounding the two poles $\omega = -\epsilon_{\mathbf{p+q}} - i\bar{\mu}$, $\omega = \epsilon_{\mathbf{p}}$ of the Green's functions in the integrand in the clockwise sense. Evaluating the residues at these poles we obtain (putting $\bar{\mu} = 0$)

$$\begin{aligned} F^0(\mathbf{q}, 0) &= \frac{1}{\beta}\sum_{\{\mathbf{p}\}}\left\{\frac{1}{(e^{\beta\epsilon_{\mathbf{p}}}+1)(\epsilon_{\mathbf{p}} + \epsilon_{\mathbf{p+q}})} - \frac{1}{(e^{-\beta\epsilon_{\mathbf{p+q}}}+1)(\epsilon_{\mathbf{p}} + \epsilon_{\mathbf{p+q}})}\right\} \\ &= -\frac{1}{\beta}\sum_{\{\mathbf{p}\}} \frac{1 - f_{\mathbf{p+q}} - f_{\mathbf{p}}}{\epsilon_{\mathbf{p+q}} + \epsilon_{\mathbf{p}}} \\ &= -\frac{1}{2\beta}\sum_{\{\mathbf{p}\}} \frac{\tanh(\frac{1}{2}\beta\epsilon_{\mathbf{p}}) + \tanh(\frac{1}{2}\beta\epsilon_{\mathbf{p+q}})}{\epsilon_{\mathbf{p+q}} + \epsilon_{\mathbf{p}}}. \end{aligned} \tag{10.2.15}$$

One can now see that the maximum value of $F^0(\mathbf{q}, 0)$ will (as in the ferromagnetic instability) occur for $\mathbf{q} = 0$, leading to a denominator of $F(\mathbf{q}, 0)$ in (10.2.12) given by

$$1 + \beta V F^0(\mathbf{q} = 0, 0) = 1 - V \sum_{\{\mathbf{p}\}} \frac{\tanh(\frac{1}{2}\beta\epsilon_{\mathbf{p}})}{2\epsilon_{\mathbf{p}}}. \qquad (10.2.16)$$

On evaluation, integrating by parts, the sum over $\mathbf{p}$ becomes

$$\sum_{\{\mathbf{p}\}} \frac{\tanh(\frac{1}{2}\beta\epsilon_{\mathbf{p}})}{2\epsilon_{\mathbf{p}}} = \mathcal{N}(0)\left\{\log(\tfrac{1}{2}\beta k_B\Theta_D) - \int_0^{\frac{1}{2}\beta k_B\Theta_D} \operatorname{sech}^2 x \log x \, dx\right\}. \qquad (10.2.17)$$

In the second term the upper limit can then be extended to ∞ to give as the condition for the divergence of $F(\mathbf{q} = 0, 0)$

$$1 = V\mathcal{N}(0) \log\left(\frac{2\gamma\beta k_B\Theta_D}{\pi}\right), \qquad (10.2.18)$$

where $\log \gamma$ is Euler's constant, equal to 0·577. . . . Thus, as T is reduced, a critical temperature T_c is reached, given by

$$T_c = 1{\cdot}14\Theta_D \, e^{-1/V\mathcal{N}(0)}, \qquad (10.2.19)$$

at which the electron pair propagator becomes singular. The notable thing about this singularity is that it occurs however weak the electron-electron attraction V. (Of course, T_c becomes very small for small V, but it is still non-zero in contrast to the case of the ferromagnetic instability which only occurs if V exceeds a critical value.) In real metals the screened Coulomb repulsion between the electrons tends to counteract the phonon-induced attraction. However, the frequency-dependence of the phonon-induced attraction extends over a much smaller frequency range because of the finite response time of the lattice which turns out to favor the attractive term. To study this effect one needs to put in the phonon propagator explicitly—but we shall not go into that question here. The theory of the actual interaction causing superconductivity and its relation to experiment is an interesting topic reviewed by Scalapino (1969).

A final point: to establish that the singularity of the pair propagator really leads to a phase transition one needs to show that the free energy has a discontinuous slope at T_c. This can be done by studying contributions to the free energy in which electron pairs and hole pairs occur by spontaneous fluctuation and then propagate by the above resonant scattering mechanism. The diagrams representing repeated scattering of pairs look as in Fig. 10.3. By evaluating this contribution to the free

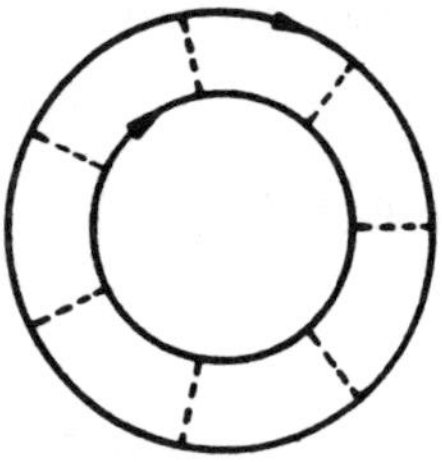

Fig. 10.3. Electron pair fluctuations leading to the superconducting instability.

energy (which is closely analogous to the paramagnon term in Sec. 7.6), one can show that the specific heat of the superconductor starts to diverge as T gets close to T_c [Thouless (1960)]. Such precursor fluctuation effects also show up in the diamagnetism of the superconductor. As T_c is approached from above it turns out that the magnetization starts to increase in a divergent fashion till at T_c the Meissner effect (perfect diamagnetism) sets in [Patton *et al.* (1969)]. Analogous effects are found in the resistive behavior of thin superconducting films above T_c [see Ambegaokar (1968)], leading to a gradual disappearance of the resistance as the temperature is lowered instead of a sharp discontinuity.

10.3. THE SUPERCONDUCTING GROUND STATE

Once the instability of the electron gas occurs, the ground state has to change its character qualitatively in order to go into a stable configuration. In the superconductor it turns out that the change occurs by a mechanism rather analogous to that in the metallic ferromagnet. In order to discuss this process we will formulate the problem in terms of a new way of writing the electron creation and annihilation operators which is specifically designed to deal with the new correlations which occur between pairs of electrons of opposite momentum and spin, in addition to describing the usual single-electron states. One of the lessons to be learned from the Cooper problem in Sec. 10.1 and from the more detailed discussion in Sec. 10.2 is that, if a pair of electrons has a non-zero center-of-mass momentum, this will simply increase the kinetic energy of the pair, since the potential energy only depends on the relative momentum in the pair. This suggests that the ground state of the electron gas will involve pairing between electrons of equal and opposite momentum **p**. The argument of Sec. 10.2 also suggests that the lowest energy state will involve pairs whose respective electrons have equal and opposite spin. This is more complicated to demonstrate explicitly and will not be established here. However, it is quite plausible on the grounds that, when a pair of plane waves, constituting a component of the Cooper pair, overlap in configuration space, the Pauli principle will tend to prevent electrons of the same spin lying in the same region of configuration space; it thus prevents them from benefiting from the effects of the attractive interaction and so reduces the resulting lowering of energy.

The creation operator generalization we use was suggested by Nambu (1960). It consists of setting up two-component creation operators which then behave as spinors. We define

$$\begin{aligned} \psi^{\dagger}_{\mathbf{p}\alpha} &= a^{\dagger}_{\mathbf{p}\uparrow} \qquad \text{for } \alpha = +1, \\ &= a_{-\mathbf{p}\downarrow} \qquad \text{for } \alpha = -1. \end{aligned} \tag{10.3.1}$$

Here α is a spinor suffix in a sub-space which is formally closely analogous to isotopic spin space in elementary-particle physics. It may now be seen that, while the operator

$$\psi^{\dagger}_{\mathbf{p},1}\psi_{\mathbf{p},1} = a^{\dagger}_{\mathbf{p}\uparrow}a_{\mathbf{p}\uparrow}$$

represents the usual number operator for electrons in state **p** with spin up, the operator

$$\psi^{\dagger}_{\mathbf{p},1}\psi_{\mathbf{p},-1} = a^{\dagger}_{\mathbf{p}\uparrow}a^{\dagger}_{-\mathbf{p}\downarrow}$$

represents instead the operator which creates a component of a Cooper pair. By introducing the Pauli matrices τ^1, τ^2, τ^3 in the new spinor sub-space we see that the operator

$$\begin{aligned} \tau_{\mathbf{p}}{}^3 &= \sum_{\alpha\beta} \psi^{\dagger}_{\mathbf{p}\alpha}\tau^3_{\alpha\beta}\psi_{\mathbf{p}\beta} = a^{\dagger}_{\mathbf{p}\uparrow}a_{\mathbf{p}\uparrow} - a_{-\mathbf{p}\downarrow}a^{\dagger}_{-\mathbf{p}\downarrow} \\ &= a^{\dagger}_{\mathbf{p}\uparrow}a_{\mathbf{p}\uparrow} + a^{\dagger}_{-\mathbf{p}\downarrow}a_{-\mathbf{p}\downarrow} - 1 \end{aligned} \tag{10.3.2}$$

(where we have used the anticommutation rules for the $a_{\mathbf{p}}, a_{\mathbf{p}}^{\dagger}$) represents the number of particles in states **p** and −**p**, while the operator

$$\tau_{\mathbf{p}}^{+} = \sum_{\alpha\beta} \psi_{\mathbf{p}\alpha}^{\dagger} \tau_{\alpha\beta}^{+} \psi_{\mathbf{p}\beta} = 2a_{\mathbf{p}\uparrow}^{\dagger} a_{-\mathbf{p}\downarrow}^{\dagger} \tag{10.3.3}$$

is the creation operator for a pair. Here we have used

$$\tau^{+} = \tau^{1} + i\tau^{2} = \begin{pmatrix} 0 & 2 \\ 0 & 0 \end{pmatrix}. \tag{10.3.4}$$

We now consider how the new representation will help to describe the problem of the gas of electrons interacting via the attractive electron–electron interaction of the form given in (10.1.2). To do this we want to restrict the interaction hamiltonian to the terms which only operate between pairs of electrons of opposite momentum and spin. In doing so we are throwing away a very large number of interaction terms which would operate between pairs of electrons with non-zero center-of-mass momentum. As discussed above, we do not want to include these pairs in the ground state of the superconductor as they will tend to increase the kinetic energy contribution to the ground state energy. The resulting reduced hamiltonian—the so-called "pairing hamiltonian"—contains all the essential features needed to describe the superconducting state: it has been shown that the neglected terms do not, if treated properly, destroy the features of the superconducting state which are described by the pairing hamiltonian. In what follows we shall therefore confine ourselves to the problem described in terms of the pairing hamiltonian. With this restriction, the hamiltonian for the interacting electron gas may be written in a very simple way directly in terms of the $\tau_{\mathbf{p}}$ operators introduced above. We have

$$\begin{aligned} H &= \sum_{\mathbf{p}} \epsilon_{\mathbf{p}}(a_{\mathbf{p}\uparrow}^{\dagger} a_{\mathbf{p}\uparrow} + a_{-\mathbf{p}\downarrow}^{\dagger} a_{-\mathbf{p}\downarrow}) - V \sum_{\mathbf{p}\mathbf{p}'} a_{\mathbf{p}\uparrow}^{\dagger} a_{-\mathbf{p}\downarrow}^{\dagger} a_{-\mathbf{p}'\downarrow} a_{\mathbf{p}'\uparrow} \\ &= \sum_{\mathbf{p}} \epsilon_{\mathbf{p}}(\tau_{\mathbf{p}}^{3} + 1) - \tfrac{1}{4} V \sum_{\mathbf{p}\mathbf{p}'} \tau_{\mathbf{p}}^{+} \tau_{\mathbf{p}'}^{-}, \end{aligned} \tag{10.3.5}$$

and using (10.3.4) this can be rewritten as

$$H = \sum_{\mathbf{p}} \epsilon_{\mathbf{p}}(\tau_{\mathbf{p}}^{3} + 1) - \tfrac{1}{4} V \sum_{\mathbf{p}\mathbf{p}'} (\tau_{\mathbf{p}}^{1} \tau_{\mathbf{p}'}^{1} + \tau_{\mathbf{p}}^{2} \tau_{\mathbf{p}'}^{2}). \tag{10.3.6}$$

It may now be seen that in this representation the pairing hamiltonian resembles quite closely the hamiltonian of an insulating Heisenberg ferromagnet. The kinetic energy term is rather like the interaction $-\mathcal{H}\cdot\sum_i \mathbf{S}_i$ with an applied magnetic field, while the interaction term represents a transverse spin–spin interaction in isospin space [compare Eq. (8.3.8)]. The main differences from the Heisenberg model are, first, that the magnetic-field-like term has a strength that depends on the momentum variable $\mathbf{p}$; it will again be convenient to measure $\boldsymbol{\epsilon}_{\mathbf{p}}$ relative to the fermi energy:

$$\epsilon_{\mathbf{p}} = \frac{\mathbf{p}^2}{2m} - \epsilon_f. \tag{10.3.7}$$

Second, the interaction term occurs between the isospin operators for particles in state $\mathbf{p}$ and a whole continuum of states $\mathbf{p}'$, restricted by the definition (10.1.2) of the interaction matrix elements, rather than near neighbor spins as in the Heisenberg model. However, the analogy is close enough to suggest a solution of the pairing problem in terms of the Weiss molecular field theory, or Hartree–Fock approximation as it becomes in the electron gas case. In this approximation the ground state is assumed to contain a non-zero isospin expectation value $\langle \boldsymbol{\tau}_{\mathbf{p}} \rangle$. In the absence of the kinetic energy term this can be chosen to point along (say) the 1-axis in isospin space as the problem is clearly rotationally invariant in the 1–2 plane. This assumption, however, implies a property of the ground state which is at first sight rather unphysical. Inserting the definition of $\langle \tau_{\mathbf{p}}{}^1 \rangle$ we have

$$\langle \tau_{\mathbf{p}}{}^1 \rangle = \langle a^\dagger_{\mathbf{p}\uparrow} a^\dagger_{-\mathbf{p}\downarrow} \rangle + \langle a_{-\mathbf{p}\downarrow} a_{\mathbf{p}\uparrow} \rangle.$$

This implies that the new ground state does not have a definite number of fermions in it, as this would lead to the vanishing of the diagonal element for a pair of diagonal creation or annihilation operators. The variable particle number would not be a problem for a system in contact with a reservoir of particles, as one normally has in the grand canonical distribution. Thus the ground state must be regarded as an admixture of states with different total numbers of particles, sharply peaked about an average number N. The average number of particles can be identified with the particle number of an isolated superconductor, and, to order $1/N$, all results obtained for the system with variable N hold for an isolated system with fixed N.

The effect of the magnetic-field-like kinetic energy term will be to pull the ground state isospin in the direction of the 3-axis. This corresponds to the mixing of pairing and empty states in the original B.C.S. variational wave function formulation of the problem:

$$\Psi_{BCS} = \prod_{\mathbf{p}} (u_{\mathbf{p}} + v_{\mathbf{p}} a^{\dagger}_{\mathbf{p}\uparrow} a^{\dagger}_{-\mathbf{p}\downarrow}) |0\rangle.$$

In order to study the nature of this ground state in more detail, we consider the equation of motion in the Nambu representation for the one-particle Green's function, defined as

$$G_{\alpha\beta}(\mathbf{p}, t) = -i\langle T[\psi_{\mathbf{p}\alpha}(t)\psi^{\dagger}_{\mathbf{p}\beta}(0)]\rangle. \tag{10.3.8}$$

Using this generalized one-particle Green's function, the averaged isospin vector in the ground state can be written as

$$\langle \boldsymbol{\tau}_{\mathbf{p}} \rangle = -\lim_{t \to 0-} i \sum_{\alpha\beta} G_{\alpha\beta}(\mathbf{p}, t) \boldsymbol{\tau}_{\alpha\beta}. \tag{10.3.9}$$

It will be convenient to write this in spinor notation as a trace over the spinor components,

$$\langle \boldsymbol{\tau}_{\mathbf{p}} \rangle = -\lim_{t \to 0-} i \,\mathrm{Tr}\, \{\boldsymbol{\tau} \mathcal{G}(\mathbf{p}, t)\}, \tag{10.3.10}$$

where $\mathcal{G}(\mathbf{p}, t)$ represents the Green's function $G_{\alpha\beta}(\mathbf{p}, t)$ in matrix notation. Using the anticommutation rules we have

$$[\psi^{\dagger}_{\mathbf{p}\alpha}\psi_{\mathbf{p}\beta}, \psi_{\mathbf{p}\gamma}] = -\delta_{\alpha\gamma}\psi_{\mathbf{p}\beta}, \tag{10.3.11}$$

from which we find

$$\begin{aligned} i\frac{\partial}{\partial t} G_{\alpha\beta}(\mathbf{p}, t) - \epsilon_{\mathbf{p}} \sum_{\gamma} \tau^{3}_{\alpha\gamma} G_{\gamma\beta}(\mathbf{p}, t) &= \delta_{\alpha\beta}\delta(t) \\ &+ \tfrac{1}{2} iV \sum_{\gamma,\mathbf{p}'} \{\tau^{1}_{\alpha\gamma}\langle T[\psi_{\mathbf{p}\gamma}(t)\tau_{\mathbf{p}'}{}^{1}(t)\psi^{\dagger}_{\mathbf{p}\beta}(0)]\rangle \\ &+ \tau^{2}_{\alpha\gamma}\langle T[\psi_{\mathbf{p}\gamma}(t)\tau_{\mathbf{p}'}{}^{2}(t)\psi^{\dagger}_{\mathbf{p}\beta}(0)]\rangle\}. \end{aligned} \tag{10.3.12}$$

The Weiss molecular field, or Hartree–Fock, approximation now consists in replacing the operator combination $\tau_{\mathbf{p}'}{}^{1}$ appearing in the right-hand side of (10.3.12) by its average value $\langle \tau_{\mathbf{p}'}{}^{1}\rangle$. Owing to the rotational invariance, discussed above, of the pairing hamiltonian in the 1–2 plane, we can without loss of generality set $\langle \tau_{\mathbf{p}'}{}^{2}\rangle = 0$. There will also be a Fock exchange term in which the $\psi_{\mathbf{p}\gamma}$ is averaged with the $\psi_{\mathbf{p}'}^{\dagger}$ occurring inside $\tau_{\mathbf{p}'}^{1}$. However, we will neglect this term as it is not important for the restricted problem of the pairing hamiltonian. For the full hamiltonian it would lead to the usual one-particle exchange term of the normal fermi system. In spinor notation the resulting average equation for the Fourier transform $\mathscr{G}(\mathrm{p}, \omega)$ may be written

$$\{\omega I - \epsilon_{\mathbf{p}}\tau^3 + \tfrac{1}{2}V\tau^1 \sum_{\mathbf{p}'} \langle \tau_{\mathbf{p}'}{}^1 \rangle\} \mathscr{G}(\mathrm{p}, \omega) = I, \tag{10.3.13}$$

where I is a unit matrix. From this we have

$$\mathscr{G}(\mathrm{p}, \omega) = \frac{I}{\omega I - \epsilon_{\mathbf{p}}\tau^3 + \Delta\tau^1}; \tag{10.3.14}$$

here we have written

$$\Delta = \tfrac{1}{2}V \sum_{\mathbf{p}'} \langle \tau_{\mathbf{p}'}{}^1 \rangle. \tag{10.3.15}$$

(10.3.14) may be rationalized by multiplying top and bottom with the matrix $\omega I + \epsilon_{\mathbf{p}}\tau^3 - \Delta\tau^1$ (and using the fact that τ^3 and τ^1 anticommute), to give

$$\mathscr{G}(\mathrm{p}, \omega) = \frac{\omega I + \epsilon_{\mathbf{p}}\tau^3 - \Delta\tau^1}{\omega^2 - (\epsilon_{\mathbf{p}}{}^2 + \Delta^2) + i\eta} \tag{10.3.16}$$

(the infinitesimal $i\eta$, with $\eta > 0$, has been added to characterize $\mathscr{G}$ as a time-ordered Green's function). The self-consistency condition (10.3.10), with (10.3.15), thus gives the solution for Δ as

$$\Delta = \tfrac{1}{2}V \sum_{\mathbf{p}'} \frac{1}{2\pi i} \int_C d\omega \ \mathrm{Tr}\, \{\tau^1 \mathscr{G}(\mathrm{p}', \omega)\}. \tag{10.3.17}$$

Here the contour C lies in the upper half-plane, and picks up the lower sign of the two zeros of the denominator of $\mathscr{G}(\mathrm{p}, \omega)$, which are the "quasiparticle" poles of the one-particle Green's function in the superconducting state:

$$E_{\mathbf{p}} = \pm\sqrt{(\epsilon_{\mathbf{p}}{}^2 + \Delta^2)}. \tag{10.3.18}$$

On performing the contour integral we are led to a self-consistent equation for Δ, the *energy gap* in the elementary excitation spectrum of a superconductor:

$$\Delta = \tfrac{1}{2}V \sum_{\mathbf{p}'} \frac{\Delta}{\sqrt{(\epsilon_{\mathbf{p}}{}^2 + \Delta^2)}}. \tag{10.3.19}$$

Replacing the sum over $\mathbf{p}'$ by an integral over $\epsilon_{\mathbf{p}'}$, and setting the density of states constant over the restricted range of integration $-k_B\Theta_D < \epsilon_{\mathbf{p}'} < k_B\Theta_D$, we finally have as the equation for Δ

$$\tfrac{1}{2}\mathcal{N}(0)V \int_{-kB\Theta D}^{kB\Theta D} \frac{d\epsilon}{\sqrt{(\epsilon^2 + \Delta^2)}} = 1. \tag{10.3.20}$$

This leads to

$$\mathcal{N}(0)V \sinh^{-1}\left(\frac{k_B\Theta_D}{\Delta}\right) = 1, \tag{10.3.21}$$

from which, for small V, we have

$$\Delta \simeq 2k_B\Theta_D\, e^{-1/\mathcal{N}(0)V}. \tag{10.3.22}$$

The above approach, which is due essentially to Anderson (1958), demonstrates the analogy between the superconducting ground state and the ferromagnetic ground state of Chap. 7. The big difference is that the symmetry breaking is here in isospin space (corresponding to symmetry breaking of the particle-conservation invariance of the original hamiltonian), as opposed to the real spin space asymmetry in the case of the ferromagnet.

Chapter 11

Strongly correlated electron systems: heavy fermions; the one–dimensional electron gas.

11.1 Heavy Fermions and Slave Bosons

In 1979 Steglich and collaborators made the discovery that an intermetallic compound of the rare earth metal cerium, $CeCu_2Si_2$, became a superconductor at $\sim 0.5^o$K. Then in 1983 Ott, Rudiger, Fisk and Smith found that the intermetallic compound UBe_{13} was superconducting at around 0.7^o K. These findings were quite surprising since it had been one of the basic results of the BCS theory of phonon-driven superconductivity that strong Coulomb repulsion, believed to be important in the narrow band f-electron compounds, would tend to suppress superconductivity and possibly inhibit it completely, as was known to happen in the transition metal palladium.

These discoveries lead to intensified studies of the rare earth compounds. Among other properties many of these compounds showed a remarkably enhanced coefficient γ of the electronic specific heat. Values of γ , of order 1 Joule/mol/deg^2 , corresponding to approximately 1000 times the effective mass at the Fermi level predicted by band structure calculations, were found. How is one to understand the enormous mass enhancements in these so-called "heavy fermion" metals?

By studying the low-temperature properties of a series of compounds in which cerium is replaced by the nonmagnetic element lanthanum, it became clear that the huge low-temperature specific heat was caused by the Kondo effect resulting from the interaction of the f electrons on the cerium exchanging with the conduction electrons in the host lattice (e.g. $LaCu_2Si_2$). To understand this phenomenon theoretically one needs to go beyond the perturbation expansion discussed in Chap. 9. To find nonperturbative solutions of the Kondo problem turned out to be a nontrivial piece of mathematical physics whose nature was first worked out by Wilson (1975). He showed that the character of the zero point fluctuations induced in the Fermi gas by coupling to a single magnetic impurity is analogous to that of the critical fluctuations of a classical system close to a second order phase transition. The analogy comes from realizing that the Kondo problem can be mapped on to a one–dimensional classical problem by means of a Wick rotation where the time axis is changed to an imaginary time axis $t \to i\tau$. Then the time course of the fluctuations in the conduction electron spin density seen by the impurity spin can be mapped on to a one-dimensional classical problem in which succeeding spin flips are coupled through interactions which are long range in the τ-variable (Anderson, Yuval, and Hamann 1970). Subsequently, exact solutions of the single impurity Kondo problem were found by Andrei (1980) and Wiegmann (1981).

Rather than go into the details of these solutions, we will instead adopt a nonperturbative approach based on the slave boson picture worked out by Coleman (1983) which becomes exact in the limit of large degeneracy $N = 2j + 1$ of the impurity with spin j. This method, as developed by Read and Newns (1983), allows the development of a mean field solution to the Kondo problem which is analogous to the BCS mean field treatment of superconductivity. This approach has the advantage that it can readily be extended to a lattice of Kondo spins where the exact solution methods cannot be applied.

We know from atomic physics and spectroscopic measurements of Ce compounds that there is a strong Coulomb repulsion between electrons attempting to occupy the Ce f orbitals. The energy of the f^2 state of Ce is approximately 10 eV above that of the f^1 state. On the other hand, the f^0 and f^1 states are quite close in energy when the Ce is in the intermetallic compound (where the charges are screened by the conduction electrons.) This leads to consideration of an Anderson lattice model for the Ce compounds:

$$\begin{aligned}\mathcal{H} &= \sum_i (v_{km} e^{i\vec{k}\cdot\vec{R}_i} d^\dagger_{im} c_k + h.c.) + \varepsilon^0_f \sum_{im} d^\dagger_{im} d_{im} + \sum_k \varepsilon_k c^\dagger_{k\sigma} c_{k\sigma} \\ &\quad + U \sum_{imm'} n_{dim} n_{dim'} \end{aligned} \tag{11.1}$$

where $d^\dagger_{im}$ creates an electron in the m'th orbital of Ce at the i'th site. $c^\dagger_{k\sigma}$ are the conduction electron Bloch state creation operators and U represents the repulsion between Ce electrons on the same site (although these are f-orbitals, we reserve the letter f for pseudo-fermion creation operators below.)

In the narrow f-band limit applicable to the cerium compounds, the ratio of the Hubbard U to the cerium f-band width becomes very large and one can take the $U \to \infty$ limit in which double occupancy of an f orbital is forbidden. This limit may be represented by the introduction of a slave boson operator $b^\dagger_i$ which measures the creation of an f^o state on site i. (This is to be contrasted with the "drone fermion" representation introduced in Chap. 9.)

In the slave boson picture we represent the electron operators, d_{im} by a product of slave boson operators and "pseudo fermion" operators, $f^\dagger_{im}$:

$$d^\dagger_{im} \to b_i f^\dagger_{im}. \tag{11.2}$$

Now to forbid double occupancy one needs to provide an additional constraint on allowable states in the Hilbert space of the problem at each site of the lattice:

$$(b^\dagger_i b_i + \sum_m f^\dagger_{im} f_{im})|\psi_i> = |\psi_i> . \tag{11.3}$$

Then ψ_i will include both f^0 states for which $< b^\dagger_i b_i >= 1$, and f^1 states for which $\sum_m f^\dagger_{im} f_{im} = 1$, but will exclude higher occupancy states.

To make progress in calculating properties of this model one may perform an expansion in $1/N$ where $N = 2j + 1$ is the degeneracy of the f- orbital of total spin + orbital angular momentum j.

This may be done in a simplified representation of (11.1) by rewriting the coupling constant v_{km} as

$$v_{km} \to \frac{v}{\sqrt{N}} = \tilde{v}. \tag{11.4}$$

Feynman diagrams for the electron propagator self-energy may then be classified in powers of $1/N$ depending on the number of vertices and sums over the f–state degeneracies in the diagram. The $N \to \infty$ limit will be represented by replacing the operators b_i by a mean field value $< b >$ where

$$z = < b_i^\dagger b_i > = < b >^2 \tag{11.5}$$

represents the average occupancy of the f^0 state and renormalizes the strength of the hybridization parameter $\tilde{v}$. The Hamiltonian may then be rewritten as

$$\mathcal{H} = \mathcal{H}_\infty + \mathcal{H}_{\text{int}} \tag{11.6}$$

where

$$\begin{aligned} \mathcal{H}_\infty &= \textstyle\sum_k \varepsilon_k c_{k\sigma}^\dagger c_{k\sigma} + \tilde{v} \sum (c_{k\sigma}^\dagger f_{im} + h.c.) + \varepsilon_f^0 \sum_{im} f_{im}^\dagger f_{im} \\ &\quad + \lambda \textstyle\sum_i (< b^\dagger b > + f_{im}^\dagger f_{im} - 1). \end{aligned} \tag{11.7}$$

Here λ is a variational parameter introduced to enforce the $U \to \infty$ constraint in the mean field limit. $\mathcal{H}_{\text{int}}$ allows for fluctuations of $b^\dagger b$ around the mean field limit:

$$\mathcal{H}_{\text{int}} = \frac{v}{\sqrt{N}} \sum (\delta b_i^\dagger f_{im} + h.c.) \tag{11.8}$$

where $\delta b_i = b_i - < b >$. Note that the electron energies $\varepsilon_k, \varepsilon_f^0$ implicitly contain the chemical potential μ ($= \varepsilon_F$ at $T = 0$) since we are working in a grand canonical ensemble so that (6) makes sense even if λ is < 0.

11.2 The Single Impurity Case

To see how the slave boson mean field solution to the Kondo problem works, we first consider the single impurity version of equation (11.7):

$$\begin{aligned} \mathcal{H}_\infty &= \mathcal{H}_{\text{conduction}} + \textstyle\sum_m \varepsilon_f^{\text{ren}} f_m^\dagger f_m + < b > \tilde{v} \sum_{km} (f_m^\dagger c_{km} + h.c.) \\ &\quad + (\varepsilon_f^{\text{ren}} - \varepsilon_f^0)(< b >^2 - 1) \end{aligned} \tag{11.9}$$

where $\varepsilon_f^{\text{ren}} = \varepsilon_f^0 + \lambda$ is now a renormalized single impurity energy shifted by the variational parameter λ relative to the bare impurity energy. The mean field solution is now found by calculating the ground state energy of (11.9) and minimizing with respect to λ (i.e., the renormalized f-state energy) and z following Newns and Read (1988). To do this we make use of the identity

$$\frac{\partial}{\partial\lambda} < \Psi(\lambda)|\mathcal{H}(\lambda)|\Psi(\lambda) > = < \Psi(\lambda)|\frac{\partial\mathcal{H}(\lambda)}{\partial\lambda}|\Psi(\lambda) > \tag{11.10}$$

where $|\Psi(\lambda) >$ is the ground state of $\mathcal{H}(\lambda)$ (the Hellman–Feynman theorem, see also Sec. 1.2). We are then led to the constraint equation

$$z = < b >^2 = 1 - \sum_m < n_{fm} > \tag{11.11}$$

and the self-consistency equation

$$-\frac{1}{\sqrt{z}} N \sum_{km} < f_m^\dagger c_{km} > = \varepsilon_f^{\mathrm{ren}} - \varepsilon_f^0. \tag{11.12}$$

Since (11.9) is a single impurity scattering problem the resulting Green's functions may be easily calculated following Sec. 4.4. Re-summing the series expansion in powers of $\tilde{v}$ for the Green's function

$$G_{ff}(t) = -i < T\{f_m(t) f_m^\dagger(0)\} > \tag{11.13}$$

gives

$$G_{ff}(\varepsilon) = [\varepsilon - \varepsilon_f - \Sigma(\varepsilon)]^{-1} \tag{11.14}$$

where the self-energy $\Sigma(\varepsilon)$ is given by

$$\Sigma(\varepsilon) = \sum_k \frac{z\tilde{v}^2}{\varepsilon - \varepsilon_f + i\epsilon} \tag{11.15}$$

For energy scales small compared with the conduction electron band width, and putting the Fermi level at the middle of the band (for convenience), the real part of $\Sigma(\varepsilon)$ goes away by symmetry and we are just left with the imaginary part

$$\Sigma(\varepsilon) \cong -i\pi z\tilde{v}^2 \rho(\varepsilon_F) \tag{11.16}$$

where $\rho(\varepsilon)$ is the conduction electron density of states $\rho(\varepsilon) = \sum_k \delta(\varepsilon - \varepsilon_k)$.

The f-level component of the one-particle density of states is given by

$$\rho_f(\varepsilon) = -\frac{1}{\pi}\mathcal{I}m G_{ff}(\varepsilon) = \frac{\Delta/\pi}{(\varepsilon - \varepsilon_f^{\mathrm{ren}})^2 + \Delta^2} \tag{11.17}$$

where Δ is a renormalized resonance level width for the f-state

$$\Delta = \pi z \tilde{v}^2 \rho(\varepsilon_F) = z\Delta_0. \tag{11.18}$$

The self-consistency condition (11.12) now involves calculation of

$$E_{\text{int}} = \sqrt{z}\tilde{v} \sum_k < f_m^\dagger c_{km} > \tag{11.19}$$

which may be obtained in terms of G_{ff} by noting that

$$G_{fc}(\varepsilon) = -i < T\{f_m(t) c_{km}^\dagger(0)\} > |_\varepsilon = G_k^0(\varepsilon)\sqrt{z}\tilde{v} G_{ff}(\varepsilon), \tag{11.20}$$

where we have taken the Fourier transform and $G_k^0(\varepsilon)$ is the conduction electron Green's function.
Hence

$$\sum_{km} < f_m c_{km}^\dagger >= -\frac{\sqrt{z}\tilde{v}}{\pi} \mathcal{I}m \int_{-\infty}^{\infty} d\varepsilon G_{ff}(\varepsilon) \sum_k G_k^0(\varepsilon). \tag{11.21}$$

At zero temperature the integral may be done by contour integration to give

$$\begin{aligned} \sum_{km} < f_m c_{km}^\dagger > &= -\tfrac{\sqrt{z}\tilde{v}}{\pi} \mathcal{I}m \textstyle\int_{-\infty}^{\varepsilon_F} d\varepsilon \frac{\rho(\varepsilon)}{\varepsilon - \varepsilon_f^{\text{ren}} + i\Delta} \\ &\cong \tfrac{\sqrt{z}\Delta_0}{\pi \tilde{v}} \ln\left(\tfrac{(\varepsilon_f^{\text{ren}})^2 + \Delta^2}{D^2}\right) \end{aligned} \tag{11.22}$$

where D, which is the energy of the Fermi level with respect to the bottom of the conduction band, measures the conduction band width, and we have treated $\rho(\varepsilon)$ as constant on energy scales large compared with Δ.

The self-consistency condition (11.12) now gives

$$\frac{1}{\sqrt{z}} N E_{\text{int}} = \varepsilon_f^{\text{ren}} - \varepsilon_f^0 = \frac{N\Delta_0}{\pi} \ln\{\frac{\sqrt{(\varepsilon_f^{\text{ren}})^2 + \Delta^2}}{D}\} \tag{11.23}$$

which may be solved in the limit that ε_f^0 is large and negative as

$$\sqrt{(\varepsilon_f^{\text{ren}})^2 + \Delta^2} \equiv T_K = De^{-\pi|\varepsilon_f^0|/N\Delta_0} \tag{11.24}$$

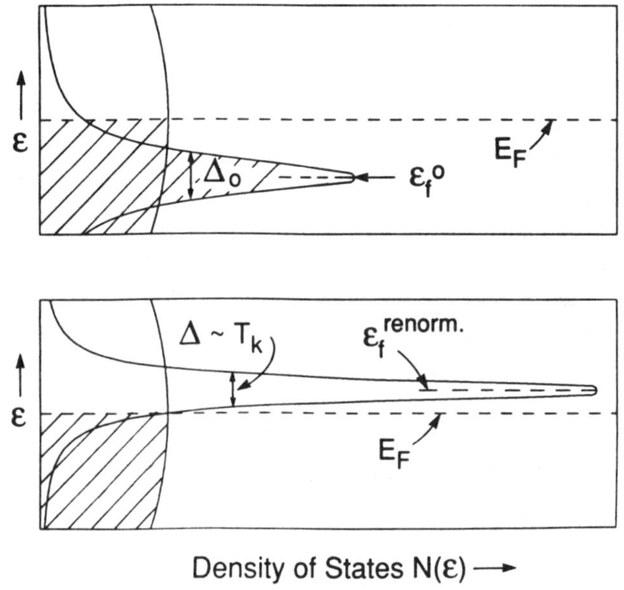

Figure 11.1: Density of f-electron states for a single impurity in the non–interacting case ($U = 0$) upper panel and renormalized case, lower panel. $\varepsilon_f^{\text{ren}}$ is renormalized so that the f occupancy is ≤ 1.

which becomes very small for large $|\varepsilon_f^0|$.

How to understand the physics of this mean field derivation of the Kondo energy scale? For $U = 0$ one would just have N impurity resonance levels of width Δ_0 centered at ε_f^0 (see Fig.(11.1)). For ε_f^0 above the Fermi level this would give a mixed valence state in which $z = n_b = < b >^2$ is of $\mathcal{O}(1)$ and Δ would be only slightly renormalized. However, as ε_f^0 becomes negative relative to the Fermi level, electrons would pour into the impurity state and the amplitude for double occupancy would become large. When U is switched on, the constraint term $(\varepsilon_f^{\text{ren}} - \varepsilon_f^0)(b^\dagger b + \sum_m f_m^\dagger f_m)$ pushes the resonance back to above the Fermi level while at the same time narrowing it. The occupancy of the f-state then becomes of order unity and $< b^\dagger b >= \Delta/\Delta_0 \cong T_K/\Delta$ very small. Physically one can think of this in a time–dependent picture as due to a delay in the amplitude for electrons to hop on to the impurity. Because of U being large, they have to wait until an f^1 electron hops off, creating a slave boson, before a succeeding electron can hop on again. This delay has the effect of reducing the hopping amplitude $\tilde{v}$ by the factor $\sqrt{z}$ to become $\sqrt{z}\tilde{v}$ so that the resonance level width is reduced from $\Delta_0 = \pi\rho(\varepsilon_F)\tilde{v}^2$ to $\Delta = z\Delta_0$. It also has the effect of shifting the f-level energy up since, in a time-dependent Hartree-Fock picture, one now has an additional oscillating factor of the form $|\psi(t) >= e^{i\lambda t}|\psi >\equiv e^{i(\varepsilon_f^{\rm ren}-\varepsilon_f^0)t}|\psi >$, which may be thought of as measuring the rate at which the boson occupancy turns on and off: $b^\dagger(t) = e^{i\mathcal{H}_\infty t}b^\dagger e^{-i\mathcal{H}_\infty t} \equiv e^{i\lambda t}b^\dagger$. Then equation (11.11) adjusts this shift of the bare f-level up to just above the Fermi level to guarantee that the occupancy of the f-level is $n_f = \sum_m < f_m^\dagger f_m >= 1- < b >^2$.

11.3 The physical electron propagator expressed in terms of the pseudo–fermion Green's function.

The mean field limit introduced above leaves out one very important piece of physics. In the mean field limit, the one-electron propagator

$$G_{f-\text{electron}} = -i < T\{c_{fm}(t)c_{fm}^\dagger(0)\} > \qquad (11.25)$$

is represented in terms of pseudo-fermions and slave bosons as

$$G_{f-\text{electron}}(t) \cong -i < b >^2 < T\{f_m(t)f_m^\dagger(0)\} > \qquad (11.26)$$

so that the density of one-electron states measured via

$$\rho_{f-\text{electron}} \cong \frac{1}{\pi}\mathcal{I}mG_{f-\text{electron}}(\varepsilon) \qquad (11.27)$$

is strongly reduced by the factor $z =< b >^2 \cong T_K/\Delta_0$ in the mean field limit. What happened to the full density of states guaranteed by the electron commutator $\{c_m, c_{m'}\} = \delta_{m,m'}$?

This would apply, for instance, in a photoemission experiment where the observation of a fast outgoing electron has guaranteed that there is a full hole (i.e., not a fraction of a hole) left behind. The f-electron photoemission cross section can then be measured via

$$\sigma(\varepsilon) = \frac{1}{\pi}\mathcal{I}mG_{f-\text{electron}}(\varepsilon) \qquad (11.28)$$

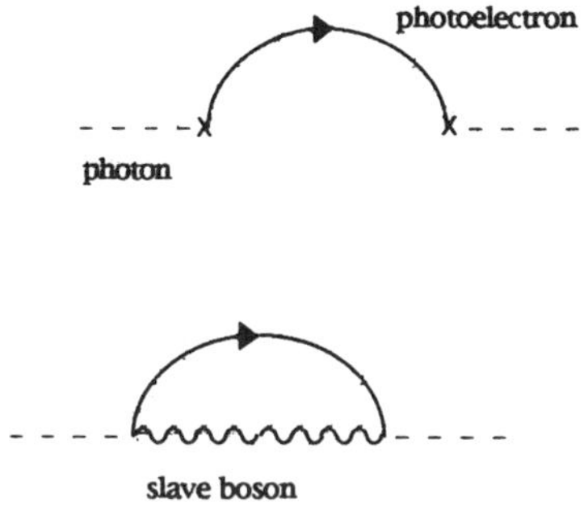

Figure 11.2: Slave boson correction term for the photoemission cross section. The x's denote the condensate amplitude $< b >$.

where ε measures the energy loss of the outgoing photoelectron caused by creating the hole (see Chap. 9), so that the sum rule reads

$$\frac{1}{\pi}\int_{-\infty}^{0} d\varepsilon G_{f-\text{electron}}(\varepsilon) \equiv \lim_{t\to 0^-} G_{f-\text{electron}}(t) \equiv < c_f^\dagger c_f > \tag{11.29}$$

which measures the f-occupancy in the ground state. As we have seen this is of $\mathcal{O}(1)$ in the Kondo limit, not of $\mathcal{O}(< b >^2)$.

The answer is seen by looking at the anti-commutator of the physical electron operators, expressed in terms of the pseudo-fermions in Eq. (11.2):

$$\{c_m, c_m^\dagger\} = \{b^\dagger f_m, f_m^\dagger b\} \equiv b^\dagger b + f_m^\dagger f_m \tag{11.30}$$

where we have made use of the the boson commutator $[b, b^\dagger] = 1$. In the mean field limit, on the other hand, one would have $\{c_m, c_m^\dagger\} \sim < b >^2$. So it is necessary to treat the b field dynamically to preserve the physical electron sum rules. This can be done approximately by including a boson fluctuation diagram in the expression for the photoemission cross section as shown in the diagram (Fig.(11.2)). The full photoemission cross section may then be written

$$\sigma(\varepsilon) \cong \frac{< b >^2}{\pi}\mathcal{I}mG_{ff}(\varepsilon) + \frac{1}{\pi}\int d\omega D(\omega) G_{ff}(\varepsilon + \omega). \tag{11.31}$$

where $D(\omega)$ denotes the boson propagator, $D(\omega) = [\omega - (\varepsilon_f^{\text{ren}} - \varepsilon_f^0) + \Pi(\omega)]^{-1}$. To work consistently to this order one also needs to include self

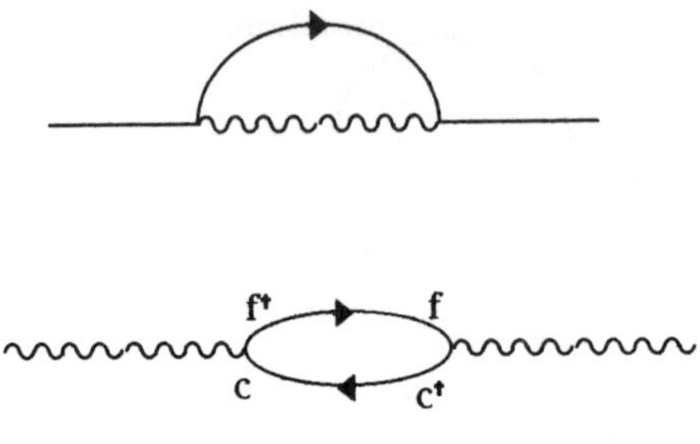

Figure 11.3: Leading order contributions to the boson and fermion self energies.

energy diagrams for the boson and pseudo-fermion propagators shown in Fig. 11.3. The net result of these corrections is that the one–particle spectral density as seen, eg, in the photoemission cross section from a Ce compound contains both a very narrow Kondo resonance just above the Fermi level with reduced weight, of order T_K/Δ_0, and a broader peak in the vicinity of the unrenormalized f-level ε_f^0 which carries most of the 1-hole spectral weight.

11.4 Instabilities of heavy fermion systems: antiferromagnetic and superconducting states.

The interest in the heavy fermion materials comes from the fact that many of the single Kondo impurity properties are also found to extend to pure crystalline intermetallic compounds such as $CuAl_3$ which have greatly enhanced specific heat coefficients γ at low temperatures. Here the theory is on much shakier grounds since the exact solution methods applicable to the single impurity problem fail for the lattice problem. Nevertheless we can still be guided by the $1/N$ expansion to get a qualitative grasp of the physics of the heavy fermion materials.

Experimentally these systems exhibit a variety of behaviors at low temperatures, depending on their detailed chemistry, which include remaining paramagnetic as $T \to 0$ (eg $CeAl_3$), becoming antiferromagnetic at low temperatures (eg $CeCu_6$), or going superconducting ($CeCu_2Si_2$, UBe_{13}, UPt_3) - see, e.g., the review by Steglich (1992) (Fig. 11.4). So the fundamental question arises as to what physical pa-

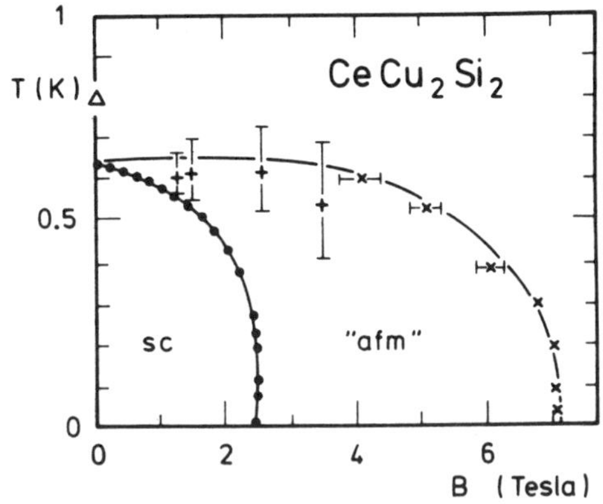

Figure 11.4: Phase diagram of $CeCu_2Si_2$ indicating superconducting and antiferromagnetic regimes (from Steglich(1992)).

rameters control these various instabilities of the heavy fermion compounds, and what is the nature of their various ground states.

To discuss these properties we now go back to reconsider the Anderson lattice model 11.1. At this point it becomes necessary to define the hybridization matrix elements between the conduction band states and the atomic f-levels more rigorously. In the simple impurity problem, one only needs to worry about the point group properties of the conduction band states since they only need to be expanded about the single impurity center. In the lattice problem on the other hand we are concerned with the space group of the conduction electron – f-electron hybrid bands which are much more dependent on the detailed band structure for the various compounds. This is complicated by the fact that the electrons need to treated relativistically in the high Z compounds so that spin–orbit coupling becomes very large.

To simplify the discussion we follow Newns and Read (1988) and consider an idealized SU(N) lattice model in which the degeneracy of the conduction bands is taken to be identical to that of the f-levels. Then, introducing the slave boson representation, the lattice model becomes

$$\mathcal{H} = \sum_{km} \varepsilon_{km} c^{\dagger}_{km} c_{km} + \varepsilon^0_f \sum_i f^{\dagger}_{im} f_{im} + \tilde{v} \sum_{im} (f^{\dagger}_{im} c_{im} b_i + \text{h.c.}) \quad (11.32)$$

where $c_{im} = (1/\sqrt{N}) \sum_{km} e^{ikR_i} c_{km}$. Introducing a Lagrange multiplier to enforce the slave boson constraint we then have in the mean field limit

$$\begin{aligned} \mathcal{H}_{\text{MF}} &= \textstyle\sum_{km} \varepsilon_{km} c^{\dagger}_{km} c_{km} + \varepsilon^0_f \sum_i f^{\dagger}_{im} f_{im} + <b> \tilde{v} \sum_{km} (f^{\dagger}_{km} c_{km} + \text{h.c.}) \\ &\quad + \lambda(\textstyle\sum_{im} f^{\dagger}_{im} f_{im} + <b>^2 - 1). \end{aligned} \quad (11.33)$$

Minimizing with respect to λ and $z = <b>^2$ as in the single impurity case we now get the constraint equation

$$z = 1 - \sum_m < f^{\dagger}_{im} f_{im} > \quad (11.34)$$

and the self-consistency condition

$$\varepsilon^{\text{ren}}_f - \varepsilon^0_f = \frac{1}{2\sqrt{z}} \tilde{v} \sum_{km} (f^{\dagger}_{km} c_{km} + \text{h.c.}). \quad (11.35)$$

We see that (11.33) now represents a set of N degenerate hybrid bands with energy spectrum

$$E^{\pm}_k = \frac{(\varepsilon^{\text{ren}}_f + \varepsilon_k)}{2} \pm \sqrt{\frac{(\varepsilon^{\text{ren}}_f - \varepsilon_k)^2}{2} + 4z\tilde{v}^2} \quad (11.36)$$

with a narrow hybridization gap as illustrated in Fig. 11.5. The density of states

$$\begin{aligned} \rho(\varepsilon) &= \textstyle\sum_k (\delta(\varepsilon - E^+_k) + \delta(\varepsilon - E^-_k)) = \int d\varepsilon \rho^0(\varepsilon) \frac{dE}{d\varepsilon} \delta(\varepsilon - E^{\pm}_k) \\ &= \rho^0(\varepsilon) \{ \tfrac{1}{2} \pm \frac{\varepsilon^{\text{ren}}_f - \varepsilon_k}{E^{\pm}_k} \} \end{aligned} \quad (11.37)$$

now has strongly enhanced peaks on each side of the gap.

To evaluate the self-consistency condition (11.35) we write it in terms of the Green's functions of $\mathcal{H}_{\text{MF}}$ (11.33) via

$$< f^{\dagger}_{im} c_m > = G_{fc}(t \to 0^+) \quad (11.38)$$

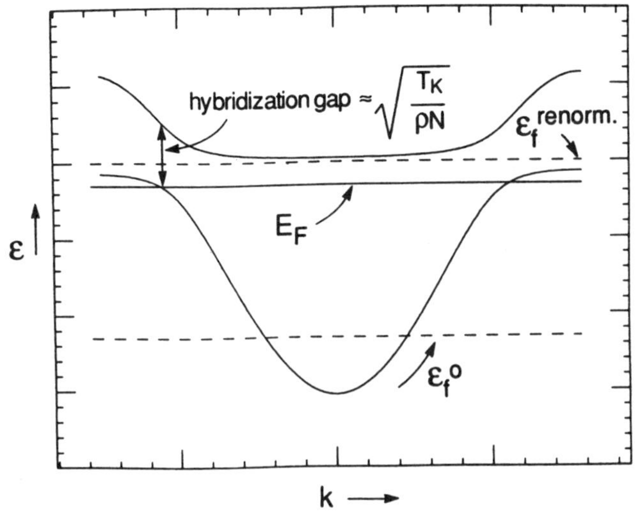

Figure 11.5: Sketch of the renormalized bands for the heavy fermion problem

and using $G_f c = \sqrt{z}\tilde{v}G_f^0 G_{cc}$ where

$$G_{cc}(\varepsilon) = \sum_k G_{cc}(k,\varepsilon) = \frac{1}{N}\sum_k \frac{1}{(\varepsilon + i\epsilon - \varepsilon_k - z\tilde{v}^2/(\varepsilon - \varepsilon_f + i\epsilon))}. \tag{11.39}$$

The evaluation of G_{cc} (11.39) now leads to the conduction electron projection of the density of states

$$\rho_{cc}(\varepsilon) = \rho_0(\varepsilon - z\tilde{v}^2/(\varepsilon - \varepsilon_f)) \tag{11.40}$$

where ρ_0 is the unperturbed conduction band density of states $(1/N)\sum_k \delta(\varepsilon - \varepsilon_k)$. If we assume that ρ_0 varies very slowly in the region of hybridization with the f–levels, then we can replace $\rho_0(\varepsilon)$ by a constant over the range $-D \leq \varepsilon \leq +D$ where D is a conduction band width parameter, and (11.40) simply introduces a hybridization gap into this constant density of conduction band states. The constraint condition (11.34) on the f–level occupancy now forces the Fermi level to be below the hybridization gap so that only one f-level (out of the possible N degenerate f- levels) is occupied per f-atom. (11.38) may now be evaluated by contour integration to give

$$\begin{aligned}\varepsilon_f^{\text{ren}} - \varepsilon_f^0 &= N\tilde{v}^2 \int_{-\infty}^{\infty} d\varepsilon G_f^0(\varepsilon) G_{cc}(\varepsilon) \\ &= \frac{N\tilde{v}^2}{2D} \int_{-D}^{\mu} d\varepsilon \frac{1}{(\varepsilon - \varepsilon_f^0)} \end{aligned} \tag{11.41}$$

leading to

$$\varepsilon_f^{\text{ren}} - \varepsilon_f^0 = N\tilde{v}^2 \ln((\mu - \varepsilon_f^0)/(-D - \varepsilon_f^0)). \tag{11.42}$$

In the limit that $\mu - \varepsilon_f^0$ is $>>$ Δ_0 where $\Delta_0 = \pi\tilde{v}^2/2D$, one finds a solution

$$n_f = \frac{1}{1 + \pi T_K / N\Delta_0} \tag{11.43}$$

with $T_K \simeq D \exp(-(\mu - \varepsilon_f^0)/N\Delta_0) << 1$. Thus in this Kondo limit the density of states becomes very large close to the Fermi level, with curvature $m/m^* = dE_k/d\varepsilon_k$ of order $(T_K/N\Delta_0)$ so that m^* can become extremely large, explaining the heaviness of the fermions.

Because of the strong correlations, the heavy fermion gas will have a tendency to order antiferromagnetically since the f spins will couple by polarizing the conduction electron spins, the so-called Ruderman-Kittel-Kasuya-Yosida (RKKY) interaction. The effective conduction electron - f spin coupling is shown to be of order $J \simeq \tilde{v}^2\rho^0$, with ρ^0 the conduction electron density of states, by the Schrieffer-Wolff transformation. From the point of view of the $1/N$ expansion, this will correspond to exchanges of pairs of bosons between a spin up particle and a spin down hole. Doniach(1977) showed that at sufficiently small values of T_K, the magnetic ordering will always win out against the RKKY ordering, since the RKKY ordering energy, $\propto J^2\rho^0$, will eventually dominate the Kondo energy, of order $T_K \propto D\exp(-1/J\rho^0)$ as J is reduced. So there is a question of how the heavy fermion metals remain paramagnetic down to a few ^{o}K before ordering antiferromagnetically, or in some cases remaining paramagnetic down to 0^oK. Read, Newns and Doniach (1984) pointed out that the large N degeneracy will tend to stabilize the paramagnetic state since the RKKY interaction energy will tend to vary as $\mathcal{O}(1/N)$ relative to the Kondo energy. So the phase boundary between antiferromagnetic and paramagnetic order will be sensitive to parameters, such as crystal field splitting, which will change the effective f-state degeneracy. (Doniach 1987).

The other important physical question is how the heavy fermions can go superconducting even though they interact via the very strong Coulomb repulsion. As may be seen from the above $1/N$ expansion, the repulsion is essentially removed by the correlations, as expressed via the slave bosons, leaving the residual RKKY interaction which is of order $(1/N)^2$. The interaction term between f-spins

$$\mathcal{H}_{\mathrm{RKKY}} = \sum_{ij} J_{\mathrm{RKKY}}(\vec{S}_{fi} \cdot \vec{S}_{fj} - \frac{1}{4} n_{fi} n_{fj}) \tag{11.44}$$

where $J_{\mathrm{RKKY}} \sim J^2 \rho^0$ can be rewritten as an attraction between fermions (see discussion in section 12.3 below)

$$\mathcal{H}_{\mathrm{RKKY}} \equiv -\frac{1}{2} \sum_{ij\sigma} J_{\mathrm{RKKY}}(f^\dagger_{i\sigma} f^\dagger_{j,-\sigma} f_{j,\sigma} f_{i,-\sigma} + f^\dagger_{i\sigma} f^\dagger_{j,-\sigma} f_{i,-\sigma} f_{j\sigma}). \tag{11.45}$$

So superconductivity may be induced by the effects of the Coulomb interactions in addition to possible electron–phonon effects as in conventional BCS superconductors.

Superconductivity in strongly correlated systems will be further discussed in Chap. 12. From the point of view of superconductors like UPt_3 it appears that the order parameter is probably d-wave –like rather than s-wave –like. However, owing to the strong spin orbit coupling, the use of separate spin and orbit quantum numbers does not work for the uranium compounds and a more complex characterization of the BCS pairing has to be made. (Norman 1992).

Finally, as pointed out by Varma (1985), the heavy fermion mass will have a tendency to cut down the transition temperature for the heavy fermion superconductors since in the usual Eliashberg formalism the mass renormalization of the fermions shows up in terms of the renormalization of the spectral weight of the quasiparticles which is reduced by the factor z as discussed in section 11.2 above. This then has the effect of reducing T_c via

$$T_c = \Theta_D \exp(-1/z\lambda), \tag{11.46}$$

where λ is the effective coupling constant for superconducting pairing and Θ_D measures the energy scale of the exchanged bosons, whether phonons or the slave bosons of the correlated electron picture. Since z is so strongly reduced in the heavy fermion materials, this could provide an explanation of the very low values of T_c found in the heavy fermion superconductors. (Why this same effect does not seem to be happening in the high T_c superconductors remains a mystery which will be further discussed in Chap. 12)

11.5 Strong correlations in the one-dimensional electron gas: spinons and holons.

The large N expansion about the slave boson mean field limit discussed above is justified for the single impurity Kondo problem by the exact solutions of the problem. This is a "zero space dimensions one time dimension" quantum problem in which the zero-point fluctuations of the impurity spin are the result of time–dependent correlations between successive events where an electron hops on and off the impurity site in a correlated fashion with time scale set by the parameter Δ_0/T_K. The one–dimensional interacting Fermi gas provides another set of exactly soluble quantum field theories where the correlations are now manifested in a two-dimensional space with one space dimension and one time dimension. The new physics which happens in these models can be traced back to classical solutions of non-linear field theories in one dimension exemplified by the Korteweg-deVries equation for water waves in a canal. Here, in addition to the usual harmonic waves – corresponding to sound waves (charge and spin density waves in the electron gas) – there appears a new class of excitation: the soliton mode.

In the classical Korteweg-deVries model these are dynamical solutions of the equations of motion in which a "bump" in the height of water in the canal propagates at constant speed (neglecting dissipation) whose magnitude depends on the height of the bump and which can pass through a second soliton going in the opposite direction without getting perturbed, apart from a time delay as the two solitons pass through each other. In the quantum system the analogous excitations are quantized and represent fermion creation and annihilation operators which now appear as soliton-like structures in the charge and spin densities of the electron gas. For the interacting system the resulting excitations have radically different properties from the usual charge and spin density electron-hole pair excitations of a Fermi liquid so that the one-dimensional electron gas has come to be named a "Luttinger liquid" following Haldane (1981). Here we give a brief introduction to the theory of the Luttinger liquid and how its properties relate to those of the Hubbard model in one dimension following H.J. Schulz (1991).

We start by considering the electron gas in 1D without interactions:

$$\mathcal{H}_0 = -t \sum_{k\sigma} c^\dagger_{k\sigma} c_{k\sigma} \tag{11.47}$$

and observe that close to 1/2 filling, the band energies $\varepsilon_k = 2t(1 - \cos k) - \mu$ are approximately linear close to the Fermi "points" (i.e., the 1-dimensional Fermi surface) $k_F = \pm\pi/a(1-x)$ where x is the doping away from 1/2 filling. Then, following Tomonaga (1950) and Luttinger (1963), the low energy physics may be described in terms of excitations about the Fermi points in terms of left- and right-going fermions by writing $\mathcal{H}_0$ (Haldane 1981) as:

$$\begin{aligned}\mathcal{H}_0 &= v_F \textstyle\sum_k (k - k_F)\{c^\dagger_{k\sigma+} c_{k\sigma+} - < n_{k\sigma+} >_0\} \\ &+ v_F \textstyle\sum_k (-k - k_F)\{c^\dagger_{k\sigma-} c_{k\sigma-} - < n_{k\sigma-} >_0\}\end{aligned} \tag{11.48}$$

Here + and − refer to right- and left- going fermions and a constant term $< n_{k\sigma\pm} >_0 \equiv \theta(k_F \mp k)$ is subtracted out to define the ground state occupancy. An essential step has been to linearize the kinetic energy about the Fermi level $k_F = \pi N_0/L$ with L the length of the system and N_0 the total number of electrons. By doing this linearization, information about the lattice spacing of the original tight binding bands has been thrown out, but can be put back by additional "umklapp" terms in the Hamiltonian.

A key step in finding analytic solutions to the 1-D electron gas is Tomonaga's observation, later refined by Luttinger, Mattis and Lieb(1965), and Haldane, that one can represent the properties of $\mathcal{H}_0$ in terms of boson operators. This is seen by looking at the commutation rules of charge and spin density operators which can be defined by linear combinations of the spin up and spin down components of

$$\rho_{q\sigma\pm} = \sum_k c^\dagger_{k+q\sigma\pm} c_{k+q\sigma\pm} \text{ for } q \neq 0. \tag{11.49}$$

It has been pointed out by Haldane that since the total momentum operator P for the 1D electron gas with linearised dispersion commutes with the Hamiltonian (taking periodic boundary conditions), the Hilbert space can be divided into a number of sectors with different quantum numbers for P. To deal with this properly he defines an algebra of number operators and their conjugate raising and lowering operators which act at $q = 0$, ie on the center of mass motion of the system. This turns out to take care of divergences which arise in the theory at $q = 0$ when interactions are turned on. Here, however, we present a simplified treatment in which these operators are left out and the divergences are instead taken into account by a cutoff procedure initially introduced by Luther and Peschel (1974).

From the anticommutation rules for the c's one then finds

$$\begin{aligned} [\rho_{q+,\sigma}, \rho_{q'+,\sigma'}] &= \delta_{qq'}\delta_{\sigma\sigma'}\frac{Lq}{2\pi}, \\ [\rho_{q-,\sigma}, \rho_{q'-,\sigma'}] &= \delta_{qq'}\delta_{\sigma\sigma'}\frac{Lq}{2\pi}, \\ [\rho_{q+,\sigma}, \rho_{q'-,\sigma'}] &= 0. \end{aligned} \tag{11.50}$$

As emphasized by Haldane (1981) care must be taken to treat the various operators in normal ordered form. This is because, since the Tomonaga–Luttinger model (11.48) does not have a finite band width, the Fermi sea becomes infinite in the limit $L \to \infty$ and one has to be careful to avoid unphysical divergences.

Using the commutation rules (11.50) it may be shown that a Hamiltonian expressed in terms of density operators

$$\mathcal{H}_0 = \frac{2\pi}{L} \sum_{q>0,\sigma} \{\rho_{q+,\sigma}\rho_{-q+,\sigma} + \rho_{-q-,\sigma}\rho_{q-,\sigma}\} \tag{11.51}$$

has the same excitation spectrum as the fermion form of the Hamiltonian (11.47). $\mathcal{H}_0$ may equivalently be defined in terms of boson creation and annihilation operators:

$$\mathcal{H}_0 = v_F \sum_q |q| a^\dagger_{q\sigma} a_{q\sigma} \tag{11.52}$$

by writing $a^\dagger_{q\sigma} = \sqrt{2\pi/L|q|}\ (\rho_{q+,\sigma} + \rho_{q-,\sigma})$.

In terms of the a's one can then write

$$\begin{aligned} \rho_{q+,\sigma} &= \sqrt{L|q|/2\pi}\ \theta(q) a^\dagger_{q\sigma} \\ \rho_{q-,\sigma} &= \sqrt{L|q|/2\pi}\ \theta(-q) a_{q\sigma}. \end{aligned} \tag{11.53}$$

where θ is the unit step function, so that the low energy dynamics of the free fermi gas in 1D (ie in the limit that the fermion energies have been linearized around the Fermi points) may be completely defined in terms of boson excitations representing charge density and spin density oscillations.

Just as for other Bose systems such as phonons (lattice vibrations) or photons (Maxwell's equations) it is also sometimes convenient to formulate the 1D electron gas in terms of charge and spin density phase fields. To simplify the discusssion we will suppress the spin suffix for the time being and focus on charge density wave excitations. Then we can define a phase field

$$\phi(x) = -\frac{i\pi}{L}\sum_{q\neq 0}\frac{1}{q}e^{-\alpha|q|x/2-iqx}[\rho_{q+}+\rho_{q-}] \tag{11.54}$$

and its conjugate momentum density

$$\Pi(x) = \frac{1}{L}\sum_{q\neq 0}e^{-\alpha|q|x/2-iqx}[\rho_{q+}-\rho_{q-}] \tag{11.55}$$

where α, which will be allowed to $\to 0$, plays the role of a cutoff to ensure convergence at small q. It then follows from the commutation rules for the ρ's that $\phi(x)$ and $\Pi(x)$ are conjugate field operators satisfying Bose commutation rules

$$[\phi(x), \Pi(x')] = \frac{i\alpha}{\alpha^2+(x-x')^2} \simeq \delta(x-x'). \tag{11.56}$$

In terms of these fields, the unperturbed boson Hamiltonian $\mathcal{H}_0$ (11.53) may now be rewritten

$$\mathcal{H}_0 = \int dx[\Pi(x)^2 + (\partial_x\phi(x))^2] \tag{11.57}$$

Now we consider Luttinger's model for the interactions between fermions. In general the potential energy between electrons can be written in terms of the Fourier transform of the electron–electron interaction potential $V(x-x')$ in the form

$$\mathcal{H}_{\text{int}} = \sum_q V_q\rho_q\rho_{-q}. \tag{11.58}$$

Luttinger observed that if the interactions in (11.58) are restricted to apply only to those which lead to scattering events with small momentum transfer (ie excluding terms which would lead to scattering from one Fermi point to the other which would involve a momentum transfer of $\mathcal{O}(2k_F)$), then the Hamiltonian for the resulting interacting electron gas

$$\mathcal{H} = \mathcal{H}_0 + \mathcal{H}_{\text{int}} \tag{11.59}$$

where

$$\mathcal{H}_{\text{int}} = \frac{2}{L} \sum_{q>0} \{V_q \rho_{q+} \rho_{-q-} + V_{-q} \rho_{-q+} \rho_{q-}\} \tag{11.60}$$

or equivalently, using(11.53)

$$\mathcal{H}_{\text{int}} = 2 \sum_{q>0} \{V_q a_q^\dagger a_{-q}^\dagger + V_{-q} a_{-q} a_q\} \tag{11.61}$$

with $\mathcal{H}_0$ given in the form (11.52), has a form analagous to that of the BCS mean field Hamiltonian (cf Chap. 10). It can then be diagonalized by a simple Bogoliubov–Valatin rotation given by

$$\tilde{\mathcal{H}} = e^{iS} \mathcal{H} e^{-iS} \tag{11.62}$$

where

$$S = \frac{2\pi}{L} \sum_{\text{all} q} \frac{\varphi(q)}{q} \rho_{q,+} \rho_{-q,-} \tag{11.63}$$

and where $\tanh(2\varphi_q) = -V_q/\pi$. The transformed Hamiltonian is now diagonal in the new boson creation and anihilation operators and may be written

$$\tilde{\mathcal{H}} = v_F \sum_q |q| \ \text{sech}(2\varphi_q) \alpha_q^\dagger \alpha_q \tag{11.64}$$

where $\alpha_q^\dagger$ and α_q are the creation and anihilation operators in the transformed system. So the effect of the Luttinger interaction is to renormalize the velocity of the density waves in the interacting system, but otherwise keep the same form of the excitation spectrum as in the non–interacting theory.

A remarkable new property of this apparently innocuous theory now makes itself felt when one looks at the physical properties involving creation and annihilation of single fermions as would be seen in a photoemission experiment for example. To see this, we calculate the single electron Green's function

$$G_\nu(x,t) = -i\theta(t) < \psi_\nu(x,t), \psi_\nu^\dagger(0,0) > \tag{11.65}$$

where $\nu = \pm$ represent the right and left moving electrons. To make use of the canonical transformation defined in (11.62) we need to express the fermion creation operators $\psi^\dagger(x)$ in terms of the boson operators $\alpha^\dagger, \alpha$, or equivalently $\rho_{\nu,q}$. This is done by the following "bosonization" expression

$$\psi_+(x) = \frac{1}{\sqrt{2\pi\alpha}} e^{ik_F x + \phi_+(x)} \tag{11.66}$$

where

$$\phi_+(x) = \frac{2\pi}{L} \sum_{q>0} \frac{e^{-\alpha q/2}}{q} [\rho_{-q+} e^{iqx} - \rho_{q+} e^{-iqx}] \tag{11.67}$$

By using the commutation relations of the $\rho_{\pm,q}$ operators, it may be shown that (11.67) satisfies the anticommutation rules of the original fermions even though it is expressed entirely in terms of boson operators. $\phi_\pm(x)$ may also be written in terms of the phase field operators $\phi(x)$ and $\Pi(x)$ as

$$\phi_\pm(x) = \frac{i}{2}[\phi(x) \pm \int_{-\infty}^{x} dx' \Pi(x')]. \tag{11.68}$$

By noting that the phase field $\phi(x)$ (11.54) has the property

$$\frac{\partial \phi(x)}{\partial x} = 2\pi\rho(x), \tag{11.69}$$

and that $\exp(i\pi \int_{-\infty}^{x} dx' \Pi(x'))$ acts as a displacement operator for $\phi(x)$, we see that the act of creating an electron (ie a "blip" in ρ at position x) by applying the operator(11.66) at x corresponds to introducing a kink into $\phi(x)$ so that $\phi(x)$ at points to the right of x are π higher than $\phi(x)$ at points to the left of x . This is the quantum analogue of a classical soliton wave.

Now that the fermion operators have been re-expressed in terms of boson operators, the fermion Green's function is seen to involve the expectation value of products of exponentials of boson operators $< \psi(x)\psi^\dagger(x') > \equiv < \exp(-iB(x)\exp(iB(x') >$. As in the x-ray problem, this may be expressed in terms of the exponential of a boson propagator

$$< \psi(x)\psi^\dagger(x') > \sim \exp(- < B(x)B(x') >) \tag{11.70}$$

which can be calculated with the help of the identities, valid for operators A and B which are linear in the boson creation and annihilation operators:

$$e^{A+B} = e^A e^B e^{-[A,B]/2} \text{ and } < e^A >= e^{<A^2>/2} \tag{11.71}$$

Applying the diagonalizing transformation (11.62) then leads to

$$G(x,t) = -i\theta(t)\frac{1}{2\pi\alpha}e^{ik_F x}\{\exp[Q_1(x,t) + Q_2(x,t)] + (t \to -t, x \to -x)\} \tag{11.72}$$

where

$$\begin{aligned} Q_1(x,t) &= \tfrac{2\pi}{L}\textstyle\sum_{q>0}(\cosh\varphi - 1 + e^{-\alpha q})^2(e^{iqx - i\epsilon_q t} - 1)/q \\ Q_2(x,t) &= \tfrac{2\pi}{L}\textstyle\sum_{q>0}(\sinh\varphi)^2(e^{-iqx - i\epsilon_q t} - 1)/q \end{aligned} \tag{11.73}$$

where $\epsilon_q = v_F q \ \mathrm{sech}\varphi_q$. It may be seen that Q_1 and Q_2 vary logarithmically in x and t so that the fermion Green's functions exhibit power law correlations which are physically quite different from the single decaying plane-wave type of behavior one expects for Fermi liquid Green's functions. A striking consequence of this power law behavior is that the number density in momentum space, obtained from $\mathcal{I}mG(k, 0^+)$, turns out to behave as:

$$n_k \simeq -\mathrm{sign}(k - k_F)|k - k_F|^\eta \ . \tag{11.74}$$

Thus the discontinuous step which occurs in the momentum distribution for a Fermi liquid at the Fermi level is replaced by a power law singularity in the Luttinger liquid. If we compare these results with the perturbation expansion of the self energy valid for Fermi liquids and discussed in Chap. 5, we see that the quasiparticle peak in the spectral function $A(k,\omega)$ of the form $-\frac{1}{\pi}\mathcal{I}mG(k_F, 0^+) = z\delta(\omega) + A_{\text{incoh}}$ where A_{incoh} represents an incoherent background, no longer occurs, so that $z = 0$ and the electron propagator becomes totally incoherent. So the Luttinger liquid has the important property that the usual Fermi liquid quasiparticles are no longer well defined excitations of the system and are replaced by new objects which may be thought of as "spinons" and "holons".

To see how this works we apply the bosonisation approach to the 1D Hubbard model for which

$$\mathcal{H}_{\text{int}} = U\sum_i n_{i\uparrow}n_{i\downarrow} \tag{11.75}$$

includes both forward scattering terms of the Luttinger type, and backscattering terms in which electrons are scattered from one Fermi point to the other. For the case of the repulsive Hubbard model ($U > 0$) one can show (Schulz 1991) that the backscattering terms renormalize to zero for small q excitations and that the renormalized Hamiltonian, valid for the small q/v_F infrared properties of the problem, may be written in terms of a pair of non-interacting Luttinger liquids in which charge and spin density wave excitations formed as $\rho_q = \frac{1}{2}(\rho_{q\sigma} + \rho_{q-\sigma})$ and $\sigma = \frac{1}{2}(\rho_{q\sigma} - \rho_{q-\sigma})$ are completely decoupled. One can then define charge and spin density phase fields via

$$\begin{aligned} \phi_\rho(x) &= \phi_\uparrow(x) + \phi_\downarrow(x) \\ \phi_\sigma(x) &= \phi_\uparrow(x) - \phi_\downarrow(x) \end{aligned} \tag{11.76}$$

where $\phi_\uparrow(x) = i\sum_q \sqrt{\frac{2\pi}{L|q|}}(e^{-iqx}a_{q\uparrow} - e^{iqx}a^\dagger_{q\uparrow})$, and similarly for spin down. The charge density part of the Hamiltonian may then be written

$$\mathcal{H}_\rho = \int dx[\frac{\pi u_\rho K_\rho}{2}\Pi_\rho^2 + \frac{u_\rho}{2\pi K_\rho}(\partial_x\phi_\rho^2)] \tag{11.77}$$

where K_ρ is an interaction–dependent compressibility and $\Pi_\rho(x)$ is conjugate to $\phi_\rho(x)$. A similar expression may be written for the spin density.

The effect of the interactions is to renormalize the speed of sound given by v_F for the noninteracting system (11.52) differently for charge and spin density waves. The low temperature thermodynamic properties of the model now can be simply understood in terms of the properties of noninteracting "jellium" charge and spin density wave modes: since acoustic phonons in one dimension have a specific heat linear in T one has

$$\gamma/\gamma_0 = \frac{1}{2}(v_F/u_\rho + v_F/u_\sigma) \tag{11.78}$$

where γ_0 is the linear coefficient of the specific heat of the noninteracting gas, $\mathcal{H}_0$, and u_ρ, u_σ are the renormalized sound velocities of the charge and spin density waves.

As discussed above for the spinless case, the non-Fermi liquid physics becomes manifest when one starts to look at physical measurements involving fermion operators such as the one–electron Green's function. In particular the power law exponent in the number density in momentum space at the Fermi level (11.74) is given by $\eta = (K_\rho + K_\rho^{-1} - 2)/4$. In the limit of large U, K_ρ takes the value $1/2$, so that $\eta = 1/8$.

Because spin and charge degrees of freedom separate completely in the long wave limit (for repulsive U), the fermion creation operators create excitations which propagate independently for the spin and charge densities. For the 1D Hubbard model close to 1/2 filling these may be visualized qualitatively as a domain wall in the local antiferromagnetic order for the spin case

$$\downarrow\uparrow\downarrow\downarrow\uparrow\downarrow\uparrow\downarrow \tag{11.79}$$

and a local vacancy in the charge case

$$\downarrow\uparrow \bigcirc \downarrow\uparrow\downarrow\uparrow\downarrow \tag{11.80}$$

These have come to be called spinon and holon excitations. Note that one needs to include umklapp terms to see these lattice-dependent effects via the Luttinger liquid picture. The result of the quantitative analysis of these 1D problems is to show that spinons and holons are well defined excitations which move independently in 1D systems.

Chapter 12

High T_c Superconductivity

12.0 Introduction

The discovery by Bednorz and Mueller in 1986 of superconductivity above 30^oK in strontium doped lanthanum cuprate ($La_{2-x}Sr_xCuO_4$) opened up a new area of the physics of strongly correlated electron systems. Although much progress, both experimental and theoretical, has been made in understanding high T_c superconductivity since 1986, there is still no consensus at the time of writing this as to the underlying mechanism causing superconductivity in the cuprate compounds. In this chapter we will present some basic theoretical ideas illustrating some of the underlying physics relevant to high T_c , but the discussion is necessarily incomplete because of the controversial nature of the field.

12.1 Effective Hamiltonian for the cuprate compounds

All the high T_c materials found to date are based on compounds in which planes of CuO_2 are separated by a number of layers of intervening perovskite structures composed of oxygen and various counter ions such as La, Y, Ca, Bi, Th, Hg, etc. The fact that compounds in which single CuO_2 layers may be spaced by 10 Å or more of intervening insulating material can still have quite high T_c 's suggests that the basic superconducting unit is a two-dimensional CuO_2 layer. (Anderson(1992) argues strongly that at least two CuO_2 adjacent layers are

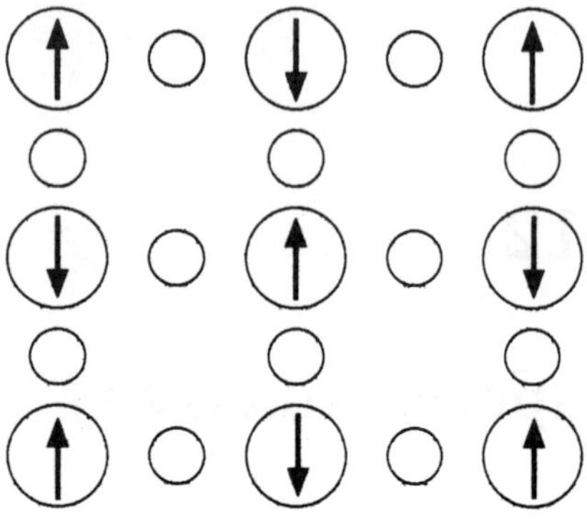

Figure 12.1: Illustration of the ground state configuration for the antiferromagnetic insulating state of a CuO_2 plane. Arrows represent d^9 spins. Intervening atoms are oxygen in a p^8 configuration.

needed for high T_c, but this claim remains controversial.)

In the insulating parent compounds from which superconductivity is produced by addition of dopants, such as La_2CuO_4, the stoichiometry leads to a d^9 configuration for the Cu^{2+} ions and filled p-shells for the O^{2-} ions. (Lanthanum forms La^{3+} ions.)

In the simplest possible picture, the holes in the Cu d^{10} shell should form a metallic band with one hole per Cu. The fact that La_2CuO_4 is an insulator with a 2eV band gap thus puts it into the class of Mott-Hubbard insulators. Since it is known from spectroscopic studies that the Cu d^8 state is about 8eV above the d^9 state, it is clear that Coulomb correlations will prevent the d^9 holes from tunneling from one Cu site to another when the band is 1/2 full. Experimentally La_2CuO_4 is an antiferromagnet, so that each d^9 state has an unpaired spin aligned opposite to the spin of its neighbors (see Fig.(12.1)).

Upon doping by replacing a few % of La with Sr^{2+}, one sees from simple valence counting that a corresponding number of oxygens will go from the O^{2-} (p^8) state to the O^-(p^7) state, i.e., will become doped with holes. A central issue is now whether these holes migrate independently of the Cu d^9 state, or whether they hybridize strongly with the

Cu states. The strongest evidence that the latter is true comes from NMR measurements of ^{63}Cu nuclei. It is found that the nuclear spin relaxation rate $1/T_1$ drops extremely rapidly as the temperature is reduced below the superconducting T_c . If the d^9 spins had remained in their antiferromagnetic configuration below T_c , one would expect $1/T_1$ to stay large as a result of the nuclear spins coupling to the antiferromagnetic spin waves. The rapid drop of $1/T_1$ below T_c shows that a "spin gap" resulting from singlet–like correlations must open up when the system becomes superconducting, which shows in turn that the d^9 configurations are very strongly perturbed by the doping–induced holes.

To represent this physics with a simple model Hamiltonian we need to include both the hybridization of the oxygen p- and copper d-orbitals and the effects of the strong correlations inhibiting the formation of d^8 states. To do this we use a three band model in which, in a tight binding picture, a p_x and a p_y orbital on each oxygen couple with a single $d_{x^2-y^2}$ on the Cu:

$$\mathcal{H} = \mathcal{H}_0 + \mathcal{H}_{\rm dp} + U \sum_i n_{di\uparrow} n_{di\downarrow}. \tag{12.1}$$

where

$$\mathcal{H} = \sum_i \{\varepsilon_d^0 d_{i\sigma}^\dagger d_{i\sigma} + \varepsilon_p^0 p_{i\nu\sigma}^\dagger p_{i\nu\sigma}\}, \tag{12.2}$$

and

$$\mathcal{H}_{\rm dp} = -t_{\rm pd}\{\sum_{i\nu\sigma} d_{i\sigma}^\dagger p_{i\nu\sigma} - \sum_{i'\nu\sigma} d_{i\sigma}^\dagger p_{i'\nu\sigma} + \text{h.c.}\}. \tag{12.3}$$

Here $d_{i\sigma}^\dagger$ represents a hole creation operator in the $d_{x^2-y^2}$ orbital of the i'th unit cell, while $p_{i\nu\sigma}^\dagger$ represents hole creation in the oxygen orbitals for each of the 2 oxygen in the unit cell. $p_{ix\sigma}$ anihilates a hole on the oxygen to the right of the Cu atom at site $\vec{R}_i$ and $p_{iy\sigma}$ acts on the oxygen above it (see Fig.(12.1)), while $p_{i'x\sigma}$ and $p_{i'y\sigma}$ act on the oxygen atoms to the left of and below the Cu atom at site i. The minus sign comes from the p–wave character of the overlap matrix elements, $t_{\rm pd}$, since the p-orbital is positive on one side and negative on the other. In this model the p_y orbital for the oxygen lying on a horizontal row of the CuO_2 plane does not bond with the Cu and neither does the p_x orbital for an oxygen in a vertical row. So solution of the tight binding Hamiltonian for $U = 0$ leads to two hybridized Cu-O bands and one non-hybridized oxygen level. In a real 3-D crystal the non-hybridized level will couple to atoms in adjacent planes and move down to lower energies, so we do not consider it further. We adopt the convention that $d^\dagger$ and $p^\dagger$ create holes (rather than electrons), so that the $U\sum_i n_{di\uparrow}n_{di\downarrow}$ term gives an energy cost to double occupancy by holes, i.e. tends to suppress the d^8 states.

To show the $U = 0$ solution explicitly we obtain the tight-binding bands of Eq.(12.10) by using the Fourier representation

$$\begin{aligned} d_{i\sigma} &= \frac{1}{\sqrt{N}}\sum_k e^{i\vec{k}\cdot\vec{R}_i} d_{k\sigma} \\ p_{ix\sigma} &= \frac{1}{\sqrt{N}}\sum_k e^{i\vec{k}\cdot(\vec{R}_i+\vec{a}_x/2)} p_{kx\sigma} \\ p_{iy\sigma} &= \frac{1}{\sqrt{N}}\sum_k e^{i\vec{k}\cdot(\vec{R}_i+\vec{a}_y/2)} p_{ky\sigma} \end{aligned} \tag{12.4}$$

and similarly for $p_{i'\nu\sigma}$ with $-\vec{a}_{x,y}/2$ in the exponent. Then we can rewrite the quadratic part of the Hamiltonian as

$$\mathcal{H}_0 + \mathcal{H}_{\rm dp} = \sum_{k\nu\nu'\sigma} \psi^\dagger_{k\nu\sigma} H_{k\nu\nu'} \psi_{k\nu'\sigma} \tag{12.5}$$

where

$$H_{k\nu\nu'} = \begin{pmatrix} \varepsilon_d^0 & it_{\rm dp}2\sin k_x/2 & it_{\rm dp}2\sin k_y/2 \\ -it_{\rm dp}2\sin k_x/2 & \varepsilon_p & 0 \\ -it_{\rm dp}2\sin k_y/2 & 0 & \varepsilon_p \end{pmatrix} \tag{12.6}$$

and $\psi^\dagger_{k\nu\sigma} = (d^\dagger_{k\sigma}, p^\dagger_{kx\sigma}, p^\dagger_{kx\sigma})$. Here $k_{x,y}$ denote $(k_{x,y}a)$, where a is the unit cell dimension. To separate out the non–bonding band it is convenient to apply a rotation matrix

$$\tilde{H}_{k\nu\nu'} = \underline{\underline{\rm R}}^T \underline{\underline{\rm H}}\ \underline{\underline{\rm R}} \tag{12.7}$$

where

$$\underline{\underline{\rm R}} = \begin{pmatrix} 1 & 0 & 0 \\ 0 & \cos\theta_k & \sin\theta_k \\ 0 & -\sin\theta_k & \cos\theta_k \end{pmatrix}. \tag{12.8}$$

Then setting $\cos\theta_k = \sin(\frac{k_x}{2})/\gamma_k$ and $\sin\theta_k = -\sin(\frac{k_y}{2})/\gamma_k$, where $\gamma_k = \sqrt{\sin(\frac{k_x}{2})^2 + \sin(\frac{k_y}{2})^2}$, one has

$$\tilde{H}_{k\nu\nu'} = \begin{pmatrix} \varepsilon_d^0 & it_{\rm pd}2\gamma_k & 0 \\ -it_{\rm pd}2\gamma_k & \varepsilon_p^0 & 0 \\ 0 & 0 & \varepsilon_p^0 \end{pmatrix} \tag{12.9}$$

so we have reduced the model to a 2–band problem.

The eigenvalues of $\tilde{H}_{k\nu\nu'}$, leaving aside the non–bonding band (which, as mentioned above will couple to other atoms surrounding the CuO_2 planes in a 3D crystal), are

$$E_{k\pm} = \frac{\varepsilon_p^0 + \varepsilon_d^0}{2} \pm R_k \tag{12.10}$$

where $R_k = \sqrt{(\frac{\Delta_0}{2})^2 + 4t_{\rm pd}^2\gamma_k^2}$ and $\Delta_0 = \varepsilon_p^0 - \varepsilon_d^0$. In the limit that energy separation, Δ_0, of the original oxygen and Cu orbitals is $\gg t_{\rm pd}$, we can expand in powers of $t_{\rm pd}/\Delta_0$ and see that the bonding and antibonding bands are each of width $8t_{\rm pd}^2/\Delta_0$. For $U = 0$ at the 1/2 filling condition the antibonding band is 1/2 full (of holes) and metallic, while the bonding band is full of electrons, i.e. empty of holes.

12.2 Effects of correlations in the CuO_2 planes.

What happens when we switch on U? The situation is now quite analogous to what happens in the mixed valent heavy fermion compounds discussed in Chap. 11. Since $U \sim 8$eV is known experimentally to be large compared with the tunneling matrix elements $t_{\rm pd}$, the energy scales are set by $t_{\rm pd}$ and we can effectively set $U = \infty$, which forbids the creation of d^8 states.

Exactly at 1/2 filling, an added electron on a Cu site forming a d^{10} configuration will migrate without involving a d^8 state by taking advantage of a Cu d^9 - O p^7 ligand–hole zero point fluctuation:

$$d^{10} - p^8 - d^9 - p^8 \longleftrightarrow d^{10} - p^7 - d^{10} - p^8 \longleftrightarrow d^9 - p^8 - d^{10} - p^8. \tag{12.11}$$

So in the limit of weak hybridization, $t_{\rm pd} << \varepsilon_p^0 - \varepsilon_d^0$, the chemical potential for adding an electron will correspond to that for converting a d^9 state to a d^{10} state, ie ε_d^0. On the other hand addition of a hole to the 1/2 filled state will involve formation of an oxygen p^7 state, since the d^8 states are forbidden, which again tunnels by using a ligand hole fluctuation:

$$d^9 - p^7 - d^9 - p^8 \longleftrightarrow d^9 - p^7 - d^{10} - p^7 \longleftrightarrow d^9 - p^8 - d^9 - p^7. \tag{12.12}$$

The chemical potential here will be that needed to convert from p^8 to p^7, ie ε_p^0. Hence there will be a "charge transfer gap" of $\Delta_0 = \varepsilon_p^0 - \varepsilon_d^0$

in the chemical potential for adding a hole versus that for adding an electron. It is this charge transfer gap which causes the 1/2 full state to be an insulator rather than a metal: creation of an electron-hole pair will cost energy of order Δ_0. As the hybridization matrix elements are increased, this gap will be diminished and eventually at large enough $t_{\rm pd}/\Delta_0$ will go to zero so that the system will be metallic.(See Zaanen, Sawatsky and Allen (1987).)

This physics can be represented quantitatively by introducing a slave boson field for the d operators:

$$d_{i\sigma} \rightarrow \tilde{d}_{i\sigma} b_i^\dagger \tag{12.13}$$

and adding a Lagrange multiplier term to the Hamiltonian to prevent double occupancy of the d-states, as in Sec. 11.1. Following Kotliar, Lee, and Read (1988) we can now solve the resulting mixed-valence problem by taking the mean field limit $b \rightarrow < b >$. Although this cannot be justified by the large N expansion since $N = 2$ in this case, we proceed to study this limit since it appears to capture some of the basic qualitative features of the charge transfer insulating state described above in pictorial terms.

The $U \rightarrow \infty$ Hamiltonian may be written

$$\begin{aligned} \mathcal{H}_\infty &= \textstyle\sum_{i\sigma} \varepsilon_d \tilde{d}_{i\sigma}^\dagger \tilde{d}_{i\sigma} + \sum_{i\nu\sigma} \varepsilon_p p_{i\nu\sigma}^\dagger p_{i\nu\sigma} \\ &+ < b > t_{\rm pd}\{\textstyle\sum_{i\nu\sigma} \tilde{d}_{i\sigma}^\dagger p_{i\nu\sigma} - \sum_{i'\nu\sigma} \tilde{d}_{i\sigma}^\dagger p_{i'\sigma} + \text{h.c.}\} \\ &+\lambda(\textstyle\sum_{k\nu\sigma} n_{\tilde{d}\nu\sigma} + < b >^2 - 1) \end{aligned} \tag{12.14}$$

where $\varepsilon_d = \varepsilon_d^0 - \mu$, $\varepsilon_p = \varepsilon_p^0 - \mu$ and λ is the Lagrange multiplier which will prevent occupancy of the d^8 state on average. $\mathcal{H}_\infty$ can now be solved in terms of tight binding bands Eq.(12.10) in which the tunneling matrix elements have been renormalized via

$$\tilde{t}_{\rm pd} =< b > t_{\rm pd} \equiv \sqrt{z} t_{\rm pd} \tag{12.15}$$

and the d–orbital energy has been shifted to

$$\varepsilon_d^{\rm ren} = \varepsilon_d + \lambda \tag{12.16}$$

We now vary the ground state energy with respect to λ (making use of the Hellman–Feynman theorem, Eq. (11.10)) to give the self consistency condition

$$z \equiv <b>^2 = 1 - \sum_{k<k_F,\sigma} u_k^2 \tag{12.17}$$

where u_k is the d–electron projection of the lower hole band eigenstates of (9), $u_k^2 = (1 + \Delta_{\rm ren}/R_k^{\rm ren})/2$, with $R_k^{\rm ren} = \sqrt{(\frac{\Delta_{\rm ren}}{2})^2 + 4zt_{\rm pd}^2\gamma_k^2}$. Variation with respect to $<b>$ then gives the second self consistency condition

$$\varepsilon_d^{\rm ren} - \varepsilon_d = \frac{t_{\rm pd}}{2\sqrt{z}} \sum_{k<k_F,\sigma} u_k v_k \gamma_k \tag{12.18}$$

where v_k is the p–electron projection $v_k^2 = (1 - \Delta_{\rm ren}/R_k^{\rm ren})/2$. Equations (12.17) and (12.18) may now be solved to give the renormalized d-band width and position in terms of the doping level of the CuO_2 band.

In the limit that $\Delta_0 = \varepsilon_p^0 - \varepsilon_d^0$ is $\gg t_{\rm pd}$, this can be done analytically by expanding in powers of $t_{\rm pd}/\Delta_0$. For doping δ holes/Cu away from 1/2 filling, the number of carriers in the lower (ie Cu rich) band determines k_F via

$$\sum_{k<k_F} 1 = 1 + \delta \tag{12.19}$$

so that the boson constraint gives

$$z = 1 - \sum_{k<k_F\sigma} u_k^2 = 1 - \sum_{k<k_F,\sigma} \frac{1}{2}(\frac{\Delta_{\rm ren}}{R_k^{\rm ren}}) \cong -\frac{\delta}{2} + \sum_{k<k_F\sigma} z\frac{4t_{\rm pd}^2\gamma_k^2}{\Delta_{\rm ren}} \tag{12.20}$$

For electron doping, δ is < 0 and $\Delta_{\rm ren} \cong \Delta_0$ so that the second term on the RHS is of order $\delta^2(t_{\rm pd}/\Delta_0)^2$ and can be neglected and one then finds $z \cong |\delta|/2$. In the limit $t_{\rm pd}/\Delta_0 \ll 1$, the band energies may be written $E_k^+ \cong \varepsilon_p^0 + zt_{\rm pd}^2 4\gamma_k^2/\Delta_0$, $E_k^- \cong \varepsilon_d^0 - zt_{\rm pd}^2 4\gamma_k^2/\Delta_0$, so the effect of correlations for electron doping is to narrow the bands by a factor δ. (In the limit of very small doping this mean field result does not make sense and the effect of spin fluctuations which accompany the propagation of the electron or hole will need to be taken into account. These will prevent the effective mass from going to ∞ and replace the band mass by a value leading to an energy scale of $\mathcal{O}(J)$ (see Trugman(1990), Nagaosa and Lee(1990).)

For hole doping, on the other hand, the self consistency equation (12.17) leads to a large shift of the d-orbital level up to just below the oxygen level so that $\Delta_{\text{ren}} \equiv \varepsilon_p - \varepsilon_d^{\text{ren}}$ becomes much smaller than $\Delta_0 = \varepsilon_p^0 - \varepsilon_d^0$ and (12.18) may be written

$$\lambda = \varepsilon_d^{\text{ren}} - \varepsilon_d^0 \cong \Delta_0 = \frac{t_{\text{pd}}}{\sqrt{z}} \sum_{k<k_F,\sigma} \gamma_k \sqrt{\frac{zt_{\text{pd}}^2 4\gamma_k^2}{\Delta_{\text{ren}}^2 + zt_{\text{pd}}^2 4\gamma_k^2}}. \tag{12.21}$$

Again in the limit $t_{\text{pd}}/\Delta_0 << 1$, we show below that neglect of the second term in the denominator leads to a solution which is consistent to lowest order in δ. This reduces to

$$\Delta_{\text{ren}} = \frac{t_{\text{pd}}^2}{\Delta_0} \sum_{k<k_F} \gamma_k^2 \tag{12.22}$$

and the self consistency equation for z may be written

$$z = -\frac{\delta}{2} + \sum_k (1 - \frac{1}{2}(1 + \frac{(\Delta_{\text{ren}}/2)}{\sqrt{(\frac{\Delta_{\text{ren}}}{2})^2 + zt_{\text{pd}}^2 4\gamma_k^2}})) \cong -\frac{\delta}{2} + 4z(\frac{t_{\text{pd}}}{\Delta_{\text{ren}}})^2 \sum_k \gamma_k^2. \tag{12.23}$$

Substituting from (12.22) we finally get

$$z \cong 2\delta(\frac{t_{\text{pd}}}{\Delta_0})^2 \sum_{k<k_F} \gamma_k^2 \tag{12.24}$$

and we can verify that the term $zt_{\text{pd}}^2 4\gamma_k^2$ in the denominator of Eq.(12.21) is of order δ times the term Δ_{ren}^2, thus justifying its neglect in solving for z to order δ. So the effect of correlations on the hole–doped samples is both to narrow the Cu–rich band by the renormalization factor $z \approx \delta\ (t_{\text{pd}}^2/\Delta_0^2)$ and to shift it up to just below the oxygen–rich band. The reason for this is, as illustrated qualitatively above, that in order to avoid the d^8 states, the additional holes need to spend a large part of their time in the oxygen p–orbitals and only visit the Cu sites when a d^{10} state becomes available. In the band picture this means that the d–projection factors u_k must be reduced by increasing the Cu hybridization with the oxygen orbitals through the reduction of Δ_{ren}.

Since the mean field Hamiltonian $\mathcal{H}_\infty$ is a free fermion model, perturbative corrections to the above mean field limit would be expected to have a Fermi liquid character (Sà de Melo and Doniach (1990)). Experimentally, the phenomenology of the normal state of the hole-doped cuprates suggests strong deviations from Fermi liquid behavior, so it is likely that perturbation theory breaks down and new physics needs to be introduced. However, a consensus on the resulting physics has yet to be reached.

12.3 Projection to a One Band Model: the Hubbard and $t - J$ Models

The result of the two band model discussed in Sec. 12.1 and Sec. 12.2 is that the ground state of the CuO_2 plane with hole doping may be described in terms of a renormalized hybrid band in which the carriers have a finite amplitude for occupying the oxygen p^7 state and, as a result of tunneling through the Cu sites, simultaneously lead to a finite amplitude for occupation of the d^{10} states. So the physical electron (and hole) propagator necessarily include the effects of slave boson fluctuations as discussed in Sec. 11.4., leading to a one particle spectral density $A_{dd}(\vec{p},\omega) = -\frac{1}{\pi}\mathcal{I}mG_{dd}(\vec{p},\omega)$ with reduced spectral weight at the Fermi level and additional weight above the Fermi level corresponding to the $(d^{10} - p^7)$ ligand hole fraction of the 1-particle state. The two hole d^8 configurations are forbidden by Bose statistics since b_i^2 acting on the vacuum state of the Bose field would give zero. At low energy scales compared with the renormalized charge transfer gap $\Delta_{pd}^{\text{ren}} = \varepsilon_p^0 - \varepsilon_d^{\text{ren}}$, one expects the physics to be dominated by dynamics of the carriers in the hybrid band. Away from 1/2- filling the carriers will behave as composite particles with Fermi-like character and effective mass enhanced by the correlation effects. One is then lead to consideration of a 1-band Hubbard model with effective hopping matrix element $t \simeq zt_{pd}^2/\Delta_{pd}$ where $z \propto$ doping concentration away from 1/2 filling in the limit of small t_{pd}/Δ_0. Interaction of carriers will now take place by exchange of slave bosons with zero'th order propagator $D_0(\omega) = 1/(\omega - \lambda + i\epsilon)$, where $\lambda = \varepsilon_d^{\text{ren}} - \varepsilon_d^0 \simeq \Delta_{pd}^0$ for hole doping. Thus we can qualitatively model the physics of this hybrid band by considering a 1-band Hubbard model in which the on-site carrier interaction is represented by a Hubbard $U_{\text{eff}} \sum_i n_{i\uparrow}n_{i\downarrow}$ term with $U_{\text{eff}} \equiv \Delta_{pd}^0$ and the bandwidth increases with doping concentration away from 1/2 filling.

As discussed in Chap. 8, the physics of the 1-band model can be represented in terms of lower and upper Hubbard bands. A simple way to do this which was introduced by Hirsch (1985) is to write $\mathcal{H}$ in terms of the Gutzwiller projection operators $d_{i\sigma}^{\dagger}P_G \equiv d_{i\sigma}^{\dagger}(1-n_{i,-\sigma})$ on to the lower and upper Hubbard bands:

$$\begin{aligned}\mathcal{H} &= -t\textstyle\sum_{<ij>}(1-n_{i,-\sigma})d_{i,\sigma}^{\dagger}d_{j,\sigma}(1-n_{j,-\sigma}) + n_{i,-\sigma}d_{i,\sigma}^{\dagger}d_{j,\sigma}n_{j,-\sigma}\\ &\quad +(1-n_{i,-\sigma})d_{i,\sigma}^{\dagger}d_{j,\sigma}n_{j,-\sigma} + n_{i,-\sigma}d_{i,\sigma}^{\dagger}d_{j,\sigma}(1-n_{j,-\sigma})\\ &\quad +\tfrac{U}{2}\textstyle\sum_i n_{i,\sigma}n_{i,-\sigma} \end{aligned} \tag{12.25}$$

and remove the cross terms involving $(1-n_{i,-\sigma})n_{j,-\sigma}$ and $(i \leftrightarrow j)$ to lowest order in t^2/U by a unitary transformation.

On writing $d_{i\sigma}^{\dagger} \to e^{i\Pi}d_{i\sigma}^{\dagger}e^{-i\Pi}$ one then finds to lowest order in t^2/U

$$\begin{aligned}\mathcal{H} &= -t\textstyle\sum_{<ij>}(1-n_{i,-\sigma})d_{i,\sigma}^{\dagger}d_{j,\sigma}(1-n_{j,-\sigma}) + n_{i,-\sigma}d_{i,\sigma}^{\dagger}d_{j,\sigma}n_{j,-\sigma}\\ &\quad +J\textstyle\sum_{<ij>}(\vec{S}_i\vec{S}_j - \tfrac{1}{4}n_{di}n_{dj}) \end{aligned} \tag{12.26}$$

with $J = t^2/U$, where $\vec{S}_i$ are the spin operators $\sum_{\mu\nu} d_{i\mu}^{\dagger}\vec{\sigma}_{\mu\nu}d_{i\nu}$, with $\vec{\sigma}_{\mu\nu}$ the Pauli matrices and n_{di} the charge operators $n_{di} = \sum_{\sigma} n_{i,\sigma}n_{i,-\sigma}$.

So the effect of the Hubbard U has been replaced by considering the lower and upper Hubbard bands with effectively $U = \infty$ (because of the projection operators, double occupancy either of particles in the lower band or of holes in the upper band is strictly forbidden). What remains is a Heisenberg type antiferromagnetic coupling. At 1/2 filling, the lower Hubbard band will therefore be completely full and may be expected to have an antiferromagnetic ground state.

Because of the Gutzwiller projection operators, this model will already have strong correlations even if J is set to zero since double occupancy is forbidden. Hence a creation operator for an electron $d_{i\sigma}^{\dagger}(1-n_{i,-\sigma})$ can only give a non-zero result if site i is unoccupied, which means the usual fermion commutation rules cannot be applied for these composite particles. Some of the resulting strong correlation physics will be discussed in Section 12.4 below.

Another way to arrive at the model of Eq. (12.26) (often referred to as the $t-J$ model) was formulated by Zhang and Rice (1988). Rather than consider the two band model, they first looked at the eigenstates of a single CuO_4 cluster. Then a hole introduced in this cluster can form an orbital among the four oxygens which has $d_{x^2-y^2}$ symmetry and hence will hybridize with the d^9 state on the central copper site.

When Coulomb interactions are included, the lowest such hybrid will be a singlet combination of $(p^8)_3 p^7_\sigma d^9_{-\sigma}$ and $(p^8)_4 d^{10}$ states which have come to be known as a "Zhang-Rice singlet." By considering tunneling of these singlets in a 1/2-filled CuO_2 lattice they were able to arrive at the $t - J$ model by an alternative route.

It should be noted that neither the 1-band Hubbard model nor the $t - J$ model really include all the physics of the CuO_2 planes since, as we have seen, the hole carriers have quite large amplitudes for being in p^7 orbitals so that their spectral function will include considerable weight at energies $\lambda = \varepsilon_p^0 - \varepsilon_d^{\text{ren}}$ above the Fermi level which would not show up in a 1-band model. Because the carriers have an appreciable amplitude of being in oxygen orbitals where Coulomb repulsion effects will be considerably weakened, the application of Gutzwiller projection operators in the 1-band model will tend to overemphasize the effects of correlation relative to what is occurring in the physical CuO_2 system.

Thus in attempting to produce quantitative estimates of the various physical properties of the CuO_2 planes, for instance by quantum Monte Carlo or exact diagonalization methods, the importance of the two band model should be kept in mind. Nevertheless, many theorists believe that much of the important physics of CuO_2 is captured by one band models (perhaps partly because they are less formidable: the so-called "looking for one's car keys under the street light" effect) and have devoted considerable theoretical effort to studying these models.

12.4 Superconductivity in the Cuprates: the d-wave BCS State

One of the characteristic features of superconductivity in the cuprate superconductors is the strong dependence of T_c on the doping concentration, x, away from 1/2-filling. In the simplest case, $La_{2-x}Sr_xCuO_4$, the onset of superconductivity occurs for $x \simeq 5\%$, below which the system behaves as an insulator. As x is increased, T_c reaches a maximum of about 38^oK around $x = 15\%$, then decreases again, reaching zero at $x \simeq 30\%$. Similar effects are seen for other cuprate compounds with much higher T_c 's ($T_c \simeq 90^o$K for $YBa_2Cu_3O_7$, and $\sim 150^o$K for $HgBa_2Ca_2Cu_3O_{8+y}$), although it is generally more difficult to cover a broad range of hole doping with these compounds.

What is the physics behind this sensitivity of T_c to doping level?

The answer appears to be a complex one, with different driving forces dominating in the overdoped and underdoped regimes.

We start by considering the 1-band $t-J$ model, Eq. (12.26), discussed in Sec. 12.3. We represent (12.26) in a slave boson representation using $(1-n_{i,-\sigma})d^\dagger_{i\sigma} \to b_i f^\dagger_{i\sigma}$ and then take the mean field limit $b_i \to <b>$ to get

$$\begin{aligned}\mathcal{H}_\infty &= -<b>^2 t \textstyle\sum_{ij} f^\dagger_{i\sigma} f_{j\sigma} - \mu \sum_i n_{fi} + J \sum_{ij}(\vec{S}_i \cdot \vec{S}_j - \frac{1}{4} n_{fi} n_{fj}) \\ &\quad + \lambda \textstyle\sum_i (<b>^2 + n_{fi} - 1) \end{aligned} \tag{12.27}$$

where μ is the chemical potential and $n_{fi} = \sum_\sigma f^\dagger_{i\sigma} f_{i\sigma}$.

On minimizing with respect to λ we then have

$$z \equiv <b>^2 = 1 - <n_f> \tag{12.28}$$

where $<n_f>$ is determined by counting the states in the band. (12.28) then gives the renormalized band width proportional to $z = (1 - <n_f>) = x$ where x is the doping away from 1/2 filling. (We revert in this section to the more chemical notation of using 'x' for doping in contrast to the symbol 'δ' used in section 12.2 as is more common in the theory literature.)

So in this mean field limit we have a band of heavy holes with effective mass increased by the factor $1/x$ interacting via the Heisenberg exchange coupling $J\sum_{ij}(\vec{S}_i \cdot \vec{S}_j - \frac{1}{4} n_{fi} n_{fj})$. Now we can rewrite this coupling in terms of pseudo fermion operators by using the identity

$$J \sum_{ij} (\vec{S}_i \cdot \vec{S}_j - \frac{1}{4} n_{fi} n_{fj}) \equiv -\frac{J}{2} \sum_{<ij>,\sigma} \{ f^\dagger_{i\sigma} f^\dagger_{j,-\sigma} f_{j,\sigma} f_{i,-\sigma} + f^\dagger_{i\sigma} f^\dagger_{j,-\sigma} f_{i,-\sigma} f_{j\sigma} \}. \tag{12.29}$$

We thus see that the antiferromagnetic repulsion between adjacent spins translates into an attraction between pseudo fermions. We can then immediately apply BCS mean field theory to find a superconducting ground state of this band of correlated pseudo fermions.

In doing so we need to be concerned that the mean field slave boson approximation has replaced the $U = \infty$ repulsion for two pseudo fermions on one site by an average reduction of the band width. Hence in applying the BCS mean field theory as a variational estimate of the ground state of (12.26) we must choose trial wave functions which do not lead to simultaneous double occupancy of a lattice site. The simplest way to do this is to build the BCS ground state out of Cooper pair creation operators of the form

$$\Psi_i^\dagger = \sum_{j\sigma} \alpha_j f_{i\sigma}^\dagger f_{j\sigma}^\dagger \tag{12.30}$$

where the sum is over nearest neighbor sites, j, to site i, and α_j are suitable coefficients.

By rewriting Eq. (12.29) in k-space

$$\mathcal{H}_{\text{int}} = -J \sum_{k,k',\sigma} [\gamma_k \gamma_{k'} + \eta_k \eta_{k'}] d_{k\sigma}^\dagger d_{-k,-\sigma}^\dagger d_{-k',-\sigma} d_{k',\sigma} \tag{12.31}$$

where $\gamma_k = \cos k_x + \cos k_y, \eta_k = \cos k_x - \cos k_y$ with k_x, k_y measured relative to the Brillouin zone, we see that the BCS gap equation becomes

$$1 = J \sum_{k<k_F} \frac{\gamma_k^2}{\sqrt{\varepsilon_k^2 + \Delta_k^2}} \tag{12.32}$$

for the extended s-wave gap

$$\Delta_k^{\text{ext.s-wave}} = \Delta_{es} \gamma_k \ , \tag{12.33}$$

or

$$1 = J \sum_{k<k_F} \frac{\eta_k^2}{\sqrt{\varepsilon_k^2 + \Delta_k^2}} \tag{12.34}$$

for the d-wave gap

$$\Delta_k^{\text{d-wave}} = \Delta_d \eta_k. \tag{12.35}$$

The single fermion energies are $\varepsilon_k = (zt\gamma_k - \mu)$ so that the BCS equation (12.34) for the d-wave solution may be written

$$1 = J \sum_{k<k_F} \frac{\eta_k^2}{\sqrt{(zt\gamma_k - \mu)^2 + \Delta_d^2 \eta_k^2}}. \tag{12.36}$$

Exactly at 1/2 filling $z = 0$ and this has a formal solution

$$\Delta_d = J \sum_{k<k_F} |\cos k_x - \cos k_y| = JA \tag{12.37}$$

where the Fermi level at 1/2 filling is bounded by the square in the Brillouin zone $k_x + k_y = \pm\pi, k_x - k_y = \pm\pi$ with the result that A is a constant of $\mathcal{O}(1)$.

Although this solution cannot be superconducting (since the band is 1/2 full it cannot carry a current) it still can represent a "resonant valence bond" or RVB alternative to the Néel antiferromagnetic ground state and will be further discussed in 12.5 below.

As the doping is increased, the band gets broader and Δ decreases monotonically with increasing x. Eventually, when $z \simeq \mathcal{O}(J/t)$, Eq. (12.36) no longer has a solution and the superconductivity disappears.

The physics of this formulation of high T_c superconductivity is that, in contrast to conventional superconductors where the Cooper pair attraction via phonon exchange is cut off by the Debye frequency, this retardation effect is now replaced by the effects of strong correlation which reduce the relative kinetic energy of the Cooper pairs by the factor $z \equiv < b >^2$. So in this mean field limit, the existence of superconductivity is essentially a result of the strong correlations, which one may think of as a kind of "Gutzwiller confinement", preventing the members of the Cooper pair from escaping from each other. Although the mean field approximations underlying the above argument are quite simplistic, they appear to provide a straightforward way to understand the observations in cuprate superconductors which show that the superconductivity goes away as the charge carriers become less correlated as a result of increased doping.

The above simple picture also provides an explanation of why a d-wave symmetry pairing dominates over an extended s-wave pairing, as is observed experimentally. For $\Delta_k^{\text{ext.s}} = \Delta_{\text{es}}\gamma_k$, the pairing wave function has lines of nodes along the 1/2-filling square $k_x + k_y = \pm\pi, k_x - k_y = \pm\pi$ in the Brillouin zone. When the band is close to 1/2 filling the Fermi surface over which the integral in (12.32) is done is then very close to the nodal lines of the extended s-wave, so this solution has a lower value of the gap Δ_{es} than does the d-wave gap function $\Delta_k^{\text{d-wave}} = \Delta_d \eta_k$. The latter has nodal lines normal to the Fermi surface along the directions $k_y = \pm k_x$. Hence the d-wave solution has a larger contribution to (12.36) over the Fermi surface and one finds $\Delta_d >> \Delta_{\text{es}}$ as a result which leads to a BCS state with lower free energy. The experimental observation of a superconducting gap with d–wave symmetry (Shen et al (1993), see Fig.(12.2)) is one of the convincing reasons behind the belief that superconductiviy in the cuprates is driven by Coulomb forces.

The above arguments give rise to a gap which monotonically increases as the doping is reduced toward 1/2-filling, in contradiction to observation for the underdoped cuprates. Doniach and Inui (1990) suggested that the reduction of T_c in underdoped samples was not due to the disappearance of the gap, but rather due to quantum fluctuations of the phase of the superconducting order parameter brought about by long range Coulomb interactions which are poorly screened in the cuprates. As the doping is reduced, the "stiffness" of the order parameter, measured by the $\int d^2x\xi_0^2|\vec{\nabla}\psi(x)|^2$ term in the Ginzburg Landau equation, becomes weaker, which reduces the phase coherence of the superconducting state, thus lowering T_c , without reducing the gap. (For an introduction to the Ginzburg Landau phenomenological theory of superconductivity see, for instance, Tinkham (1975).) A similar argument, though differing in details, has been put forward by Emery and Kivelson (1995). Recently photoemission measurements (Shen et al 1996) have shown evidence that a pairing gap of d-wave symmetry still exists in underdoped samples even though the system is not superconducting, thus providing support for the "phase winding" argument for the disappearance of T_c at low doping. Physically, quantum zero-point winding of the phase $\varphi(x)$ of the superconducting order parameter $\psi(x) = e^{i\varphi(x)}|\psi(x)|$ corresponds to localization of the Cooper pairs since phase and charge are conjugate variables. Thus the disappearance of superconductivity at low doping may be thought of as a localization phenomenon in which a superconductor/insulator transition occurs by formation of a "Cooper crystal" of localized Cooper pairs. In reality the underdoped cuprate materials are spatially inhomogeneous on the atomic scale because of the Sr atoms (or oxygen vacancies) so that the insulating state may be quite disordered. The detailed nature of this state remains unclear at this time.

Although the above discussion of applying BCS theory to the $t - J$ model probably gets some of the physics right - notably the dominance of the d-wave pairing state - it has some significant limitations when Fermi–liquid style correlation effects are included. As in the heavy fermion case discussed in Section 11.4, strong correlation–induced enhancement of the effective mass of the holes tends to suppress T_c . Calculations of T_c based on use of renormalized hole Green's functions lead to an Eliashberg type of integral equation (see Schrieffer 1983) which has to be solved numerically. Studies by Monthoux and Pines

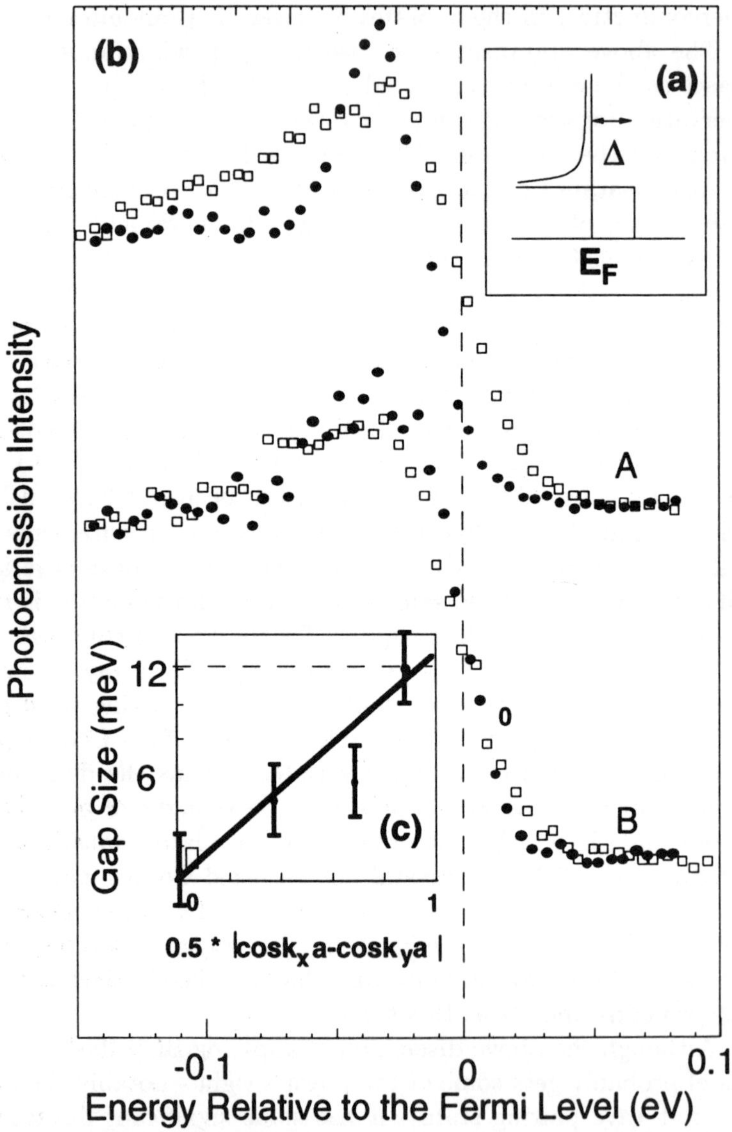

Figure 12.2: d-wave character of the superconducting energy gap for BSCCO as measured by Angularly Resolved Photoemission Spectroscopy. ARPES spectra are taken above and below T_c (78°K) at points in the Brillouin zone where the gap is expected to be large (A) and is expected to vanish (B) for the $d_{x^2-y^2}$ state.
(**Inset**) Comparison of the experimental data with d–wave theory. The horizontal dashed line is what would be expected for an s–wave superconductor (from Shen et al (1995)).

(1992) and Dickinson and Doniach (1993) lead to the need for unrealistically large values of J in order to account for the observed T_c . At the same time photoemission measurements on the high T_c materials show that the hole spectral density is very broad so that the whole idea of the simple quasiparticle picture consistent with Fermi liquid theory is called into question for the cuprate materials, and understanding some new kind of strong correlation physics is probably important for achieving a satisfactory theory of the details of BCS pairing. Ideas on what this will look like are briefly outlined in the following section.

12.5 Strong Correlation Effects in two–dimensional Fermi Systems

Experimentally the cuprate materials show some rather anomalous properties when not in the superconducting state. (See Anderson 1992 for a review.) There are indications that usual ideas of Fermi liquid physics are not a good description of the effects of strong correlation in the cuprates and new physics needs to be understood. One approach is to learn from the strong correlation physics of 1-dimensional Fermi systems where we have a very firm theoretical understanding based on exact solutions, as discussed in Sec. 11.5.

Is there an analog to the one dimensional fermion→ boson mapping in a 2D system? In 1D we saw that a fermion creation operator may be represented as a kink–anti kink, or soliton-like excitation in the phase of the charge and spin density fields. The simple extension of a kink to 2D is a domain wall between two domains of an ordered system (such as a ferromagnet, for example). However, the higher dimensionality also gives rise to a new kind of topological entity: a vortex. These objects are familiar in type–II superconductors in a magnetic field where they represent a topological singularity in which the phase of the order parameter at a point $\vec{x}$ increases as the angle subtended by the vector joining $\vec{x}$ to the center of the vortex at $\vec{x}_0$:

$$\phi(\vec{x} - \vec{x}_0) = \arctan(y/x) \tag{12.38}$$

where x and y are the cartesian coordinates of $\vec{x} - \vec{x}_0$. Is there an analogous quantity in a 2-D electron liquid?

The answer is yes for a 2D electron gas in a strong magnetic field where, in semiconducting films at low electron density, the fractional quantum Hall effect (FQHE) is a manifestation of a new kind of quantum correlated ground state which is well represented by the Laughlin wave function (see Prange and Girvin 1990).

A doped CuO_2 plane exhibits properties which can be considered analogous to those of excitations in a quantum Hall liquid, but in the absence of an applied magnetic field. Laughlin (1988) has hypothesized that a 2D electron gas could form a ground state in which a "statistical gauge field" could mimic the effects of a physical magnetic field and stabilize a quantum Hall ground state. However, an electron gas in a magnetic field exhibits a net diamagnetic moment which changes sign upon time reversal, a so-called time-reversal- breaking or T-breaking state. (A ferromagnetic state is a familiar example.) Experimentally, the cuprate superconductors do not show evidence of T-breaking (Spielman *et al* 1992) so Laughlin's hypothesis does not appear to work for the cuprate superconductors. Nevertheless there appear to be qualitative reasons that the excitations from the physical ground state of the doped $t - J$ model, for example, have properties analogous to those of excitations in the Laughlin state, so we proceed to give a very elementary account of the theory of statistical gauge fields in 2D systems.

As mentioned above, the mean field equations for the BCS ground state have a well defined solution (12.37) at 1/2-filling even though this is not a current–carrying state. It was observed by Affleck and Marston (1988) that, by using the transformation $f^{\dagger}_{\text{B-sublattice}} \rightarrow f_{\text{B-sublattice}}$ for a bipartite lattice, this same BCS-type state in which the order parameter Δ is proportional to $< f^{\dagger}_{i\sigma} f^{\dagger}_{j,-\sigma} >$ where i and j are nearest neighbors, is identical to a Hartree Fock (i.e., mean field) state in which bond averages

$$\chi_{ij} = \sum_{\sigma} < f^{\dagger}_{i\sigma} f_{j,\sigma} > \tag{12.39}$$

are non zero. Here the $< >$ is with respect to a mean field average of the J term in the $t - J$ hamiltonian (since the t term averages to zero at 1/2-filling)

$$\mathcal{H}_{\infty} = \frac{J}{2} \sum_{<ij>} \chi_{ij} f^{\dagger}_{i\sigma} f_{j\sigma} + \text{h.c.} \tag{12.40}$$

By assuming that χ_{ij} has a periodicity of $\sqrt{2}a$, corresponding to the symmetry of translation across diagonals on the original square lattice of side a, they were able to show that a solution in which $\chi_{ij} = |\chi| e^{i\pi/4}$ with directions (ij) as indicated in Fig. 12.3, provides an energy minimum

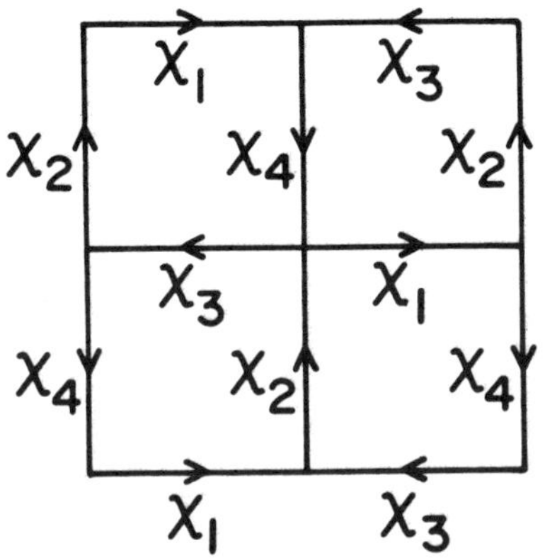

Figure 12.3: Definition of vector potentials for the flux phase (after Affleck and Marston 1988).

(as a function of the χ's) which is competitive with the Néel Hartree Fock state in which *site* as opposed to *bond* averages are taken:

$$\begin{aligned} < S_{zi} > &\equiv < f_{i\uparrow}^\dagger f_{i\uparrow} > - < f_{i\downarrow}^\dagger f_{i\downarrow} > \\ &= 1 \text{ for } i \text{ on the A sublattice} \\ &= -1 \text{ for } j \text{ on the B sublattice} \end{aligned} \tag{12.41}$$

Now (12.40) represents electrons moving in a pair of tight binding bands with double the unit cell, i.e., 1/2 the Brillouin zone, of the original lattice. Just as in the BCS case (Chapter 10) the Hamiltonian can be written in Nambu, or iso–spin, form as

$$\mathcal{H}_\infty = \sum_k{}' \psi_k^\dagger h_k \psi_k \tag{12.42}$$

where

$$\psi_k = \begin{bmatrix} f_k \\ f_{k+Q} \end{bmatrix} \tag{12.43}$$

is a 2-component spinor, Q is the vector $[\pi, \pi]$, $\sum_k'$ is a sum over 1/2 the Brillouin zone and

$$h_k = \frac{J}{2}\{(\cos k_x + \cos k_y)\tau_3 + (\cos k_x - \cos k_y)\tau_2\} \tag{12.44}$$

where $\tau_3 = \begin{bmatrix} 1 & 0 \\ 0 & -1 \end{bmatrix}$, $\tau_2 = \begin{bmatrix} 0 & i \\ -i & 0 \end{bmatrix}$.

The new bands, which are eigenstates of (12.42) have energies $\varepsilon_{k\pm} = \pm\sqrt{(\cos k_x)^2 + (\cos k_y)^2}$ and the Fermi surface is now a pair of Fermi "points" in 2-dimensions with zeros at $k_{F1} = (\pi/2, \pi/2)$ and $k_{F2} = (\pi/2, -\pi/2)$. In the vicinity of these points the bands form cones. In the ground state the lower bands each contain 1/2 a spin $\uparrow$ electron and 1/2 a spin$\downarrow$ electron per site.

This new state may be regarded as what would happen to a tight binding model if placed in a huge external magnetic field

$$\mathcal{H} = J \sum_{<ij>} e^{iA_{ij}} f^\dagger_{i\sigma} f_{j\sigma} \tag{12.45}$$

where the vector potential $A_{ij} \equiv \int_i^j d\vec{l} \cdot \vec{A}(\vec{l})$ increments are exactly $\pi/4$ and the integration directions are as shown in Fig.(12.3).

Then the total magnetic flux per plaquette is obtained from Stokes' theorem as

$$\Phi = \int d\sigma H_z \equiv \int_C d\vec{l} \cdot \vec{A}_l = \pi \text{ per plaquette.} \tag{12.46}$$

Note that the sign of Φ alternates as one goes along an axis, but since the flux is π, this does not alter the eigenstates. So we see that the effect of the electron correlations as estimated in the mean field limit is to replace the original coupling $J \sum (\vec{S}_i \cdot \vec{S}_j - \frac{1}{4} n_i n_j)$ by a fictitious magnetic field, or "statistical gauge field." This formulation of the effects of correlations in terms of gauge fields is very natural in 2-dimensions and may be regarded as providing an extension to two dimensions of the bosonization ideas for 1-D physics discussed in Section 11.4.

This flux phase state is in the general class of "resonant valence bond" states in which the usual Néel antiferromagnetism is replaced by a superposition of singlet combinations of spins on different sites. However, considerable calculational effort in recent years has shown that such states generally have higher energy than the Néel state and in practise states of this type, though close to the Néel state in energy, have not (so far) been found in a physical system.

Nevertheless, the close analogy to the FQHE for the flux phase - i.e., the idea that in the mean field limit the strong particle correlations can be represented by fictitious magnetic fields - has lead to the study of what effects doping would have on the above flux phase.

To see this, we put back in the hopping term in the $t - J$ model by reintroducing the slave bosons as a device to mimic the Gutzwiller projection operators. Then we can write:

$$\mathcal{H}_{tJ} = -t \sum_{<ij>} b_i^\dagger b_j f_{i\sigma}^\dagger f_{j\sigma} + \frac{J}{2} \sum_{<ij>\sigma\sigma'} \{ f_{i\sigma}^\dagger f_{j\sigma'}^\dagger f_{j\sigma} f_{i\sigma'} - \frac{1}{4} b_i^\dagger b_j^\dagger b_j b_i \} \quad (12.47)$$

where we have made use of the constraint

$$b_i^\dagger b_i + n_{fi} = 1. \quad (12.48)$$

Now we can look at a generalized mean field solution in which, following Zou, Levy and Laughlin (1992) we set

$$\chi_{ij} = < \sum_\sigma (f_{i\sigma}^\dagger f_{j\sigma} - 2t b_i^\dagger b_j) > = |\bar{\chi}| e^{i a_{ij}} \quad (12.49)$$

and add a Lagrange multiplier term $\sum(b_i^\dagger b_i + n_{fi} - 1)$ to enforce the slave boson constraint.

The effect of doping the above $\Phi = \pi$ phase away from 1/2 filling now introduces a new degeneracy into the problem. By analogy with FQHE, we can think of the $\Phi = \pi$ phase as one in which each pseudo–fermion carries a flux tube of π around with it. If the number of particles is reduced, by adding x holes per site, one may expect $\Phi = \pi(1 - x)$ on average, i.e. the system will now be in an average effective magnetic field leading to $\chi_{ij} = |\chi| e^{i a_{ij}}$ which now has a much larger unit cell. For fluxes Φ which are now rational fractions, $\Phi = 2\pi p/q$, where p and q are integers, one can then define new "magnetic" band structures with unit cells containing q lattice sites. These new Hofstadter bands (Hofstadter(1976)) will be the tight binding analog of Landau levels for free electrons in a magnetic field. Hence the energy levels $\varepsilon(\vec{k})$ will now have series of gaps, rather than going to Fermi points as in the $\Phi = \pi$ case. However, there still remains an overall degeneracy given by the direction of the fictitious magnetic field. This corresponds to the magnetization direction in an Ising ferromagnet and will lead to a global "chiral" symmetry breaking (as formulated by Wen, Wilczek and Zee (1990)), which will be T-breaking as discussed above.

To make this pseudo magnetic ground state self consistent Zou, Levy and Laughlin (1992) show that dynamic interactions between "holons" represented by the $b_i^\dagger$ creation operators and the "spinons" represented by the pseudo–fermions $f_{i\sigma}^\dagger$ via the hopping term

$$\mathcal{H}_{\text{hop}} = t \sum_{<ij>} b_i^\dagger b_j (f_{i\sigma}^\dagger f_{j\sigma} - \bar{\chi}_{ij}) \tag{12.50}$$

may be represented by introducing an effective dynamic gauge field

$$\mathcal{H}_{\text{hop}} = t \sum_{<ij>} |\chi| e^{i a_{ij}} b_i^\dagger b_j - \frac{J}{2} \sum_{<ij>} |\chi| e^{i a_{ij}} f_{i\sigma}^\dagger f_{j\sigma} \tag{12.51}$$

where the gauge increments $a_{ij} = \int_j^i d\vec{l} \cdot \vec{a}(\vec{x})$ are now expressed in terms of a fluctuating $\vec{a}$ field rather than the mean field values used above. This is done by a Hubbard Stratanovich transformation which is derived using the functional integral (or Feynman path integral) formulation of quantum field theory. The reader is referred to the book by Negele and Orland (1985) for details of this approach.

Finally, one can then show that the effects of the lowest order fluctuation corrections to the propagator for the statistical gauge field (Fig.(12.4)) may be represented in the limit of small doping by adding a Chern Simons term (see Wen, Wilczeck and Zee(1990)) to the effective Lagrangean for the holons:

$$\mathcal{H}_{CS} = \int \frac{d^2x}{2\pi} \epsilon_{\alpha\beta\gamma} a_\alpha(x) (\partial_\beta a_\gamma(x) - \partial_\gamma a_\beta(x)) \tag{12.52}$$

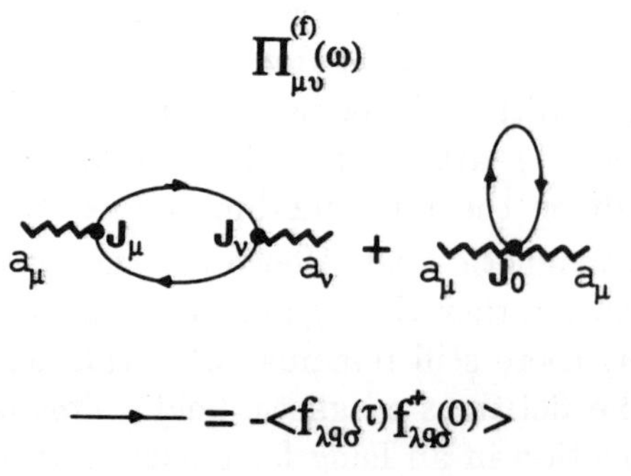

Figure 12.4: lowest order contributions to the response function for the statistical gauge field a_α.

where $\partial_\alpha = \partial/\partial x_\alpha$ and $\epsilon_{\alpha\beta\gamma}$ is the antisymmetric unit tensor.

As a result the holons interact with each other as though they each had a flux tube of π attached to them , so that interchanging two holons will lead to a factor $e^{i\pi/2}$ in the wave function of the system. This is half way between Bose statistics, where a pair interchange leads to a factor of 1, and Fermi statistics where an interchange leads to a factor of $e^{i\pi}$. For this reason they are fractional statistics particles referred to as "semions". The holons therefore carry with them a topological winding of the gauge field $\vec{a}(x)$ and so are qualitatively analogous to vortices, although with $\int_C d\vec{l} \cdot \vec{a} = \pi$ rather than 2π as would be the case for the analogous integral for a superconducting vortex.

As mentioned above, the cuprate superconductors do not exhibit T-breaking experimentally. Nevertheless many of their normal state properties may be interpreted using the holon picture so that it seems likely that some aspects of holon physics will be useful for understanding the physics of the cuprates.

Appendix 1

Second Quantization for Fermions and Bosons

In elementary wave mechanics one deals with many-particle systems by forming appropriately symmetrized many-body wave functions from the single-particle wave functions and then using complete sets of such functions as basis functions for the expansion of arbitrary (many-body) wave functions. This procedure is unnecessarily cumbersome and can be simplified by putting the theory into an equivalent form called "second quantization." This is based on the "occupation number representation," where one works with state vectors which specify the single-particle states which are occupied, and introduces creation and annihilation operators which change by one the occupancy of the single-particle states. The symmetry requirements are then easily satisfied by imposing commutation rules on the creation and annihilation operators.

We consider first fermions and suppose that a system with one fermi particle has a complete set of orthonormal basis states $\varphi_k(r)$. The coordinate r stands for x, y, z, α, where $\alpha = 1$ or 2 is a spin coordinate, and the allowed values of the quantum number k ($k = 1, 2, 3, \ldots$) may, in the case of extended states (e.g. plane waves), be determined by imposing box boundary conditions. A system with two identical particles has states

$$\varphi_{kl}(r_1, r_2) = \frac{1}{\sqrt{2!}}\begin{vmatrix} \varphi_k(r_1) & \varphi_k(r_2) \\ \varphi_l(r_1) & \varphi_l(r_2) \end{vmatrix}, \tag{A.1.1}$$

and a system with three particles has states

$$\varphi_{klm} = \frac{1}{\sqrt{3!}}\begin{vmatrix} \varphi_k(r_1) & \varphi_k(r_2) & \varphi_k(r_3) \\ \varphi_l(r_1) & \varphi_l(r_2) & \varphi_l(r_3) \\ \varphi_m(r_1) & \varphi_m(r_2) & \varphi_m(r_3) \end{vmatrix}. \tag{A.1.2}$$

For $k > l > m$ these last form an orthonormal basis for a function space Φ_3 of antisymmetric functions in (r_1, r_2, r_3). We may proceed similarly for the general case Φ_N (N = any integer).

We now introduce an abstract vector space V with basis vectors v which is the direct sum of vector spaces $V_0, V_1, V_2, \ldots$. V_0 is one-dimensional with normal base $|0\rangle$ or **0**. For $N \geqslant 1$ we make V_N isomorphic with Φ_N. The creation operators $a_k^\dagger$ take V_N into V_{N+1} according to

$$a_k^\dagger|0\rangle \Leftrightarrow \varphi_k; \tag{A.1.3}$$

if $\mathrm{v} \Leftrightarrow \varphi_{lmn\ldots}$, then

$$a_k^\dagger \mathrm{v} \Leftrightarrow \varphi_{klmn\ldots} \qquad \text{if } k \neq l, m, n \ldots,$$
$$= 0 \text{ otherwise.}$$

Since φ_{klmn} is antisymmetric in its suffixes, this scheme is well defined and consistent provided the anticommutator $\{a_k^\dagger, a_l^\dagger\} = 0$ (in particular, $a_k^{\dagger 2} = 0$). Scalar products in V_N are defined in terms of the images in Φ_N, and we also take $(V_N, V_{N'}) = 0$ if $N \neq N'$, and $\langle 0|0\rangle = 1$. A basis vector $\varphi_{klmn\ldots}$ (or its image in V) is said to correspond to a state with $k, l, m, n \ldots$ occupied. Specifying which states are occupied determines a basis vector up to a sign.

The annihilation (or destruction) operator a_k is defined as follows. Suppose $\mathrm{v} \in V$ is a basis vector such that k is not occupied, i.e., $a_k^\dagger \mathrm{v} \neq 0$, but is otherwise arbitrary. Then

$$a_k \mathrm{v} = 0, \qquad a_k(a_k^\dagger \mathrm{v}) = \mathrm{v}. \tag{A.1.4}$$

Since any basis vector in V_N can be written as v or $a_k^\dagger \mathrm{v}$ (up to a sign), this defines a_k.

By considering the effect of $(a_k^\dagger a_k + a_k a_k^\dagger)$ on the above vectors v and $a_k^\dagger \mathrm{v}$ we get $\{a_k, a_k^\dagger\} = 1$. Similarly, for $k \neq l$, we consider the effect of $(a_k a_l^\dagger + a_l^\dagger a_k)$ on w, $a_k^\dagger \mathrm{w}$, $a_l^\dagger \mathrm{w}$ and $a_k^\dagger a_l^\dagger \mathrm{w}$, where w is an arbitrary basis vector with k and l unoccupied, and we find $\{a_k, a_l^\dagger\} = 0$. We can similarly show that $\{a_k, a_l\} = 0$. Summarizing, we have the anticommutation rules

$$\{a_k^\dagger, a_l^\dagger\} = 0, \qquad \{a_k, a_l\} = 0, \qquad \{a_k, a_l^\dagger\} = \delta_{kl}. \tag{A.1.5}$$

$a_k^\dagger a_k$ is the number operator for the mode k; clearly $a_k^\dagger a_k \mathrm{v} = 0$ if v is such that k is not occupied, and $a_k^\dagger a_k \mathrm{v} = \mathrm{v}$ if v is such that k is occupied. Thus $a_k^\dagger a_k$ has eigenvalues zero and one, as required for fermions. Since $a_k^\dagger a_k^\dagger \mathrm{v} = 0$, two fermions cannot occupy the same state.

We now show that $a_k^\dagger$ and a_k are hermitian conjugates. We have to verify that $A = B$, where

$$A = (a_k^\dagger \mathbf{u}, \mathbf{v}), \qquad B = (\mathbf{u}, a_k \mathbf{v}),$$

and **u**, **v** are basis vectors. Both quantities vanish unless k is unoccupied in **u** and k is occupied in **v**. In this latter case we can put $\mathbf{v} = \pm a_k^\dagger \mathbf{w}$, where **u**, **w**, have k unoccupied and are either orthogonal or identical; thus

$$A = \pm\delta_{\mathbf{u},\mathbf{w}}, \qquad B = \pm(\mathbf{u}, a_k a_k^\dagger \mathbf{w}) = \pm\delta_{\mathbf{u},\mathbf{w}},$$

which completes the proof.

Operators in Φ_N, the space of antisymmetrical functions of $r_1, r_2, \ldots, r_N$, are transformed as follows.

Consider first single-particle operators. These are of the form $\Sigma_{i=1}^N \hat{T}^{(i)}$, where (because of the completeness of the $\varphi_k(r)$)

$$\hat{T}^{(i)}\varphi_k(r_i)\psi(\text{rest of } r\text{'s}) = \sum_l \varphi_l(r_i)\psi(\text{rest of } r\text{'s})\, T_{lk}, \tag{A.1.6}$$

with

$$T_{lk} = (\varphi_l, \hat{T}\varphi_k) = \int dr_1 \varphi_l^*(r_1)\hat{T}^{(1)}\varphi_k(r_1). \tag{A.1.7}$$

($\int dr_1$ denotes $\int dx_1\, dy_1\, dz_1\, \Sigma_\alpha$.) Writing

$$\hat{X}_{lk}^{(i)}\varphi_{k'}(r_i)\psi(\text{rest of } r\text{'s}) = \delta_{kk'}\varphi_l(r_i)\psi(\text{rest of } r\text{'s}), \tag{A.1.8}$$

we have

$$\hat{T}^{(i)} = \sum_{lm} \hat{X}_{lm}^{(i)} T_{lm}. \tag{A.1.9}$$

Now consider

$$\Phi = \sum_i \hat{X}_{lm}^{(i)} \left\{ \sum_P \epsilon_P \prod_{j=1}^N \varphi_j(r_{Pj}) \right\}, \tag{A.1.10}$$

where $\{\ldots\}$ is a determinantal wave function in Φ_N. [P is a permutation of $1, \ldots, N$; $\epsilon_P = +1(-1)$ if P is even (odd); and without loss of generality we have labeled the states so that $1, \ldots, N$ are occupied, $N+1, \ldots,$ are unoccupied.] We have

$$\begin{aligned}\Phi &= \sum_P \epsilon_P \sum_i \hat{X}_{lm}^{(Pi)} \varphi_i(r_{Pi}) \prod_{j(\neq i)} \varphi_j(r_{Pj}) \\ &= \sum_P \epsilon_P \sum_i \delta_{mi} \varphi_l(r_{Pi}) \prod_{j(\neq i)} \varphi_j(r_{Pj}) \\ &= \sum_P \epsilon_P \varphi_l(r_{Pm}) \prod_{j(\neq m)} \varphi_j(r_{Pj}) \text{ or } 0, \end{aligned} \tag{A.1.11}$$

depending on whether $m \in 1, \ldots, N$ or not. Thus Φ is the determinantal wave function in which φ_m has been replaced by φ_l (or is zero). Hence the image of the operator $\Sigma_i \hat{X}_{lm}^{(i)}$ is just $a_l^\dagger a_m$, and [see (A.1.9)] the image of $\Sigma_i \hat{T}^{(i)}$ is

$$\sum_{lm} T_{lm} a_l^\dagger a_m = \int dr_1 \psi^\dagger(r_1) \hat{T}^{(1)} \psi(r_1), \tag{A.1.12}$$

where

$$\psi(r) = \sum_k \varphi_k(r) a_k, \qquad \psi^\dagger(r) = \sum_k \varphi_k^*(r) a_k^\dagger. \tag{A.1.13}$$

$\psi(r)$, $\psi^\dagger(r)$ are linear combinations of the creation and annihilation operators known as *field operators*.

For two-particle operators ($N \geqslant 2$)

$$\sum_{i \neq i'} \hat{V}^{(i,i')} \qquad (i, i' \in 1, \ldots, N) \tag{A.1.14}$$

we have, similarly,

$$\hat{V}^{(i,i')} \varphi_k(r_i) \varphi_{k'}(r_{i'}) \psi(\text{rest of } r\text{'s}) = \sum_{ll'} \varphi_l(r_i) \varphi_{l'}(r_{i'}) \psi(\text{rest of } r\text{'s}) V_{ll'kk'}, \tag{A.1.15}$$

where

$$V_{ll'kk'} = \int \int \varphi_l^*(r_1) \varphi_{l'}^*(r_2) \hat{V}^{(1,2)} \varphi_k(r_1) \varphi_{k'}(r_2)\, dr_1\, dr_2. \tag{A.1.16}$$

Let

$$\hat{X}_{ll'mm'}^{(i,i')} \varphi_n(r_i) \varphi_{n'}(r_{i'}) \psi(\text{rest of } r\text{'s}) = \delta_{mn} \delta_{m'n'} \varphi_l(r_i) \varphi_{l'}(r_{i'}) \psi(\text{rest of } r\text{'s}); \tag{A.1.17}$$

then

$$\hat{V}^{(i,i')} = \sum_{ll'mm'} \hat{X}^{(i,i')}_{ll'mm'} V_{ll'mm'}. \tag{A.1.18}$$

Consider

$$\Phi = \sum_{i \neq i'} \hat{X}^{(i,i')}_{ll'mm'} \left\{ \sum_P \epsilon_P \prod_j \varphi_j(r_{Pj}) \right\}$$

$$= \sum_P \epsilon_P \sum_{i \neq i'} \hat{X}^{(Pi,Pi')}_{ll'mm'} \varphi_i(r_{Pi}) \varphi_{i'}(r_{Pi'}) \prod_{j \neq i,i'} \varphi_j(r_{Pj})$$

$$= \sum_P \epsilon_P \sum_{i \neq i'} \varphi_l(r_{Pi}) \varphi_{l'}(r_{Pi'}) \delta_{mi} \delta_{m'i'} \prod_{j \neq i,i'} \varphi_j(r_{Pj})$$

$$= \sum_P \epsilon_P \varphi_l(r_{Pm}) \varphi_{l'}(r_{Pm'}) \prod_{j \neq m,m'} \varphi_j(r_{Pj}) \text{ or } 0, \tag{A.1.19}$$

depending on whether $m \neq m'$ and both are in $1, \ldots, N$ or not. Thus we have the determinant where φ_m has been replaced by φ_l *and* $\varphi_{m'}$ has been replaced by $\varphi_{l'}$ (or we have zero). It follows from this that the image of

$$\sum_{i \neq i'} \hat{X}^{(i,i')}_{ll'mm'} \quad \text{is} \quad a_{l'}^\dagger a_l^\dagger a_m a_{m'}$$

(note the order), and the image of $\sum_{i \neq i'} \hat{V}^{(i,i')}$ is

$$\sum_{ll'kk'} \hat{V}_{ll'kk'} a_{l'}^\dagger a_l^\dagger a_k a_{k'} = \int\int dr_1\, dr_2\, \psi^\dagger(r_2) \psi^\dagger(r_1) \hat{V}^{(1,2)} \psi(r_1) \psi(r_2). \tag{A.1.20}$$

In the case of bosons Φ_N is the space of symmetrical functions of $r_1, \ldots, r_N$. The basis functions are

$$\Psi_{k_1,\ldots,k_N}(r_1,\ldots, r_N) = \frac{1}{\sqrt{N!}} \sum_P \varphi_{k_1}(r_{P1}) \ldots \varphi_{k_N}(r_{PN}), \tag{A.1.21}$$

where the state labels $k_1, \ldots, k_N$ need not all be different. We image this state by $a_{k_1}^\dagger \ldots a_{k_N}^\dagger |0\rangle$ in V_N. It is evident that the $a_k^\dagger$'s commute. Unfortunately the state (A.1.21) is not normalized (in Φ_N, and so also in V_N). We have

$$(\Psi, \Psi) = \frac{1}{N!} \int dr_1 \ldots dr_N \left\{ \sum_P \varphi_{k_1}^*(r_{P1}) \ldots \right\} \left\{ \sum_Q \varphi_{k_1}(r_{Q1}) \ldots \right\},$$

(where P, Q are permutations of $1, \ldots, N$)

$$= \frac{1}{N!} \sum_P \int ds_1 \ldots ds_N \left\{ \varphi^*_{k_1}(s_1) \ldots \right\} \left\{ \sum_{Q'} \varphi_{k_1}(s_{Q'1}) \ldots \right\},$$

where $s_i = r_{Pi}$, and $Q' = P^{-1}Q$ (goes through all permutations). The integrand is now independent of P, so that Σ_P gives just $N!$ Hence, writing $Q'' = (Q')^{-1}$,

$$(\Psi, \Psi) = \int ds_1 \ldots ds_N \left\{ \varphi^*_{k'}(s_1) \ldots \right\} \left\{ \sum_{Q''} \varphi_{kQ''1}(s_1) \ldots \right\}.$$

Now suppose that in $k_1, \ldots, k_N$ the state r occurs n_r times. For each Q'' the integral is zero unless $k_{Q''i} = k_i$ $(i = 1, \ldots, N)$, when it is unity. But the number of such permutations Q'' is $n_1!n_2! \ldots$, and hence $(\Psi, \Psi) = n_1!n_2! \ldots$. Thus the state $(n_1!n_2! \ldots)^{-1/2}\Psi$ is normalized. We call its image in V_N $|n_1, n_2, \ldots\rangle$, and the image of the unnormalized state Ψ $|Un_1, n_2, \ldots\rangle$. Hence

$$\Psi \Leftrightarrow (a_1{}^\dagger)^{n_1} \ldots |0\rangle = |Un_1, \ldots\rangle = (n_1!)^{1/2} \ldots |n_1, \ldots\rangle.$$

Thus

$$a_1{}^\dagger |Un_1, \ldots\rangle = |U(n_1 + 1), \ldots\rangle$$

implies

$$a_1{}^\dagger |n_1, \ldots\rangle = \sqrt{(n_1 + 1)}\, |n_1 + 1, \ldots\rangle, \tag{A.1.22}$$

and similarly for $a_r{}^\dagger$. We define the operator a_1 by

$$a_1 |Un_1, \ldots\rangle = n_1 |U(n_1 - 1), \ldots\rangle,$$

giving

$$a_1 |n_1, \ldots\rangle = \sqrt{n_1}\, |n_1 - 1, \ldots\rangle, \tag{A.1.23}$$

and similarly for a_r.

We soon find that $a_k, a_k{}^\dagger$ satisfy the commutation rules

$$[a_k{}^\dagger, a_l{}^\dagger] = 0, \qquad [a_k, a_l] = 0, \qquad [a_k, a_l{}^\dagger] = \delta_{kl}, \tag{A.1.24}$$

and that a_k, $a_k^\dagger$ are hermitian conjugates. The number operator is $a_r^\dagger a_r$ and is such that

$$a_r^\dagger a_r |n_1, \ldots\rangle = n_r |n_1, \ldots\rangle; \qquad \text{(A.1.25)}$$

the spectrum of eigenvalues n_r may be shown to run over the positive integers and zero. The total-number operator is

$$\hat{N} = \sum_r a_r^\dagger a_r. \qquad \text{(A.1.26)}$$

By considering the effect of the terms $\Sigma_i \hat{T}^{(i)}$ in the hamiltonian on the *unnormalized* basis functions Ψ we get

$$\sum_i \hat{T}^{(i)} \Leftrightarrow \sum_{kl} T_{kl} a_k^\dagger a_l. \qquad \text{(A.1.27)}$$

Similarly

$$\sum_{i \neq i'} \hat{V}^{(i,i')} \Leftrightarrow \sum_{ll'kk'} V_{ll'kk'} a_{l'}^\dagger a_l^\dagger a_k a_{k'}. \qquad \text{(A.1.28)}$$

The argument is the same as for fermions, except that we put $\epsilon_P = 1$ for all P.

Appendix 2

Time Correlation Functions and Green's Functions

In a liquid or gas in thermal equilibrium the random impacts of the molecules produce irregular driving forces, manifested for example as the Brownian motion of colloidal particles suspended in a liquid. If the Brownian particles are driven by an applied force (such as an electric field, if the particles are charged), the same molecular impacts produce frictional resistive forces, described macroscopically through quantities such as the electrical resistance of the medium. Since the random and systematic parts of the microscopic forces have the same physical origin, we may expect a mathematical relation to exist between them. In its general form, this relation is known as the *fluctuation-dissipation theorem* [Kubo (1957)]. Special cases of the relation have been known for a long time in classical physics: the earliest example is Einstein's relation (1905) between the diffusion coefficient and the friction constant in Brownian motion, and another example is Nyquist's formula (1928) relating the random current fluctuations (the intrinsic noise) in a metallic resistor to the resistivity of the metal.

In quantum-mechanical terms, the fluctuations of a system in thermal equilibrium may be characterized by time correlation functions of the type $\langle A(t)B(0)\rangle$, where A, B are operators, or by the Fourier transforms of these correlation functions which give the *fluctuation spectrum*. The linear response to an applied driving force (electrical conductivity, magnetic susceptibility, etc.) is given by a retarded Green's function of the type

$$-i\theta(t)\langle[A(t), B(0)]\rangle.$$

We now derive the general relation between these quantities. For the sake of generality we work with a canonical ensemble at non-zero temperature T, treating the case $T = 0$ as a limiting form of the general result.

We suppose the system has hamiltonian H, with a complete set of eigenstates $|n\rangle$ such that $H|n\rangle = E_n|n\rangle$. Consider

$$\begin{aligned}\langle A(t)B(0)\rangle &= Z^{-1}\,\mathrm{Tr}\,\{e^{-\beta H}\, e^{iHt}\, A\, e^{-iHt} B\} \\ &= Z^{-1} \sum_n \langle n|e^{-\beta H}\, e^{iHt}\, A\, e^{-iHt} B|n\rangle.\end{aligned} \tag{A.2.1}$$

To extract the time dependence we insert a complete set $|m\rangle$ $(\Sigma_m |m\rangle\langle m| = 1)$, and obtain

$$\begin{aligned}\langle A(t)B(0)\rangle &= Z^{-1} \sum_{mn} \langle n|B\, e^{-\beta H}|m\rangle\langle m|e^{iHt} A\, e^{-iHt}|n\rangle \\ &= Z^{-1} \sum_{mn} e^{-\beta E_m} \langle n|B|m\rangle\langle m|A|n\rangle\, e^{i(E_m-E_n)t}.\end{aligned} \tag{A.2.2}$$

[This type of analysis was first given by H. Lehmann (1954) in quantum field theory.] The time Fourier transform of $\langle A(t)B(0)\rangle$ is thus

$$\begin{aligned}J_1(\omega) &= \int_{-\infty}^{\infty} \langle A(t)B(0)\rangle\, e^{i\omega t}\, dt \\ &= 2\pi Z^{-1} \sum_{mn} e^{-\beta E_m} \langle n|B|m\rangle\langle m|A|n\rangle\delta(E_m - E_n + \omega).\end{aligned} \tag{A.2.3}$$

In the same way we find

$$\begin{aligned}J_2(\omega) &= \int_{-\infty}^{\infty} \langle B(0)A(t)\rangle\, e^{i\omega t}\, dt \\ &= 2\pi Z^{-1} \sum_{mn} e^{-\beta E_n} \langle n|B|m\rangle\langle m|A|n\rangle\delta(E_m - E_n + \omega).\end{aligned} \tag{A.2.4}$$

Since the δ-function is non-zero only for $E_n = E_m + \omega$, we have

$$\begin{aligned}J_2(\omega) &= 2\pi Z^{-1} \sum_{mn} e^{-\beta(E_m+\omega)} \langle n|B|m\rangle\langle m|A|n\rangle\delta(E_m - E_n + \omega) \\ &= e^{-\beta\omega} J_1(\omega).\end{aligned} \tag{A.2.5}$$

$J_1(\omega)$ is the *spectral density function* associated with the time correlation function $\langle A(t)B(0)\rangle$. At $T = 0$ only the terms for which $m = 0$ survive in (A.2.3), where E_0 is the ground state energy. Thus in this limit $J_1(\omega)$ is non-zero only for positive frequencies $\omega > 0$. Similarly $J_2(\omega)$ is non-zero only for $\omega < 0$.

Now consider the retarded Green's function

$$\begin{aligned} G^R(t) &= -i\theta(t)\langle[A(t), B(0)]\rangle \\ &= -i\theta(t)\langle A(t)B(0) - B(0)A(t)\rangle. \end{aligned} \tag{A.2.6}$$

This can be expressed in terms of $J_1(\omega)$ as

$$\begin{aligned} G^R(t) &= -i\theta(t) \int_{-\infty}^{\infty} \{J_1(\omega') - J_2(\omega')\}\, e^{-i\omega' t} \frac{d\omega'}{2\pi} \\ &= -i\theta(t) \int_{-\infty}^{\infty} (1 - e^{-\beta\omega'}) J_1(\omega')\, e^{-i\omega' t} \frac{d\omega'}{2\pi}. \end{aligned} \tag{A.2.7}$$

The Fourier transform is

$$\begin{aligned} G^R(\omega) &= -i \int_{-\infty}^{\infty} (1 - e^{-\beta\omega'}) J_1(\omega') \frac{d\omega'}{2\pi} \int_0^{\infty} e^{i(\omega - \omega' + i\eta)t}\, dt \\ &= \int_{-\infty}^{\infty} (1 - e^{-\beta\omega'}) \frac{J_1(\omega')}{\omega - \omega' + i\eta} \frac{d\omega'}{2\pi}, \qquad (\eta = 0+). \end{aligned} \tag{A.2.8}$$

Hence, assuming $J_1(\omega)$ is real (which is true if A and B are the same operator or hermitian conjugates),

$$\begin{aligned} \operatorname{Im} G^R(\omega) &= -\pi \int_{-\infty}^{\infty} (1 - e^{-\beta\omega'}) J_1(\omega') \delta(\omega - \omega') \frac{d\omega'}{2\pi} \\ &= -\tfrac{1}{2}(1 - e^{-\beta\omega}) J_1(\omega), \end{aligned}$$

or (for $\omega \neq 0$)

$$J_1(\omega) = -\frac{2}{1 - e^{-\beta\omega}} \operatorname{Im} G^R(\omega). \tag{A.2.9}$$

This is one form of the fluctuation-dissipation theorem.

The time-ordered Green's function

$$G^T(t) = -i\langle T[A(t)B(0)]\rangle \tag{A.2.10}$$

can also be expressed in terms of the spectral density $J_1(\omega)$. (T here is a boson T-product.) We have

$$G^T(t) = -i\theta(t)\int_{-\infty}^{\infty} J_1(\omega')\, e^{-i\omega' t}\frac{d\omega'}{2\pi} - i\theta(-t)\int_{-\infty}^{\infty} J_2(\omega')\, e^{-i\omega' t}\frac{d\omega'}{2\pi},$$

with Fourier transform

$$\begin{aligned} G^T(\omega) &= -i\int_{-\infty}^{\infty} J_1(\omega')\frac{d\omega'}{2\pi}\int_{0}^{\infty} e^{i(\omega-\omega'+i\eta)t}\, dt \\ &\quad -i\int_{-\infty}^{\infty} J_2(\omega')\frac{d\omega'}{2\pi}\int_{-\infty}^{0} e^{i(\omega-\omega'-i\eta)t}\, dt \\ &= \int_{-\infty}^{\infty}\left\{\frac{J_1(\omega')}{\omega-\omega'+i\eta} - \frac{J_2(\omega')}{\omega-\omega'-i\eta}\right\}\frac{d\omega'}{2\pi} \\ &= \int_{-\infty}^{\infty} J_1(\omega')\left\{\frac{1}{\omega-\omega'+i\eta} - \frac{e^{-\beta\omega'}}{\omega-\omega'-i\eta}\right\}\frac{d\omega'}{2\pi}. \end{aligned} \qquad \text{(A.2.11)}$$

From (A.2.8) and (A.2.11) we see that, if $J_1(\omega)$ is real,

$$\operatorname{Re} G^R(\omega) = \operatorname{Re} G^T(\omega) = \mathscr{P}\int_{-\infty}^{\infty}(1-e^{-\beta\omega'})\frac{J_1(\omega')}{\omega-\omega'}\frac{d\omega'}{2\pi}, \qquad \text{(A.2.12)}$$

and

$$\begin{aligned} \operatorname{Im} G^R(\omega) &= -\tfrac{1}{2}(1-e^{-\beta\omega})J_1(\omega), \\ \operatorname{Im} G^T(\omega) &= -\tfrac{1}{2}(1+e^{-\beta\omega})J_1(\omega). \end{aligned} \qquad \text{(A.2.13)}$$

The relation between $G^R(\omega)$ and $G^T(\omega)$ is thus

$$\operatorname{Re} G^R(\omega) = \operatorname{Re} G^T(\omega), \qquad \text{(A.2.14)}$$

$$\operatorname{Im} G^R(\omega) = \tanh(\tfrac{1}{2}\beta\omega).\ \operatorname{Im} G^T(\omega). \qquad \text{(A.2.15)}$$

At $T = 0$ ($\beta = \infty$), (A.2.15) reduces to

$$\operatorname{Im} G^R(\omega) = \frac{\omega}{|\omega|}\operatorname{Im} G^T(\omega). \qquad \text{(A.2.16)}$$

In the case of metallic conduction $G^R(\omega) = i\omega\sigma$, where σ is the conductivity [compare Eq. (5.7.18)], and $J_1(\omega)$ is the Fourier transform of the current–current correlation function $\langle j(t)j(0)\rangle$. Eq. (A.2.9) then gives

$$J_1(\omega) = -\frac{2}{1-e^{-\beta\omega}}\,\omega\sigma, \tag{A.2.17}$$

which is a quantum form of the Nyquist relation [Callen and Welton (1951)].

It is important to note that the spectral density function $J_1(\omega)$ also determines the *temperature Green's function* discussed in Chap. 2. To see this, consider

$$\begin{aligned} G^T(\sigma) &= \langle T[A(\sigma)B(0)]\rangle \\ &= \theta(\sigma)\langle A(\sigma)B(0)\rangle + \theta(-\sigma)\langle B(0)A(\sigma)\rangle, \end{aligned} \tag{A.2.18}$$

where σ is a temperature variable. Proceeding as in the case of the real-time correlation functions, we find

$$\langle A(\sigma)B(0)\rangle = Z^{-1}\sum_{mn} e^{-\beta E_m}\langle n|B|m\rangle\langle m|A|n\rangle\, e^{\sigma(E_m-E_n)}, \tag{A.2.19}$$

$$\langle B(0)A(\sigma)\rangle = Z^{-1}\sum_{mn} e^{-\beta E_n}\langle n|B|m\rangle\langle m|A|n\rangle\, e^{\sigma(E_m-E_n)}. \tag{A.2.20}$$

From these expressions it is immediately seen that, for $0<\sigma<\beta$, $G^T(\sigma) = G^T(\sigma-\beta)$, which demonstrates the periodicity property of the temperature Green's function.

Thus, expanding $G^T(\sigma)$ in the range $0\leqslant\sigma\leqslant\beta$ as a Fourier series by means of the reciprocal formulas (2.2.15) and (2.2.16) of Chap. 2, we obtain for the Fourier coefficients

$$\begin{aligned} G^T(\bar{\mu}) &= \frac{1}{\beta}\int_0^\beta d\sigma\, e^{-i\bar{\mu}\sigma}G^T(\sigma) \\ &= \frac{1}{\beta}\int_0^\beta d\sigma\, e^{-i\bar{\mu}\sigma}\langle A(\sigma)B(0)\rangle \\ &= \frac{1}{\beta Z}\sum_{mn}\langle n|B|m\rangle\langle m|A|n\rangle\frac{e^{-\beta E_n}-e^{-\beta E_m}}{E_m-E_n-i\bar{\mu}}. \end{aligned} \tag{A.2.21}$$

Here $\bar{\mu} = 2\pi\mu/\beta$, μ is an integer, and we have used Eq. (A.2.19). Comparing this form with (A.2.3) and (A.2.4), we see that $G^T(\bar{\mu})$ can be expressed in terms of $J_1(\omega)$ and $J_2(\omega)$ as

$$G^T(\bar{\mu}) = \frac{1}{\beta} \int_{-\infty}^{\infty} \frac{J_1(\omega') - J_2(\omega')}{i\bar{\mu} + \omega'} \frac{d\omega'}{2\pi}.$$

This can be written

$$G^T(\bar{\mu}) = -\frac{1}{\beta} \int_{-\infty}^{\infty} (1 - e^{-\beta\omega'}) \frac{J_1(\omega')}{-i\bar{\mu} - \omega'} \frac{d\omega'}{2\pi}, \qquad \text{(A.2.22)}$$

showing that $G^T(\bar{\mu})$ can be obtained from $J_1(\omega)$.

Comparing this result with the expression (A.2.8) for the retarded Green's function $G^R(\omega)$, we see that the function $G^R(\omega)$ can be obtained formally from $-\beta G^T(\bar{\mu})$ by replacing the imaginary variable $-i\bar{\mu}$ by the complex variable $\omega + i\eta$. Since $G^T(\bar{\mu})$ is known only at a discrete set of points on the imaginary axis, this procedure amounts to an *analytic continuation* of the temperature Green's function into the whole complex plane [for a discussion of uniqueness, see Baym and Mermin (1961)]. Finally, the *time-ordered* real-time Green's function at finite temperature can now be obtained via Eqs. (A.2.9) and (A.2.11).

Bibliography

It is of course not possible to give a complete list of references in a work of this kind. We give references to (a) some general texts recommended for further reading, (b) original articles cited in the text, or otherwise of interest in relation to the text, and (c) review articles developing particular topics in greater detail, in which many further references will be found.

General Texts

A. Messiah (1961) *Quantum Mechanics* (Amsterdam: North-Holland).

A.A. Abrikosov, L.P. Gor'kov, and L.E. Dzyaloshinskii (1963) *Methods of Quantum Field Theory in Statistical Physics* (translated and edited by R.A. Silverman) (Englewood Cliffs, New Jersey: Prentice-Hall).

C. Kittel (1963) *Quantum Theory of Solids* (New York: Wiley).

D. Pines and P. Nozières (1966) *The Theory of Quantum Liquids.* Vol. 1, *Normal Fermi Liquids* (New York: W.A. Benjamin).

A.L. Fetter and J.D. Walecka (1971) *Quantum Theory of Many-Particle Systems* (New York: McGraw-Hill).

Chapter 1

General (*lattice dynamics in the harmonic approximation*)

A.A. Maradudin, E.W. Montroll, G.H. Weiss, and I.P. Ipatova (2nd ed., 1971) *Theory of Lattice Dynamics in the Harmonic Approximation, Solid State Physics* (ed. Seitz, Turnbull, and Ehrenreich, Supplement 3) (New York: Academic Press).

Sec. 1.3 (*neutron scattering cross-section*)

G. Placzek (1952) *Phys. Rev. 86*, 377.

L. van Hove (1954) *Phys. Rev. 95*, 249, 1374.

Sec. 1.6 (*Fourier transforms*)

E.C. Titchmarsh (1937) *Introduction to the Theory of Fourier Integrals* (Oxford: Clarendon Press), Sec. 1.3.

Sec. 1.7 (*phonon dispersion by neutron scattering*)

B.N. Brockhouse (1964) in *Phonons and Phonon Interactions* (ed. T.A. Bak) (New York: W.A. Benjamin), p. 221.

(*anharmonic lattice*)
R.A. Cowley (1963) *Advances in Phys. 12*, 421.
D.H. Martin (1965) *Advances in Phys. 14*, 39.
(*phonons and lattice imperfections*)
A.A. Maradudin (1964) in *Phonons and Phonon Interactions* (ed. T.A. Bak) (New York: W.A. Benjamin), p. 424.

Chapter 2

General (*finite-temperature Green's functions*)
D.N. Zubarev (1960) *Uspekhi Fiz. Nauk. 71*, 71 (English transl.: *Soviet Physics Uspekhi 3*, 320).
L.P. Kadanoff and G. Baym (1962) *Quantum Statistical Mechanics* (New York: W.A. Benjamin).
Sec. 2.2 (*Bloch equation*)
F. Bloch (1932) *Z. Physik 74*, 295.
(*Fourier series for finite-temperature Green's functions*)
A.A. Abrikosov, L.P. Gor'kov, and I.E. Dzyaloshinskii (1959) *Ž. Èksper. Teoret. Fiz.36*, 900 (English transl.: *Soviet Physics JETP 9*, 636).
E.S. Fradkin (1959) *Ž. Èksper. Teoret. Fiz. 36*, 1286 (English transl.: *Soviet Physics JETP 9*, 912).
P.C. Martin and J. Schwinger (1959) *Phys. Rev. 115*, 1342.

Chapter 3

General (*texts on quantum field theory*)
A.I. Akhiezer and V.B. Berestetskii (1965) *Quantum Electrodynamics* (revised English ed.) (New York: Interscience).
N.N. Bogoliubov and D.V. Shirkov (1959) *Introduction to the Theory of Quantized Fields* (New York: Interscience).
(*the Feynman–Dyson expansion*)
R.P. Feynman (1949) *Phys. Rev. 76*, 749, 769.
F.J. Dyson (1949) *Phys. Rev. 75*, 486, 1736.
(*time-labeled form of many-body perturbation theory*)
N.M. Hugenholtz (1957) *Physica 23*, 481, 533.
J. Goldstone (1957) *Proc. Roy. Soc. (London) A239*, 267.
(*functional integral formalism*)
D.R. Hamann (1970) *Phys. Rev. B2*, 1373.
Sec. 3.1 (*adiabatic theorem*)
M. Gell-Mann and F. Low (1951) *Phys. Rev. 84*, 350.

(*Wick's theorem*)
G.C. Wick (1950) *Phys. Rev. 80*, 268.
Sec. 3.2 (*linked-cluster theorem*)
K.A. Brueckner (1955) *Phys. Rev. 100*, 36.
J. Goldstone (1957) *loc cit.*
Sec. 3.3 (*Wick's theorem for thermodynamic averages*)
T. Matsubara (1955) *Progr. Theoret. Phys. (Kyoto) 14*, 351.
M. Gaudin (1960) *Nuclear Phys. 15*, 89.
Sec. 3.4 (*$T \to 0$ limit for interacting fermions*)
W. Kohn and J.M. Luttinger (1960) *Phys. Rev. 118*, 41.
J.M. Luttinger and J.C. Ward (1960) *Phys. Rev. 118*, 1417.

Chapter 4

General (*theory of normal fermi systems*)
V. Ambegaokar (1963) Green's Functions in Many-Body Problems, in *Astrophysics and the Many-Body Problem* (Brandeis Summer Institute 1962) (New York: W.A. Benjamin), p. 321.
Sec. 4.3 (*Green's functions and T matrix*)
A. Messiah (1961) *op. cit.*, Chap. XIX.
Sec. 4.4 (*short-range potential model and virtual states*)
G.F. Koster and J.C. Slater (1954) *Phys. Rev. 96*, 1208.
J. Friedel (1958) *Nuovo Cimento, Suppl. 7*, 287.
P.A. Wolff (1961) *Phys. Rev. 124*, 1030.
A.M. Clogston (1962) *Phys. Rev. 125*, 439.
Secs. 4.5–4.7 (*Friedel sum rule*)
J. Friedel (1952) *Philos. Mag. 43*, 153; (1958) *loc. cit.*
C. Kittel (1963) *op. cit.*, Chap. 18.
O.P. Sinha (1971) *Canad. J. Physics 49*, 61.
See also: V. Ambegaokar (1963) *loc.cit.*
For experiments on V_3Ga see:
A.M. Clogston and V. Jaccarino (1961) *Phys. Rev. 121*, 1357.
(*nuclear resonance in dilute alloys*)
N. Bloembergen and T.J. Rowland (1953) *Acta Metallurgica 1*, 731.
T.J. Rowland (1960) *Phys. Rev. 119*, 900.
W. Kohn and S.H. Vosko (1960) *Phys. Rev. 119*, 912.

Chapter 5

General (*irreversible processes*)
Review: G.V. Chester (1963) *Rep. Progr. Phys. XXVI*, 411 (and references given there).

(*transport equations*)

L. van Hove (1955) *Physica 21,* 517.

L.P. Kadanoff and G. Baym (1962) *op. cit.*

S. Fujita (1966) *Introduction to Non-equilibrium Quantum Statistical Mechanics* (Philadelphia and London: Saunders).

G.V. Chester, *loc. cit.*

Sec. 5.1 (*quantum-mechanical recurrence theorem*)

P. Bocchieri and A. Loinger (1957) *Phys. Rev. 107,* 337.

I. Percival (1961) *J. Math. Phys. 2*, 235.

Sec. 5.2 (*random impurity distribution*)

W. Kohn and J.M. Luttinger (1957) *Phys. Rev. 108*, 590; (1958) *109*, 1892.

Sec. 5.3 (*Dyson's equation*)

F.J. Dyson (1949) *Phys. Rev. 75*, 1736.

Sec. 5.5 (*Kubo formula*)

R. Kubo (1957) *J. Phys. Soc. Japan 12*, 570; (1958) *Lectures in Theoretical Physics (Boulder) 1,* 120.

Secs. 5.6, 5.7 (*evaluation of response function for transport*)

S.F. Edwards (1958) *Philos. Mag. 3*, 1020.

A.A. Abrikosov and L.P. Gor'kov (1958) *Ž. Èksper. Teoret. Fiz. 35*, 1558 (English transl.: *Soviet Physics JETP 8*, 1090).

See also:

G. Rickayzen (1962) Scattering by Impurities in Metals, in *The Many Body Problem*, Bergen International School of Physics (New York: W.A. Benjamin), p. 85.

A.A. Abrikosov, L.P. Gor'kov, and I.E. Dzyaloshinskii (1963) *op. cit.*, Sec. 39.

(*anomalous skin effect*)

A.B. Pippard (1965) *The Dynamics of Conduction Electrons* (London: Blackie), p. 58.

Chapter 6

General

R. Brout and P. Carruthers (1963) *Lectures on the Many-Electron Problem* (New York: Interscience).

Introduction (*fermi liquid theory*)

L.D. Landau (1956) *Ž. Èksper. Teoret. Fiz. 30*, 1058; (1957) *32*, 59; (1958) *35*, 97 (English transl.: *Soviet Physics JETP 3*, 920; *5*,101; *8*, 70).

J.M. Luttinger (1960) *Phys. Rev. 119*, 1153.
See also: Pines and Nozières (1966) *op. cit.*

Sec. 6.1 (*T*/log *T specific heat*)

J. Bardeen (1936) *Phys. Rev. 50*, 1098.

Sec. 6.4 (*ground state energy of dense electron gas*)

M. Gell-Mann and K.A. Brueckner (1957) *Phys. Rev. 106*, 364.

(*random phase approximation*)

D. Bohm and D. Pines (1953) *Phys. Rev. 92*, 609.

(*dielectric formulation of ground state energy*)

P. Nozières and D. Pines (1958) *Phys. Rev. 109*, 762.

Sec. 6.5 (*Thomas–Fermi approximation*)

L.H. Thomas (1927) *Proc. Cambridge Philos. Soc. 23*, 542.
E. Fermi (1928) *Z. Physik 48*, 73.

(*Friedel oscillations*)

J. Friedel (1958) *loc. cit.*

(*Kohn effect*)

W. Kohn (1959) *Phys. Rev. Lett. 2*, 393.
E.J. Woll, Jr., and W. Kohn (1962) *Phys. Rev. 126*, 1693.
Review: S.K. Joshi and A.K. Rajagopal (1968) Lattice Dynamics of Metals, in *Solid State Physics* (ed. Seitz, Turnbull, and Ehrenreich) (New York: Academic Press), Vol. 22, p. 284.

(*dielectric response function in r.p.a.*)

J. Lindhard (1954) *Kgl. Danske Vid. Selsk., Mat.-Fys. Medd. 28*, No. 8.

Sec. 6.6 (*correlation energy of dense electron gas*)

W. Macke (1950) *Z. Naturforsch. 5a*, 192.
M. Gell-Mann and K.A. Brueckner (1957) *loc. cit.*
L. Onsager, L. Mittag, and M.J. Stephen (1966) *Ann. Physik 18*, 71.
D.F. DuBois (1959) *Ann. Physics 7*, 174.
W.J. Carr, Jr., and A.A. Maradudin (1964) *Phys. Rev. 133*, A371.

(*improvements over r.p.a.*)

J. Hubbard (1967) *Phys. Lett. 25A*, 709.
K.S. Singwi, M.P. Tosi, R.H. Land, and A. Sjölander (1970) *Phys. Rev. B1*, 1044.

(*Wigner crystal*)

E. Wigner (1938) *Trans. Faraday Soc. 34*, 678.
W.J. Carr, Jr. (1961) *Phys. Rev. 122*, 1437.

Chapter 7

General

C. Herring (1966) *Exchange Interactions among Itinerant Electrons*

(Vol. IV of *Magnetism*, ed: G.T. Rado and H. Suhl) (New York: Academic Press).
W. Marshall (ed.) (1967) *Theory of Magnetism in Transition Metals* (Proc. Intern. School of Physics "Enrico Fermi," Course XXXVII) (New York: Academic Press).
Introduction (*excitonic insulator*)
W. Kohn (1968) Metals and Insulators, in *Many-Body Physics* (ed. DeWitt and Balian) (New York: Gordon and Breach), p. 372.
B.I. Halperin and T.M. Rice (1968) The Excitonic State at the Semiconductor–Semimetal Transition, in *Solid State Physics* (ed. Seitz, Turnbull, and Ehrenreich) (New York: Academic Press), Vol. 21, p. 115.
Sec. 7.1 (*Hubbard model*)
J. Hubbard (1963) *Proc. Roy. Soc. (London) A276*, 238; (1964) *277*, 237; *281*, 401.
(*classification of phase transitions*)
L.D. Landau and E.M. Lifshitz (1959) *Statistical Physics* (London: Pergamon), p. 430 ff.
Sec. 7.3 (*r.p.a. evaluation of susceptibility*)
P.A. Wolff (1960) *Phys. Rev. 120*, 814.
T. Izuyama, D.-J. Kim, and R. Kubo (1963) *J. Phys. Soc. (Japan) 18*, 1025.
Sec. 7.4 (*spin density waves*)
A.W. Overhauser (1962) *Phys. Rev. 128*, 1437.
D.R. Penn (1966) *Phys. Rev. 142*, 350.
Sec. 7.5 (*neutron scattering*)
L. van Hove (1954) *loc. cit.*
S. Doniach (1967) *Proc. Phys. Soc. (London) 91*, 86.
R.D. Lowde and C.G. Windsor (1970) *Advances in Phys. 19*, 813.
Secs. 7.5, 7.6 (*properties of* He_3)
S. Doniach (1968) *Proc. 11th Intern. Conf. on Low Temp. Physics* (*St. Andrews 1968*), p. 76.
(*electrical resistivity of Pd alloys*)
P. Lederer and D.L. Mills (1968) *Phys. Rev. 165*, 837.
M.J. Rice (1968) *J. Appl. Phys. (U.S.A.) 39*, 958.
A.I. Schindler and B.R. Coles (1968) *J. Appl. Phys. (U.S.A.) 39*, 956.
(*paramagnon specific heat*)
N.F. Berk and J.R. Schrieffer (1966) *Phys. Rev. Lett. 17*, 433.
S. Doniach and S. Engelsberg (1966) *Phys. Rev. Lett. 17*, 750.
W. Brenig and H.J. Mikeska (1967) *Phys. Lett. 24*, 332.

E. Fawcett, E. Bucher, W.F. Brinkman, J.P. Maita, and J.H. Wernick (1969) *J. Appl. Phys. (U.S.A.) 40*, 1097.

Sec. 7.7 (*Goldstone modes*)

J. Goldstone (1961) *Nuovo Cimento 19*, 154.

Review: T.W.B. Kibble (1966) *Proc. Oxford Intern. Conf. on Elem. Particles 1965* (Didcot, England: Rutherford High Energy Lab.), Pt. I, p. 19.

(*band theory of ferromagnetism*)

F. Englert and M.M. Antonoff (1964) *Physica 30*, 429.

E.P. Wohlfarth (1968) *J. Inst. Math. Appl. 4*, 359.

A. Blandin (1968) in *Theory of Condensed Matter* (Vienna: International Atomic Energy Agency), p. 691.

Sec. 7.8 (*Anderson model*)

P.W. Anderson (1961) *Phys. Rev. 124*, 41.

J.R. Schrieffer and D.C. Mattis (1965) *Phys. Rev. 140*, A1412.

(*Wolff model*)

P.A. Wolff (1961) *Phys. Rev. 124*, 1030.

(*functional integral formalism*)

W.E. Evenson, J.R. Schrieffer, and S.Q. Wong (1970) *J. Appl. Phys. (U.S.A.) 41*, 1199.

D.R. Hamann (1970) *Phys. Rev. B2*, 1373.

(*measurements of local moments*)

A.M. Clogston, B.T. Matthias, M. Peter, H.J. Williams, E. Corenzwit, and R.C. Sherwood (1962) *Phys. Rev. 125*, 541.

Chapter 8

Introduction, Secs. 8.1, 8.2 (*metal-insulator transition*)

N.F. Mott (1949) *Proc. Phys. Soc. (London) A62*, 416; (1961) *Philos. Mag. 6*, 287.

J. Hubbard (1964) *loc. cit.*, paper III.

Review articles:

D. Adler (1968) Insulating and Metallic States in Transition Metal Oxides, in *Solid State Physics* (ed. Seitz, Turnbull, and Ehrenreich) (New York: Academic Press), Vol. 21, p. 1.

N.F. Mott (1969) *Philos. Mag. 20*, 1.

S. Doniach (1969) *Advances in Phys. 18*, 819.

Sec. 8.3 (*superexchange*)

P.W. Anderson (1950) *Phys. Rev. 79*, 350; (1963) Exchange in Insulators: Super-exchange, Direct Exchange, and Double Exchange, in *Magnetism*

(Vol. I, ed. G.T. Rado and H. Suhl) (New York: Academic Press), p. 25.

(*theory of phase transition in spin system*)

N.N. Bogoliubov and S.V. Tyablikov (1959) *Dokl. Akad. Nauk SSSR 126*, 53.

(*Holstein–Primakoff transformation*)

T. Holstein and H. Primakoff (1940) *Phys. Rev. 58*, 1098.

(*spin-wave interactions*)

F.J. Dyson (1956a) *Phys. Rev. 102*, 1217.

(*thermodynamics of Heisenberg ferromagnet*)

F.J. Dyson (1956b) *Phys. Rev. 102*, 1230.

M. Wortis (1965) *Phys. Rev. 138*, A1126.

(*spin waves in Heisenberg antiferromagnets*)

A.B. Harris, D. Kumar, B.I. Halperin, and P.C. Hohenberg (1971) *Phys. Rev. B3*, 961.

(*spin waves:* review article)

F. Keffer (1966) *Handbuch der Physik* (Berlin: Springer) Vol. XVIII/2, p. 1.

Chapter 9

Introduction

J. Kondo (1964) *Progr. Theoret. Phys. (Kyoto) 32*, 37.

J.W. Loram, T.E. Whall, and P.J. Ford (1970) *Phys. Rev. B2*, 857.

Sec. 9.1

A.I. Akhiezer and V.B. Berestetskii (1965) *op. cit.*

Sec. 9.2 (*x-ray singularity in photoemission*)

G.D. Mahan (1967) *Phys. Rev. 163*, 612.

B. Roulet, J. Gavoret, and P. Nozières (1969) *Phys. Rev. 178*, 1072.

P. Nozières, J. Gavoret, and B. Roulet (1969) *Phys. Rev. 178*, 1084.

P. Nozières and C.T. DeDominicis (1969) *Phys. Rev. 178*, 1097.

K.D. Schotte and U. Schotte (1969) *Phys. Rev. 182*, 479; *185*, 509.

(*Baker–Hausdorff theorem*)

W. Magnus (1954) *Comm. Pure Appl. Math. 7*, 649.

(*lifetime effects*)

S. Doniach and M. Šunjić (1970) *J. Phys. C (Solid State Phys.) 3*, 285.

(*experiments*)

C. Nordling, E. Sokolowski, and K. Siegbahn (1958) *Ark. Fys. 13*, 483.

L.G. Parratt (1959) *Rev. Modern Phys. 31*, 616.

Sec. 9.3 (*x-ray absorption and emission spectra*)
P. Nozières and C.T. DeDominicis (1969) *loc. cit.*
K.D. Schotte and U. Schotte (1969) *loc. cit.*
G.A. Ausman and A.J. Glick (1969) *Phys. Rev. 183*, 687.
Sec. 9.4 (*Kondo hamiltonian*)
T. Kasuya (1956) *Progr. Theoret. Phys. (Kyoto) 16*, 45, 58.
K. Yosida (1957) *Phys. Rev. 106*, 893.
J. Kondo (1964) *loc. cit.*
(*Schrieffer–Wolff transformation*)
J.R. Schrieffer and P.A. Wolff (1966) *Phys. Rev. 149*, 491.
(*relation between Kondo and x-ray problems*)
G. Yuval and P.W. Anderson (1970) *Phys. Rev. B1*, 1522.
Sec. 9.5 (*drone-fermion representation*)
D.C. Mattis (1965) *Theory of Magnetism* (New York: Harper and Row), p. 78.
H.J. Spencer (1968) *Phys. Rev.171*, 515.
(*self-consistent solutions for T matrix*)
Y. Nagaoka (1965) *Phys. Rev. 138*, 1112.
A.A. Abrikosov (1965) *Physics 2*, 5.
H. Suhl (1965) *Physics 2*, 39.
J. Zittartz and E. Müller-Hartmann (1968) *Z. Physik 212*, 380.
See also: P.W. Anderson, G. Yuval, and D.R. Hamann (1970) *Phys. Rev. B1*, 4464.
(*resistivity of dilute alloys*)
M.D. Daybell and W.A. Steyert (1967) *Phys. Rev. Lett. 18*, 398.
J.W. Loram, T.E. Whall, and P.J. Ford (1970) *loc. cit.*
Review articles:
J. Kondo (1969) Theory of Dilute Magnetic Alloys, in *Solid State Physics* (ed. Seitz, Turnbull, and Ehrenreich) (New York: Academic Press), Vol. 23, p. 183.
A.J. Heeger (1969) Localized Moments and Non-moments in Metals, in *Solid State Physics* (ed. Seitz, Turnbull, and Ehrenreich) (New York: Academic Press), Vol. 23, p. 283.

Chapter 10
General
J.R. Schrieffer (1964) *Theory of Superconductivity* (New York: W.A. Benjamin).
G. Rickayzen (1965) *Theory of Superconductivity* (New York: Interscience).

R.D. Parks (ed.) (1969) *Superconductivity* (2 Vols.) (New York: Dekker).

Sec. 10.1 (*Cooper pairs*)

L.N. Cooper (1956) *Phys. Rev. 104*, 1189.

(*B.C.S. theory*)

J. Bardeen, L.N. Cooper, and J.R. Schrieffer (1957) *Phys. Rev. 108*, 1175.

Sec. 10.2 (*two-particle Green's function*)

L.P. Kadanoff and P.C. Martin (1961) *Phys. Rev. 124*, 670.

V. Ambegaokar (1969) The Green's Function Method, in R.D. Parks (ed.) *op. cit.*, Vol. 1, p. 284.

(*multiparticle tunneling*)

B.N. Taylor and E. Burstein (1963) *Phys. Rev. Lett. 10*, 14.

J.R. Schrieffer and J.W. Wilkins (1963) *Phys. Rev. Lett. 10*, 17.

Review: J.M. Rowell (1969) Tunneling and Superconductivity, in *Superconductivity* (ed. P.R. Wallace) (New York: Gordon and Breach), Vol. 2, p. 721.

(*electron–electron interaction in superconductors*)

G.M. Eliashberg (1960) *Ž. Èksper. Teoret. Fiz. 38*, 966 (English transl.: *Soviet Physics JETP 11*, 696).

P. Morel and P.W. Anderson (1962) *Phys. Rev. 125*, 1263.

Review: D.C. Scalapino (1969) The Electron-phonon Interaction and Strong-coupling Superconductors, in R.D. Parks (ed.) *op. cit.*, Vol. 1, p. 449.

(*precursor fluctuation effects*)

D.J. Thouless (1960) *Ann. Physics 10*, 553.

B.R. Patton, V. Ambegaokar, and J.W. Wilkins (1969) *Solid State Comm.* 7, 1287.

V. Ambegaokar (1969) Critical Currents and the Onset of Resistance in Superconductors, in *Superconductivity* (ed. P.R. Wallace) (New York: Gordon and Breach), Vol. 1, p. 117.

Sec. 10.3 (*superconducting ground state*)

P.W. Anderson (1958) *Phys. Rev. 112*, 1900.

L.P. Gor'kov (1958) *Ž. Èksper. Teoret. Fiz. 34*, 735 (English transl.: *Soviet Physics JETP* 7, 505).

Y. Nambu (1960) *Phys. Rev. 117*, 648.

Chapter 11

11.6 References

P.W. Anderson, G. Yuval, and D.R. Hamann (1970) *Phys. Rev. B1*, 4464.

N. Andrei (1980) *Phys. Rev. Letts.* vol. 45, no. 5, p. 379–82.
P. Coleman (1983) *Phys. Rev. B29*, 5255.
S. Doniach (1977) *Physica 91B*, 231.
S. Doniach (1987) *Phys. Rev. 53*, 1814.
F.D.M. Haldane (1981) *J. Phys. C14*, 2585.
A. Luther and I. Peschel (1974) *Phys. Rev. B9*, 2911.
J.M. Luttinger (1963) *J. Math. Phys. 15*, 609.
D.C. Mattis and E.H. Lieb (1965) *J. Math. Phys. 6*, 304.
D.M. Newns and N. Read (1988) *Advances in Physics 36*, 799.
M.R. Norman (1992) *Physica C194*, 203.
H.R. Ott, H. Rudiger, Z. Fisk, and J.L. Smith (1983) *Phys. Rev. Lett. 50*, 1595.
N. Read and D.M. Newns (1983) *J. Phys. C16*, 3273.
N. Read, D.M. Newns, and S. Doniach (1984) *Phys. Rev. B30*, 3841–3844.
H.J. Schulz (1991) *Int. J. Modern Physics B5*, 57.
F. Steglich, J. Aarts, C.D. Bredl, W. Lieke, D. Meschede, W. Franz, and H. Schäfer (1979) *Phys. Rev. Letts. 43*, 1892.
F. Steglich (1992) in *Electron Correlation and Disorder Effects in Metals* (ed. S.N. Behera) (Singapore: World Scientific Publishing Co).
S. Tomonaga (1950) *Prog. Theor. Physics (Kyoto) 5*, 544.
C.M. Varma (1985) *Comments on Solid State Physics 11*, 221–43.
P.B. Wiegmann (1981) *J. Phys. C14*, 1463.
K.G. Wilson (1975) *Rev. Mod. Phys. 47*, 773.

Chapter 12

12.6 References

I. Affleck and J.B. Marston (1988) *Phys. Rev. B37*, 3774.
P.W. Anderson (1992) *Science*, vol. 256, p. 1526–31.
P.H. Dickinson and S. Doniach (1993) *Phys. Rev. B47*, 11, 447.
S. Doniach and M. Inui (1990) *Phys. Rev. B41*, 6668–6678.
V. Emery and S. Kivelson (1995) *Nature 374*, 434–437
J.E. Hirsch (1985) *Phys. Rev. Letts. 54*, 1317.
D.R. Hofstadter (1976) *Phys. Rev. B14*, 2239–2249.
G. Kotliar, P.A. Lee, and N. Read (1988) *Physica C* 153–155,538.
R.B. Laughlin (1988) *Science 242*, 525.
P. Monthoux and D. Pines (1992) *Phys. Rev. Letts. 69*, 961.

N. Nagaosa and P.A. Lee (1990) *Phys. Rev. Letts. 64*, 2450–2453.

J.W. Negele and H. Orland (1988) *Quantum Many-Particle Systems* (Redwood City, Calif.: Addison-Wesley).

R.E. Prange and S.M. Girvin (1990) *The Quantized Hall Effect*, 2nd Edition (Berlin: Springer-Verlag).

C.A.R. Sà de Melo and S. Doniach (1990) *Phys. Rev. B41*, 6633–40.

J.R. Schrieffer (1983) *Theory of Superconductivity*, Revised Printing, (Reading, Mass: Benjamin/Cummings Pub. Co.).

Z.-X. Shen, D.S. Dessau, B.O. Wells, D.M. King, W.E. Spicer, A.J. Arko, D.S. Marshall, L.W. Lombardo, A. Kapitulnik, P. Dickinson, S. Doniach, J. Dicarlo, T. Loeser, and C.H. Park (1993) *Phys. Rev. Letts. 70*, 1553.

Z.X. Shen, W.E. Spicer, D.M. King, D.S. Dessau, and B.O. Wells (1995) *Science 267*, 343–350.

Z.X. Shen, D.S. Marshall, D.S. Dessau, A.G. Loeser, C.H. Park, A.Y. Matsuura, J.N. Eckstein, I. Bozovic, P. Fourier, A. Kapitulnik, and W.E. Spicer (1996) *Phys. Rev. Letts. 76*, 4841–4844.

S. Spielman, J.S. Dodge, L.W. Lombardo, C.B. Eom, M.M. Fejer, T.H. Geballe, and A. Kapitulnik (1992) *Phys. Rev. Letts. 68*, 3472–3475.

M. Tinkham (1975) *Introduction to Superconductivity* (New York: McGraw-Hill).

S.A. Trugman (1990) *Phys. Rev. B41*, 892–895.

X.G. Wen, F. Wilczek, and A. Zee (1990) *Phys. Rev. B39*, 11,413.

J. Zaanen, G.A. Sawatzky, and J.W. Allen (1985) *Phys. Rev. Letts. 55*, 418–421

F.C. Zhang and T.M. Rice (1988) *Phys Rev B37*, 3759.

Z. Zou, J.L. Levy and R.B. Laughlin (1992) *Phys. Rev. B45*, 993.

Appendix 1

(second quantization)

F.A. Berezin (1966) *The Method of Second Quantization* (trans. from the Russian) (New York: Academic Press).

J. de Boer (1965) *Studies in Statistical Mechanics* (ed. J. de Boer and G. E. Uhlenbeck) (Amsterdam; North-Holland) Vol. III, p. 215.

V. Fock (1932) *Z. Physik 75*, 622.

P. Jordan and O. Klein (1927) *Z. Physik 45*, 751.

P. Jordan and E.P. Wigner (1928) *Z. Physik 47*, 631.

(Lehmann representation)

H. Lehmann (1954) *Nuovo Cimento 11*, 342.

Appendix 2

(*fluctuation-dissipation theorem*)

A. Einstein (1905) *Ann. Physik 17*, 549.

H. Nyquist (1928) *Phys. Rev. 32*, 110.

H.B. Callen and T.A. Welton (1951) *Phys. Rev. 83*, 34.

R. Kubo (1957) *J. Phys. Soc. (Japan) 12*, 570.

(*classical two-particle correlation function*)

L.S. Ornstein and F. Zernike (1914) *Proc. Sect. Sci. K. med. Akad. Wetensch. 17*, 793.

(*analytic properties of thermodynamic Green's functions*)

G. Baym and N.D. Mermin (1961) *J. Math. Phys. 2*, 232.

See also: A.L. Fetter and J.D. Walecka (1971) *op. cit.*, Chap. 9.

Historical Note on George Green[1]

George Green, who gave his name to the functions discussed in this book, was born in 1793 in Nottingham, England. His father was a baker who later purchased an estate at a nearby village and became a prosperous miller. We know very little about Green's schooling and education. According to his brother-in-law, "his mathematical abilities were such that he quickly outstripped his teachers and in consequence his schooling ceased at an early age, whereupon he assisted his father first as a baker and then as a miller." He was clearly almost entirely self-taught; fortunately there was an excellent subscription library in Nottingham, offering access to books such as Laplace's "Mécanique Céleste" and to the *Philosophical Transactions of the Royal Society.*

In 1828 George Green published his first and greatest paper, "An Essay on the Application of Mathematical Analysis to the Theories of Electricity and Magnetism." It was printed and published in Nottingham with the aid of private subscription. The paper—of quite astonishing originality—attracted no notice, and Green decided to give up mathematics. However, one of the subscribers, Sir Edward Ffrench Bromhead, Baronet, of Thurlby Hall, Lincolnshire, became Green's patron and encouraged him to pursue his studies. His father's death in 1829 left Green independent and comparatively affluent, and in 1833 (having sold the business) he entered Gonville and Caius College, Cambridge. The mathematical honors examination (the Tripos) was at that time highly competitive, the candidates with first-class honors (the Wranglers) being placed in order of merit. Then, as now, performance in the Tripos and ability in research were not synonymous, and in 1837 Green was returned as fourth Wrangler, disappointing his friends who had expected the highest honors. After graduating Green remained in Cambridge and continued with his researches. In 1839 he was elected to a College Fellowship, but he was

[1] For the biographical material we consulted: A Biography of George Green, Mathematical Physicist of Nottingham and Cambridge 1793-1841, by H. Gwynedd Green, M.A. (Reprinted from: Studies and Essays in the History of Science and Learning in Honor of George Sarton, 1946.) We thank Dr. F. C. Powell of Gonville and Caius College, Cambridge, for showing us this reference.

already in failing health and soon returned to his village where he died in 1841. Green's death was ignored by the scientific world; a local newspaper published an appreciation ending with the opinion: "Had his life been prolonged, he might have stood eminently high as a mathematician."

In the 1828 Essay, Green established the integral identities, equivalent to the divergence theorem, which bear his name, and used them to make a systematic investigation of the properties of the potential. (The term *potential function* is due to him, and he was the first to apply this concept—already used by Lagrange, Laplace and Poisson in the study of gravitational attractions—to the theory of electricity and magnetism.) One of Green's results was a formula for the potential at any point P inside a closed region, expressed in terms of its boundary values on the surface and the charge density induced, by a unit charge at P, on an earthed sheet coinciding with the surface. This was the first example of the use of a Green's function—here used to establish a fundamental result in potential theory. It was Riemann who first attached Green's name to this function. Later the concept was generalized to denote a much wider class of solutions of inhomogeneous differential equations with point sources, and the physical interpretation was extended to denote, quite generally, the response of any system to a standard input.

Green published nine further papers, dealing with the equilibrium of fluids, the attraction of ellipsoids, the motion of waves in canals and the reflection and refraction of sound and light. In 1837, he used an asymptotic method, equivalent to the "W.K.B.J." approximation in wave mechanics, to investigate the propagation of tidal waves in a canal of slowly varying cross-section. In his last paper he developed a theory of propagation of light in crystalline media.

Recognition of Green's work on the potential came in 1845, when the Essay was discovered by William Thomson (later Lord Kelvin), who showed it to the French mathematicians and had it reprinted in *Crelle's Journal* (now *Journal für die reine und angewandte Mathematik*).[2] In the meantime all Green's general theorems had been rediscovered by Thomson, by Gauss in Germany and by Chasles and Sturm in France. It is not altogether surprising that Green's work remained neglected during his lifetime. Quite apart from the circumstance of the Essay's private publication in a provincial town, English mathematics in the 1820's was only just beginning to emerge from a century of isolation, the legacy of

[2] A complete edition of Green's papers was published in 1871: Mathematical Papers of the late George Green, edited by N.M. Ferrers (London: Macmillan).

the Newton–Leibniz controversy. In Green's time there were no major mathematicians in Cambridge, and he received his early inspiration from the French analysts, particularly Poisson. His own writings greatly influenced the work of Kelvin and his contemporaries, and Green can be said to stand first in line of the great British school of natural philosophers which flourished in the second half of the nineteenth century and included Kelvin, Stokes, Rayleigh, Clerk Maxwell, and J.J. Thomson.

Index